STRATHCLYDE UNIVERSITY LIBRARY
30125 00076529 6

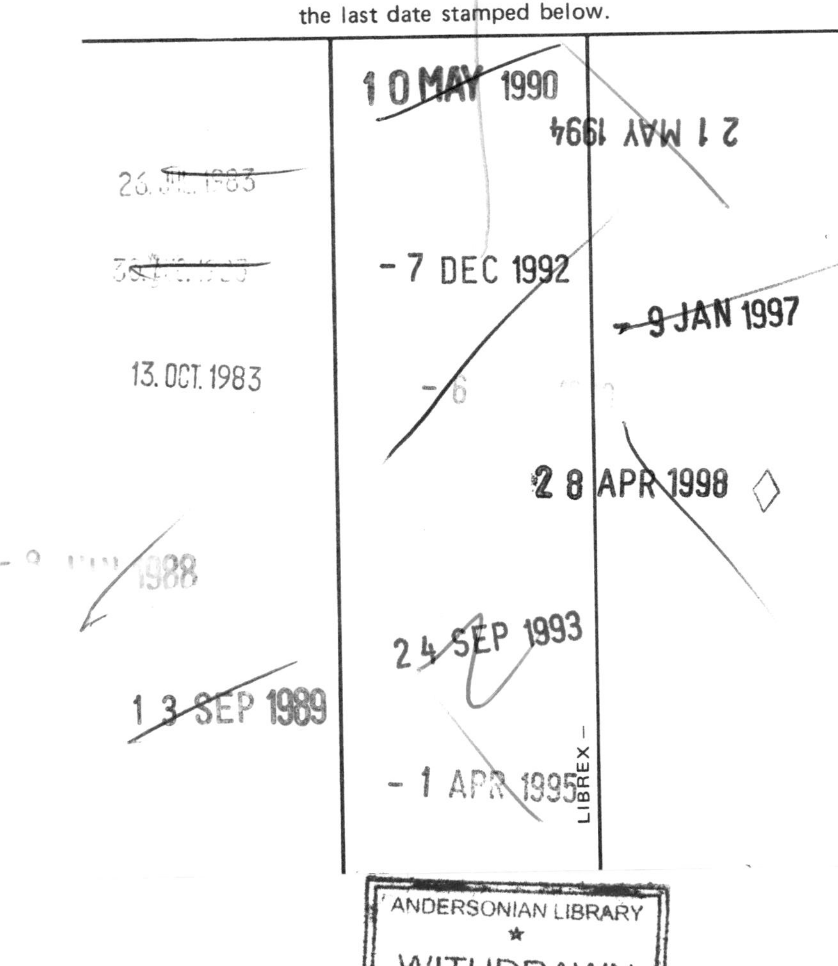

Gas Cleaning for Air Quality Control

Industrial and Environmental Health and Safety Requirements

CHEMICAL PROCESSING AND ENGINEERING

An International Series of Monographs and Textbooks

Volume 1: Chemical Process Economics, Second Edition, Revised and Expanded
by John Happel and Donald G. Jordan

Volume 2: Gas Cleaning for Air Quality Control: Industrial and Environmental Health and Safety Requirements,
edited by Joseph M. Marchello and John J. Kelly

IN PREPARATION

Petroleum Economics and Engineering: An Introduction, *by H. K. Abdel-Aal and Robert Schmelzlee*

Thermodynamics of Fluids: An Introduction to Equilibrium Theory, *by K.C. Chao and R.A. Greenkorn*

Chemical Reactions as a Separation Mechanism—Sulfur Removal, *edited by Billy L. Crynes*

Continuum Mechanics of Viscoelastic Fluids, *by Ronald Darby*

Control of Air Pollution Sources, *by J. M. Marchello*

Gas-Solids Handling in the Process Industry, *edited by J. M. Marchello and Albert Gomezplata*

Computers in Process Control, *by Ulrich Rembold, Mahesh Seth, and Jeremy S. Weinstein*

Solvent Extraction in Hydrometallurgical Processing, *by G. M. Ritcey and A. W. Ashbrook*

Petroleum Refining: Technology and Economics, *by James H. Gary and Glenn E. Handwerk*

ADDITIONAL VOLUMES IN PREPARATION

Gas Cleaning for Air Quality Control

INDUSTRIAL AND ENVIRONMENTAL HEALTH AND SAFETY REQUIREMENTS

edited by Joseph M. Marchello
Division of Mathematical and Physical Sciences and Engineering
University of Maryland
College Park, Maryland

and

John J. Kelly
Chemical Engineering Department
University College
Dublin, Ireland

MARCEL DEKKER, INC. New York

MARCEL DEKKER, INC.

270 Madison Avenue, New York, New York 10016

LIBRARY OF CONGRESS CATALOG CARD NUMBER: 73-78559

ISBN: 0-8247-6079-4

Current printing (last digit):
10 9 8 7 6 5 4 3 2 1

PRINTED IN THE UNITED STATES OF AMERICA

CONTENTS

LIST OF CONTRIBUTORS

PAUL G. GORMAN, Midwest Research Institute, Kansas City, Missouri

JOHN J. KELLY, Chemical Engineering Department, University College, Dublin, Ireland

JOSEPH M. MARCHELLO, Division of Mathematical and Physical Sciences and Engineering, University of Maryland, College Park, Maryland

GRADY B. NICHOLS, Southern Research Institute, Birmingham, Alabama

SABERT OGLESBY, Jr., Southern Research Institute, Birmingham, Alabama

LARRY J. SHANNON, Midwest Research Institute, Kansas City, Missouri

A. EUGENE VANDEGRIFT, Midwest Research Institute, Kansas City, Missouri

FOREWORD

This Chemical Engineering and Processing Series will be a welcome addition to the literature in that it provides new material on a series of topics important to the chemical engineering aspects of the process industries.

Today we are in great need of carrying new scientific and engineering developments through to the application state. Unfortunately chemical engineering had a lull in attention to new information via university research having to do with the real world of the process industries. It is believed this situation has turned around both in research and teaching and this series of books should assist in the educational effort in this direction. The bridge between science and engineering application is a continuing challenge to chemical engineers and has been taken up by the editors of this series.

One important characteristic of the book or series is the background and ability of the authors and editors, their discernment of what is important and what should be left out, their dedication to the task of communicating the essence of ideas to the reader, and their preparation of material from the viewpoint of the user overrides most other considerations. I wish to commend Marcel Dekker, Inc., for the leadership they have chosen and the editors in turn for prospective lists of authors.

I wish the editors and authors well in their efforts to increase the use of technical knowledge for the benefit of process industries and hence the people of our nation.

DONALD L. KATZ

PREFACE

Gas-solid systems is an interdisciplinary subject requiring application of physics, chemistry, mathematics, and engineering. Many practical problems have been solved by engineers in the metal, chemical, food, and pharmaceutical industries. The objective of this book is to bring together the latest theories and practice on the control of particulate emissions to provide an integrated view and reference source of the field.

Control of fine particle emissions to the atmosphere is a subject of concern from the standpoint of reducing material losses and in terms of improving the quality of the environment. This book is directed toward gas cleaning as required and practiced by industry to meet air pollution control requirements. The basic aspects of particle removal and experience and design of separation systems are presented with the aim of providing the reader with the background to better handle this growing problem.

The chapters of this book lead to a concluding summary of particle control practice in Chapter 6, which presents equipment costs, comparisons, and applications. Chapter 1 deals with air quality regulations, characteristics of emission sources, and sampling. Chapters 2, 3, 4 and 5 present detailed discussions of the design and operation of major types of equipment for reducing emissions to the atmosphere.

The contributing authors possess wide and extensive levels of experience in control of particulate emissions. Some have years of experience in industry with both equipment manufacturers and users. Others have experience in federal, state, and local government regulatory activities. This book draws upon their collective experience and summarizes it for the convenience of future workers.

Joseph M. Marchello

John J. Kelly

Gas Cleaning for Air Quality Control

Industrial and Environmental Health and Safety Requirements

Chapter 1

INTRODUCTION

Joseph M. Marchello

Division of Mathematical and
Physical Sciences and Engineering
University of Maryland
College Park, Maryland

I. GAS CLEANING

A variety of man's activities result in the emission to the atmosphere of solid and liquid particles and reactive gases that subsequently participate in the formation of particles. The cumulative effect of such activities poses a threat to the health and welfare of the public and has been the basis for regulations requiring that gaseous emissions be cleaned prior to release (1). The removal of particulate matter from gases is generally known as gas cleaning.

Control of particulate emissions may be divided into the control of material present in gas streams inside equipment or building enclosures and the control of materials arising in open or unconfined areas. Strategies for control include basic changes in procedures and materials which eliminate or alter emissions and the use of add-on or auxiliary equipment to remove pollutants prior to release of the carrier gas to the atmosphere.

Changes in hooding, materials handling, and general housekeeping can sometimes significantly reduce the emission of particulate matter into the atmosphere. Process changes involving new techniques and the use of alternate materials may also offer important ways of abating emissions.

In most cases, pollution control equipment is selected to control the quantity of particulate matter. Emission regulations usually stipulate the maximum allowable mass emission rate but frequently also contain a visibility limitation on the plume expressed as equivalent opacity or Ringlemann number. Particle concentration, size distribution and optical properties establish the opacity of emissions. Thus mass and visibility regulations are interrelated for many process sources. Either kind of regulation establishes the efficiency for which the control equipment must be designed, based on given inlet conditions.

Equipment for the control of particulate emissions (2) includes dry inertial and centrifugal collectors such as cyclones, low- and high-energy scrubbers, electrostatic precipitators, cleanable fabric filters, and mist eliminators. Gas incinerators (afterburners) are sometimes used for removal of combustible particles.

The primary design consideration in gas cleaning is to establish what control devices will be able to meet the emission regulations for a given source. Specifications for a particular application require information on the pollutant and carrier gas characteristics such as: particle and gas concentration (average and range), average particle size and size distribution; particle shape, density, packing characteristics, and resistivity; gas flow rate, temperature, and moisture content, corrosivity, and flammability.

The estimated level of control, on a weight basis, of major sources of particulate matter for new installations is summarized in Table 1 based on sales of equipment of all types to various industries in 1967 (3). Potential weight efficiencies are generally over 90% and in some cases over 99%. There has been a steady improvement in the collection efficiencies over the years, but some operational changes can have marked effects on collector system performance. For example, the removal of sulfur trioxide from coal burning power plant effluents significantly lowers the performance of electrostatic precipitators. This results from the removal of sulfur, leading to the formation of a more resistive fly ash.

As discussed in Chap. 6, costs of particulate control equipment (2,4) vary widely. Costs cover a range of values because of local conditions, the nature of the particles and the gas stream, equipment size (gas volume), and design collection efficiency.

The high mass efficiencies for particle collection can lead to deceptive conclusions about emissions because of the large tonnages involved. Weight percentages that pass through the control equipment collectively represent large quantities of material escaping to the atmosphere on a total mass basis.

TABLE 1

Average Efficiency (Weight Basis) Of Newly Installed Industrial Air Pollution Control Equipment For Removal Of Particulates (3)

Industry description	Average operating efficiency %	Expected efficiency at optimum conditions %
Coal-fired electric utilities	93	98
Coal-fired industrial boilers	90	95
Crushed stone/sand and gravel	94	96
Agriculture operations	92	94
Iron and Steel		
Ore crushing	70	90
Materials handling	95	97
Sinter plants	95	97
Coke ovens	70	90
Blast furnaces	97	99
Basic oxygen	97	99
Electric arc	97	99
Scarfing	95	97
Cement	97	99
Woodpulp	96	97
Lime	92	94
Clay	96	97
Primary nonferrous		
Aluminum	85	90
Copper	97	98
Zinc	97	98
Lead	97	99
Phosphate rock	97	98
Fertilizer	96	98

TABLE 1 (continued)

Industry description	Average operating efficiency %	Expected efficiency at optimum conditions %
Asphalt	98	99
Ferroalloys	90	95
Iron foundries	85	90
Secondary nonferrous		
Copper	96	99
Aluminum	96	99
Lead	96	99
Zinc	96	99
Coal cleaning	99	99
Petroleum	99	99
Acids	94	96

Moreover, efficiencies for the finest particles, which play a key role in air pollution effects, are significantly less than for larger particles. For some types of collectors, theory indicates that particle size removal efficiency passes through a minimum in the size range between 0.1 and 1.0 μm, which is a particularly important range so far as visibility, health effects, and weather modification are concerned. In the case of high-temperature particulate sources such as combustion of fossil fuel, there appears to be a concentration of metals such as cadmium, chromium, and lead in the smaller size fraction--possibly by vaporization and condensation of metal fumes.

II. AIR QUALITY REGULATIONS

At the beginning of the industrial revolution manufacturers had little or no conscience about the atmospheric discharge of noxious effluents. However, modern industrialization, growth of cities, increased vehicular traffic on the roads, crowded living conditions, and waste problems have joined together in recent years to magnify the air pollution problem.

Atmospheric particulate matter is produced from natural sources and from man-made sources such as fuel combustion in power plants, industrial processes of diverse nature, and internal-combustion engines. Formation processes can be classified as primary, which means that particles are introduced into the atmosphere in particulate form, or secondary, which refers to the formation of particles from gases and vapors.

The most easily observed effect of air pollution is the reduction in visibility produced by the scattering of light from the surface of airborne particles. The degree of light obstruction is related to particle size, aerosol density, distance, and other factors. The destruction of metals, coatings, fabrics, and vegetation is mainly due to acid mists, oxidants of various kinds, and particulate products of combustion and industrial processing.

Knowledge of sources of air pollution in a community and the quantities of the various pollutants emitted to the air provides the basic framework for air conservation activities. Through an emission inventory, information relating to the quantities of the various pollutants released, the relative contribution of pollutants from different source categories, and the geographical distribution of pollutant emissions within the study area may be obtained. The results of an emission survey may be used effectively in metropolitan planning, pollution abatement activities, sampling programs, and diffusion models for predicting atmospheric levels of pollutants.

National primary and secondary ambient air quality standards have been set by the Environmental Protection

Agency, EPA (5). Primary ambient air quality standards define levels of air quality judged to allow an adequate margin of safety to protect the public health. National secondary ambient air quality standards define levels judged to protect the public welfare from adverse effects associated with the presence of air pollutants in the ambient air. Part 410, Chapter IV, Title 42, Code of Federal Regulations contains the standards for air pollutants and the reference methods for their measurement.

The national *primary* ambient air quality standards for particulate matter are:

a. 75 $\mu g/m^3$--annual geometric mean;

b. 260 $\mu g/m^3$--maximum 24-hr concentration not to be exceeded more than once per year.

The national *secondary* ambient air quality standards for particulate matter are:

a. 60 $\mu g/m^3$--annual geometric mean, as a guide to be used in assessing implementation plans to achieve a 24-hr concentration;

b. 150 $\mu g/m^3$--maximum 24-hr concentration not to be exceeded more than once per year.

The national ambient standards are continually under review by EPA and may be changed and extended in the future. In addition to particulate matter, the present standards cover sulfur dioxide, nitrogen oxides, hydrocarbons, carbon monoxide, and photochemical oxidants. Polycyclic hydrocarbons, trace elements, and other specific chemicals may be included in the standards in the future.

Under the Clean Air Amendments of 1970 (7) the States were required to submit plans which provide for implementation, maintenance, and enforcement of these standards. The States may adopt more stringent standards. Recommendations by EPA (6) to the States for possible inclusion in their implementation plans cover emissions of particulates. These, together with the standards for performance for several new stationary sources, are as follows:

	Required for new sources[a]	Recommended for implementation plans
Fuel burning equipment	(a) 0.10 lb/million Btu, (b) 30% opacity	0.30 lb/million Btu
Incinerators	0.08 grains/scf corrected to 12% CO_2	0.20 lb/100 lbs charged
Portland cement plants	Kiln: (a) 0.30 lb/ton of feed, (b) 10% opacity Clinker cooler: (a) 0.10 lb/ton of feed, (b) 10% opacity	Use of process weight formula, ranges between 0.1 and 1.0 lb/ton of feed
Sulfuric acid	(a) 0.15 lb of mist/ton of acid produced, (b) 10% opacity	--

[a]Mass emission limitations provide control for all particles. Opacity restrictions limit emission of fine particles which have high light scattering capability per unit mass.

The preparation, adoption, submission, and implementation of plans to meet the national standards are covered in Part 420, Chapter IV of the Code of Federal Regulations. It provides for a statewide system of permits for construction and operation of stationary sources, a strategy to attain and maintain ambient air standards, and emergency procedures during episodes to reduce emissions.

III. PARTICULATE EMISSIONS

A. SOURCES AND EMISSION RATES

Midwest Research Institute, under contract with the Office of Air Programs, completed a survey in 1971 of particulate emissions and control (3). Some of the results are summarized in Tables 2-4. In rural areas, 80% by weight of the total emissions are estimated to arise from natural sources. In urban areas, man-made contributions to particulate pollution are usually several times greater than those from natural sources. The national goal of lowering urban ambient concentrations from the present levels of approximately 120 $\mu g/m^3$ to a level of 60 $\mu g/m^3$ will require close control over all man-made sources.

The objective of the study was to identify, characterize, and quantify the particulate air pollution emissions resulting from stationary sources in the continental United States. Emissions from each source or industry were determined from: (1) emission factors for an uncontrolled source based on a unit of production; (2) the material processed per year; (3) the average or expected efficiency of control equipment; and, (4) the percentage of production capacity equipped with control devices.

As shown in Table 2, the largest sources of particulates are natural dusts and forest fires. These sources account for an estimated 85% of the national atmospheric primary particulate loading on a mass basis and are a substantial portion of background levels. However, their effect on the population is much less than for man-made sources because of the concentration of the latter in urban areas.

Fugitive or unconfined sources directly related to man's activities result from agriculture, mining, construction and transportation. A number of unconfined sources are contained

TABLE 2

Sources Of Particulate Pollution (12):
United States--1968

Source	Emissions tons/yr	% by wt
Natural dusts	63,000,000	44.7
Forest fires	56,400,000	40.1
Major stationary industrial sources	18,000,000	12.9
Transportation	1,200,000	0.8
Incineration	930,000	0.7
Other Sources	1,300,000	0.8
TOTAL	140,830,000	100.0

in the data in Table 3. For example, some crushing and grinding and materials-handling operations are unconfined and are important functions within a number of the processes listed.

Information on industrial particle emissions and properties, based on mass measurements of particles collected from stack effluents is presented in Tables 3 and 4. Other measurement and rating methods for particle emissions, such as light scattering, toxicity, and the potential of gaseous emissions to form aerosols in the atmosphere, may be needed in the future (12).

Most, if not all, of the industrial sources listed in Table 3 may be controlled by the installation of presently available control equipment. Estimates of the efficiency of control equipment and the extent to which it is presently used are given in Table 3. In some cases, such as the combustion of coal by electric utilities, over 90% collection efficiency is now being routinely achieved. Yet by virtue of the large amounts of coal burned, total particulate emissions remain a major source of total emissions.

Not reflected in Table 3 is the relative contribution that the emissions make to the long-lived suspended particulate levels in the atmosphere. A portion of the industrial emissions will deposit near the emission point in a short time. Available information on particle size distribution, outlet microgram loadings, and chemical composition for the sources is listed in Table 4.

Forecasts of future particulate emission levels have been made by several investigators (3,12). These are based on: (1) changes in production capacity; (2) improvements in control devices; and, (3) increased application of control devices based on legislative or regulatory enforcement. With industrial emission sources may be expected to increase from the 1968 level of 18 million tons per year to over 50 million tons per year by the year 2000. Legislative active action, requiring the best existing control equipment on all sources could reduce emissions to about eight million tons per year by the year 2000. Legislative action and improvements in control devices and practices could reduce emissions to as low as 2.5 million tons by the year 2000.

B. FINE PARTICLES

The environment in a stack or process usually differs greatly from that in the atmosphere. Stack environments are often characterized by high temperatures, high moisture content, high particle number concentrations, the presence of cocontaminants, and a short time between particle formation and emission. Atmospheric conditions are usually characterized by low temperature, long exposure to ambient conditions, variable humidities, and particle number concentrations so low that coagulation is negligible.

In a typical urban atmosphere most of the particles,

TABLE 3

Major Industrial Sources Of Particulate Pollutants (3,4): United States--1968

Source	Annual Tonnage P
1. Fuel combustion	
A. Coal	
1. Electric utility	
a. Pulverized	258,400,000 tons of coal
b. Stoker	9,900,000 tons of coal
c. Cyclone	28,700,000 tons of coal
2. Industrial boilers	
a. Pulverized	20,000,000 tons of coal
b. Stoker	70,000,000 tons of coal
c. Cyclone	10,000,000 tons of coal
B. Fuel oil	
1. Electric utility	7.18×10^9 gal
2. Industrial	
a. Residual	7.51×10^9 gal
b. Distillate	2.36×10^9 gal
C. Natural gas & LPG	
1. Electric utility	3.14×10^6 mil. scf
2. Industrial	9.27×10^6 mil. scf
2. Crushed stone, sand & gravel	
A. Crushed stone	681,000,000
B. Sand & gravel	918,000,000
3. Operations related to agriculture	
A. Grain elevators	177,000,000 tons grain handled
B. Cotton gins	11,000,000 bales
C. Feed mills	
1. Alfalfa mills	1,600,000 tons dry meal
2. Mills other than alfalfa	8,364,000 tons
4. Iron and steel	
A. Ore crushing	82,000,000 tons of ore
B. Materials handling	131,000,000 tons of steel
C. Pellet plants	50,000,000 tons of pellets
D. Sinter plants	51,000,000 tons of sinter
1. Sintering process	
2. Crushing, screening, etc.	
E. Coke manufacture	
1. Beehive	1,300,000 tons of coal
2. By-product	90,000,000 tons of coal
3. Pushing and Quenching	91,300,000 tons of coal
F. Blast furnace	88,800,000 tons of iron
G. Steel furnaces	
1. Open hearth	65,800,000 tons of steel
2. Basic oxygen	48,000,000 tons of steel
3. Electric arc	16,800,000 tons of steel
H. Scarfing	131,000,000 tons of steel

Emission factor lb/ton e_f	Efficiency[b] of control C_c	Application[a] of control C_t	Net[c] control C_c C_t	Emissions tons/yr E
16A=190[d] lb/ton of coal	0.92	0.97	0.89	2,710,000
13A=146 lb/ton of coal	0.80	0.87	0.70	217,000
3A=35 lb/ton of coal	0.91	0.71	0.64	182,000
	TOTAL from Electric utility coal			3,109,000
16A=170[e] lb/ton of coal	0.85	0.95	0.81	322,000
13A=133 lb/ton of coal	0.85	0.62	0.52	2,234,000
3A=31 lb/ton of coal	0.82	0.91	0.75	39,000
	TOTAL from industrial coal			2,595,000
0.010 lb/gal	0	0	0	36,000
0.023 lb/gal	0	0	0	87,000
0.015 lb/gal	0	0	0	18,000
	TOTAL from fuel oil			141,000
15 lb/mil. scf	0	0	0	24,000
18 lb/mil. scf	0	0	0	84,000
	TOTAL from natural gas & LPG			108,000
	TOTAL from utility and industrial fuel combustion			5,953,000
17	0.80	0.25	0.20	4,554,000
0.1	--	--	0	46,000
	TOTAL from crushed stone,sand & gravel			4,600,000
	0.80	0.40	0.28	1,700,000*
12 lb/bale	0.80	0.40	0.32	45,000
50 lb/ton dry meal	0.85	0.50	0.42	23,000
1% of production	0.85	0.50	0.42	49,000
	TOTAL from listed agricultural operations			1,817,000
2 lb/ton of ore	0	0	0	82,000
10 lb/ton of steel	0.90	0.35	0.32	446,000
--	--		--	80,000*
20 lb/ton of sinter	0.90	1.0	0.90	51,000
22 lb/ton of sinter	0.90	1.0	0.90	56,000
200 lb/ton of coal	0	0	0	130,000
2 lb/ton of coal	0	0	0	90,000
0.46 lb/ton of coal	--	--	--	21,000
130 lb/ton of iron	0.99	1.0	0.99	58,000
17 lb/ton of steel	0.97	0.41	0.40	337,000
40 lb/ton of steel	0.99	1.0	0.99	10,000
10 lb/ton of steel	0.99	0.79	0.78	18,000
3 lb/ton of steel	0.90	0.75	0.68	63,000
	TOTAL from iron and steel			1,442,000

TABLE 3 (Continued)

Source	Annual tonnage P
5. Cement	
A. Wet process	43,600,000 tons of cement
1. Kilns	
2. Grinders, dryers, etc.	
B. Dry process	31,000,000 tons of cement
1. Kilns	
2. Grinders, dryers, etc.	
6. Forest products	
A. Wigwam burners	27,500,000 tons of waste
B. Pulp mills	
1. Kraft process	24,300,000 tons of pulp
a. Recovery furnace	
b. Lime kilns	
c. Dissolving tanks	
2. Sulfite process	2,500,000 tons of pulp
a. Recovery furnace	833,000 tons of pulp
3. NSSC process	3,500,000 tons of pulp
a. Recovery furnace	1,167,000 tons of pulp
b. Fluid-bed reactor	525,000 tons of pulp
4. Bark boilers	--
C. Particleboard, etc.	--
7. Lime	
A. Crushing, screening	28,000,000 tons or rock
B. Rotary kilns	16,200,000 tons of lime
C. Vertical kilns	1,800,000 tons of lime
D. Materials handling	18,000,000 tons of lime
8. Primary nonferrous metals	
A. Aluminum	
1. Grinding of bauxite	13,000,000 tons of bauxite
2. Calcining of hydroxide	5,840,000 tons of alumina
3. Reduction cells	
a. H. S. Soderberg	800,000 tons of aluminum
b. V. S. Soderberg	700,000 tons of aluminum
c. Prebake	1,755,000 tons of aluminum
4. Materials handling	3,300,000 tons of aluminum
B. Copper	
1. Ore crushing	170,000,000 tons of ore
2. Roasting	575,000 tons of copper
3. Reverb. furnace	1,437,000 tons of copper
4. Converters	1,437,000 tons of copper
5. Materials handling	1,437,000 tons of copper
C. Zinc	
1. Ore crushing	18,000,000 tons of ore

Emission factor lb/ton e_f	Efficiency[b] of control C_c	Application[a] of control C_t	Net[c] control C_c C_t	Emissions tons/yr E
167 lb/ton of cement	0.94	0.94	0.88	435,000
25 lb/ton of cement	0.94	0.94	0.88	65,000
167 lb/ton of cement	0.94	0.94	0.88	310,000
67 lb/ton of cement	0.94	0.94	0.88	124.000
	TOTAL from cement			934,000
10 lb/ton of waste	0	0	0	132,000
150 lb/ton of pulp	0.92	0.99	0.91	164,000
45 lb/ton of pulp	0.95	0.99	0.94	33,000
5 lb/ton of pulp	0.90	0.33	0.30	42,000
268 lb/ton of pulp	0.92	0.99	0.91	10,000
24 lb/ton of pulp	0.92	0.99	0.91	1,000
533 lb/ton of pulp	0.70	1.00	0.70	42,000
--	--	--	--	82,000*
--	--	--	--	74.000*
	TOTAL from forest products			580,000
24 lb/ton of rock	0.80	0.25	0.20	264,000
180 lb/ton of lime	0.93	0.87	0.81	224,000
7 lb/ton of lime	0.97	0.40	0.39	4,000
5 lb/ton of lime	0.95	0.80	0.76	11,000
				573,000
	TOTAL from lime			
6 lb/ton of bauxite	--	--	0.80	8,000
200 lb/ton of alumina	--	--	0.90	58,000
144 lb/ton of aluminum	0.40	1.0	0.40	35,000
84 lb/ton of aluminum	0.64	1.0	0.64	10,000
63 lb/ton of aluminum	0.64	1.0	0.64	20,000
10 lb/ton of aluminum	0.90	0.35	0.32	11,000
	TOTAL from primary aluminum			142,000
2 lb/ton of ore	0	0	0	170,000
168 lb/ton of Cu	0.85	1.0	0.85	7,000
206 lb/ton of Cu	0.95	0.85	0.81	28,000
235 lb/ton of Cu	0.95	0.85	0.81	33,000
10 lb/ton of Cu	0.90	0.35	0.32	5,000
	TOTAL from primary copper			243,000
2 lb/ton of ore	0	0	0	18,000

TABLE 3 (Continued)

Source	Annual tonnage P
2. Roasting	
a. Fluid-bed	765,000 tons of zinc
b. Ropp, multi-hearth	153,000 tons of zinc
3. Sintering	612,000 tons of zinc
4. Distillation	612,000 tons of zinc
5. Materials handling	1,020,000 tons of zinc
D. Lead	
1. Ore crushing	4,500,000 tons of ore
2. Sintering	467,000 tons of lead
3. Blast furnace	467,000 tons of lead
4. Dross reverb. furnace	467,000 tons of lead
5. Materials handling	467,000 tons of lead
9. Clay	
A. Ceramic	
1. Grinding	4,722,000 tons
2. Drying	7,870,000 tons
B. Refractories	
1. Kiln-fired	
a. Calcining	688,000 tons
b. Drying	1,032,000 tons
c. Grinding	3,440,000 tons
2. Castable	550,000 tons
3. Magnesite	120,000 tons
4. Mortars	
a. Grinding	120,000 tons
b. Drying	120,000 tons
5. Mixes	249,000 tons
C. Heavy clay products	
1. Grinding	4,740,000 tons
2. Drying	7,110,000 tons
10. Fertilizer and Phosphate rock	
A. Phosphate rock	41,300,000 tons of rock
1. Drying	
2. Grinding	
3. Materials handling	
4. Calcining	8,260,000 tons
B. Fertilizers	
1. Amonium nitrate	2,800,000 tons of granules
2. Urea	1,000,000 tons of granules
3. Phosphates	
a. Rock pulverizing	17,000,000 tons of rock
b. Acid-rock reaction	4,370,000 tons of P_2O_5
c. Granulation and drying etc.	18,100,000 tons of granules
d. Materials handling	--
e. Bagging	9,000,000 tons of granules
4. Ammonium sulfate	2,700,000 tons of granules

Emission factor lb/ton e_f	Efficiency[b] of control C_c	Application[a] of control C_t	Net[c] control C_c C_t	Emissions tons/yr E
2,000 lb/ton of Zn	0.98	1.0	0.98	15,000
333 lb/ton of Zn	0.85	1.0	0.85	4,000
180 lb/ton of Zn	0.95	1.0	0.95	3,000
--	--	--	--	15,000*
7 lb/ton of Zn	0.90	0.35	0.32	2,000
	TOTAL from primary zinc			57,000
2 lb/ton of ore	0	0	0	4,000
520 lb/ton of lead	0.95	0.90	0.86	17,000
250 lb/ton of lead	0.85	0.98	0.83	10,000
20 lb/ton of lead	--	--	0.50	2,000
5 lb/ton of lead	0.90	0.35	0.32	1,000
	TOTAL from primary lead			34,000
	TOTAL from primary nonferrous metals			476,000
76 lb/ton	0.80	0.75	0.60	72,000
70 lb/ton	0.80	0.75	0.60	110,000
200 lb/ton	0.80	0.80	0.64	25,000
70 lb/ton	0.80	0.80	0.64	13,000
76 lb/ton	0.80	0.80	0.64	47,000
225 lb/ton	0.90	0.85	0.77	14,000
250 lb/ton	0.80	0.70	0.56	7,000
76 lb/ton	0.80	0.75	0.60	2,000
70 lb/ton	0.80	0.75	0.60	2,000
76 lb/ton	0.80	0.75	0.60	4,000
76 lb/ton	0.80	0.75	0.60	72,000
70 lb/ton	0.80	0.75	0.60	99,000
	TOTAL from clay			467,000
12 lb/ton	0.94	1.0	0.94	14,000
2 lb/ton	0.97	1.0	0.97	1,000
2 lb/ton	0.90	0.25	0.22	30,000
40 lb/ton	0.95	1.0	0.95	8,000
--	--	--	--	28,000*
--	--	--	--	10,000*
6 lb/ton of rock	0.80	1.0	0.80	10,000
48 lb/ton of P_2O_5	0.95	0.95	0.90	9,000
195 lb/ton	0.95	0.95	0.90	169,000
		--	--	18,000*
--	--	--	--	4,000*
--	--	--	--	27,000*
	TOTAL from fertilizers and Phosphate rock			328,000

TABLE 3 (Continued)

Source	Annual Tonnage P
11. Asphalt	
A. Paving material	251,000,000 tons of material
1. Dryers	
2. Secondary sources	
B. Roofing material	6,264,000 tons of asphalt
1. Blowing	
2. Saturator	
12. Ferroalloys	
A. Blast furnace	591,000 tons of ferroalloy
B. Electric furnace	2,119,000 tons of ferroalloy
C. Materials handling	2,710,000 tons of ferroalloy
13. Iron foundries	
A. Furnaces	18,000,000 tons of hot metal
B. Materials handling	
1. Coke, limestone, etc.	
2. Sand	10,500,000 tons of sand
14. Secondary nonferrous metals	
A. Copper	
1. Material preparation	
a. Wire burning	300,000 tons insulated wire
b. Sweating furnaces	64,000 tons scrap
c. Blast furnaces	287,000 tons scrap
2. Smelting and refining	1,170,000 tons scrap
B. Aluminum	
1. Sweating furnaces	500,000 tons scrap
2. Refining furnaces	1,015,000 tons scrap
3. Chlorine fluxing	136,000 tons Cl used
C. Lead	
1. Pot furnaces	53,000 tons scrap
2. Blast furnaces	119,000 tons scrap
3. Reverb. furnaces	554,000 tons scrap
D. Zinc	
1. Sweating furnaces	
a. Metallic scrap	52,000 tons of scrap
b. Residual scrap	210,000 tons of scrap
2. Distillation furnace	233,000 tons Zn recovered
15. Coal cleaning	
A. Thermal dryers	73,000,000 tons dried
16. Carbon black	
A. Channel process	71,000
B. Furnace process	
1. Gas	156,000
2. Oil	1,180,000

Emission factor lb/ton e_f	Efficiency[b] of control C_c	Application[a] of control C_t	Net[c] control C_c C_t	Emissions tons/yr E
32 lb/ton of material	0.97	0.99	0.96	161,000
8 lb/ton of material	0.97	0.99	0.96	40,000
4 lb/ton of asphalt	--	--	0.50	3,000
--	--	--	--	14,000*
	TOTAL from asphalt			218,000
410 lb/ton ferroalloy	0.99	1.00	0.99	1,000
240 lb/ton ferroalloy	0.80	0.50	0.40	150,000
10 lb/ton ferroalloy	0.90	0.35	0.32	9,000
TOTAL from ferroalloys				160,000
16 lb/ton of metal	0.80	0.33	0.27	105,000
5 lb/ton of metal	0.80	0.25	0.20	37,000
0.3 lb/ton of sand	0	0	0	1,000
TOTAL from iron foundries				143,000
275 lb/ton of wire	0	0	0	41,000
15 lb/ton of scrap	0.95	0.20	0.19	--
50 lb/ton of scrap	0.90	0.75	0.68	2,000
70 lb/ton of scrap	0.95	0.60	0.57	17,000
TOTAL from secondary copper				60,000
32 lb/ton of scrap	0.95	0.20	0.19	6,000
4 lb/ton of scrap	0.95	0.60	0.57	1,000
1,000 lb/ton Cl used	--	--	0.25	51,000
TOTAL from secondary aluminum				58,000
0.8 lb/ton of scrap	0.95	0.95	0.90	--
190 lb/ton of scrap	0.95	0.95	0.90	1,000
100 lb/ton of scrap	0.95	0.95	0.90	3,000
TOTAL from secondary lead				4,000
12 lb/ton of scrap	0.95	0.20	0.19	--
30 lb/ton of scrap	0.95	0.20	0.19	3,000
45 lb/ton of zinc	0.95	0.60	0.57	2,000
TOTAL from secondary zinc				5,000
TOTAL from secondary nonferrous metals				127,000
--	--	1.0	--	94,000*
2,300	0	0	0	82,000
--	--	1.00	--	5,000*
--	--	1.00	--	6,000*
	TOTAL from carbon black			93,000

TABLE 3 (Continued)

Source	Annual tonnage P
17. Petroleum	
A. FCC units	1.19×10^9 bbl of feed
18. Acids	
A. Sulfuric	
1. New acid	
a. Chamber	1,000,000 tons of 100% H_2SO_4
b. Contact	27,000,000 tons of 100% H_2SO_4
2. Spent-acid concentrators	11,200,000 tons of spent acid
B. Phosphoric	
1. Thermal process	1,020,000 tons of P_2O_5

* See specific industry section of Volume I for method of calculating quantity emitted.

a Application of control is defined as that fraction of the total production which has controls.

b Efficiency of control is defined as the average fractional efficiency of the control equipment, prorated on the basis of production capacity.

c Net control is defined as the overall level of control, and is the product of the application of control multiplied by the efficiency of control.

d,e Average ash content of coal used, determined by phone survey:

Type boiler	(d) Elec. util.	(e) Industrial
Pulverized	11.9%	10.6%
Stoker	11.2%	10.2%
Cyclone	11.8%	10.3%

Emission factor lb/ton e_f	Efficiency[b] of control C_c	Application[a] of control C_t	Net[c] control C_c C_t	Emissions tons/yr E
--	--	1.0	--	45,000*
5 lb/ton of 100% H_2SO_4	--	0	0	2,000
2 lb/ton of 100% H_2SO_4	0.95	0.90	0.85	4,000
30 lb/ton of spent acid	0.95	0.85	0.80	8,000
134 lb/ton of P_2O_5	0.97	1.0	0.97	2,000
	TOTAL from acids			16,000
TOTAL FROM MAJOR INDUSTRIAL SOURCES				18,056,000

TABLE 4

Selected Effluent Characteristics of Particulate Pollutant Sources[a]

Source	Particulate[b]			Carrier gas[c]	
	Particle size	Outlet grain loading	Chemical composition	Carrier-gas flow rate	Chemical composition
I. Electric utility power plants					
a. Coal fired	Typical distribution (pulverized unit): 81<40, 65<20 42<10, 25<5 (particle size varies with type of unit)	0.2-5.6 (Dependent on type of unit)	Fly ash: SiO_2: 17-64 Fe_2O_3: 2-36 Al_2O_3: 9-58 CaO: 0.1-22 MgO: 0.1-5 Na_2O: 0.2-4	a. 44-558 b. 351-643	CO_2, O_2, N_2, SO_2 SO_3 and NO_x
b. Oil fired	90<1	0.01-0.2	Carbon, ash, NiO, V_2O_3, Al_2O_3, sulfates, and a wide variety of minor components	a. 47-12,400 b. 348-780	CO_2, O_2, N_2, SO_2 SO_3 and NO_x
II. Kilns					
a. Cement					
1. Rotary	Mass median: 8.5 σ = 4.1	Dry process 1-17[d]	CaO: 39-50 SiO_2: 9-19 Fe_2O_3: 2-11	Dry process a. 34-300[d] b. 94-614[d]	Typical analysis: CO_2: 17-25 O_2: 1-4

		Wet process	Al_2O_3: 2-8	Wet process	CO: 0-2
		1-14[d]	K_2O + Na_2O: 0.9-8	a. 72-445[d]	N_2: 75-80
			MgO: 1.3-2.5	b. 166-570[d]	
2. Vertical		0.8[d]		a. 73.6[d]	
b. Lime					
1. Rotary	Mass median:44 σ = 13.7	2-223	$CaCO_3$: 23-61 CaO: 6-66 Na_2CO_3: 1-4 $MgCO_3$: 1-19 Fe_2O_3: 3 Al_2O_3: 3	a. 57-204[d]	CO_2, O_2, N_2, H_2O, SO_2
2. Vertical	10<10, 50<30 (one kiln)	0.3-1.0		a. 33.5[d]	
c. Bauxite (rotary)	25-40<10	2.2[d]	Al_2O_3<99 SiO_2: 0.005-0.015 Fe_2O_3: 0.005-0.02 Na_2O: 0.04-0.80	a. 29.5[d]	CO_2, O_2, N_2, H_2O, SO_2
d. Magnesite (rotary)	50<10 (one kiln)				
e. Zinc ore (rotary)		5.7-17.3[d]		a. 4.7-7.4[d]	
f. Nickel ore (rotary)		12		a. 26.5[d]	
III. Dryers					
a. Rotary					
1. Grain			Chaff		

TABLE 4 (continued)

Source	Particulate[b]			Carrier gas[c]	
	Particle size	Outlet grain loading	Chemical composition	Carrier-gas flow rate	Chemical composition
2. Cement	40-70<10	13-40		b. 26-64[d]	
3. Phosphate rock		7.63	Typical Composition P_2O_5: 32.5 SiO_2: 11.0 Al_2O_3: 2.0 MgO: 0.7 CaO: 45.5 Fe_2O_3: 0.8		CO_2, O_2, N_2, H_2O, SO_2
4. Titanium dioxide	0.5-1	1-5			
5. Ammonium nitrate	Large, unstable agglomerates				
6. Fertilizer superphosphate	Mass median:85 σ = 6.3	0.7-4.0	Fertilizer, fluorides Ca, Mg, P, Fe, and Al compounds	a. 16.5 (one unit)	Fluorides, NH_3, SO_x, CO_2, N_2, O_2
7. Asphalt stone (Hot mix paving plant)	Mass median:17.8 σ = 5.1	20-70	Stone dust, fly ash, soot, unburned oil	a. 7.7-46 b. 3.9-24.6	CO_2, NO_x, N_2, O_2, CO, SO_2
8. Bauxite		20-40		a. 11-150[d]	
9. Fuller's earth		1.9		a. 17.6[d]	

10. Limestone	Mass median:7 σ = 9.7	4.3-11		a. 14.7[d]	
11. Dolomite		6.1-30.4		b. 36-43[d]	
12. Clay				a. 23.8	
13. Coal			Coal, dust, fly ash	a. 20-135	CO_2, N_2, O_2, CO, SO_2
b. Spray					
1. Detergent	50<40	3			
2. Fertilizer					
a. Urea					
IV. Sinter machines					
a. Iron ore	Mass median:100 σ = 5.4	0.2-5.0	Fe_2O_3: 45-50 SiO_2: 3-15 CaO: 7-25 MgO: 1-10 Al_2O_3: 2-8 C: 0.5-5 S: 0-2.5 Fluorides	a. 30-460 b. 148-230[d]	O_2: 10-20 CO_2: 4-10 CO: 0-6 SO_2: 0-0.4 N_2: 64-86 Fluorides
b. Lead ore		0.87-6.6	Pb: 40-65 Zn: 10-20 S: 8-12	a. 140 b. 130 (one plant)	O_2, N_2, CO_2, CO, H_2O, SO_2
c. Zinc ore	100<10	0.44-5	Zn: 5-18 Pb: 45-55 Cd: 2-8 S: 8-13	b. 140	O_2, N_2, CO_2, CO, H_2O, SO_2

TABLE 4 (continued)

Source	Particulate[b]			Carrier gas[c]	
	Particle size	Outlet grain loading	Chemical composition	Carrier-gas flow rate	Chemical composition
V. Roaster					
a. Copper ore		6.6	Cu: 9 S: 10 Fe: 26	a. 4.6 b. 47.3	O_2, N_2, CO_2, SO_2,
b. Zinc ore					
1. Ropp				a. 25-30	O_2, N_2, CO_2, SO_2, H_2O
2. Multiple hearth				a. 5-6	O_2, N_2, CO_2, SO_2, H_2O
3. Suspension				a. 10-15	O_2, N_2, CO_2, SO_2, H_2O
4. Fluid-bed				a. 6-10	O_2, N_2, CO_2, SO_2, H_2O
c. Pyrites		0.5-1.0[d]		a. 6.5-7.7[d]	
VI. Furnaces					
a. Blast					
1. Iron ore	15-90<74	4-30	Fe: 36-50 FeO: 12-47 SiO_2: 8-30 Al_2O_3: 2-15 MgO: 0.2-5 C: 3.5-15 CaO: 3.8-28 Mn: 0.5-1.0 P: 0.03-0.2 S: 0.2-0.4	a. 40-140[d] b. 60-138	CO: 21-42 Av 26 CO_2: 7-19 Av 18 H_2: 1.7-5.7 Av 3.1 CH_4: 0.2-2.3 N_2: 50-60

2. Lead	0.03-0.3	2-6.6	PbO, ZnO CdO, Pb_3O_4 Coke dust	a. 6-190 b. 180	Typical analysis: CO_2: 4.7 O_2: 15 CO: 1.3 SO_2: 0.14 N_2: balance
3. Copper		6.6	Cu: 4.4 Zn: 12.5 S: 7.3	a. 21.2 b. 76.5	
4. Secondary lead		2-12		a. 2.1	CO_2, O_2, CO, N_2, SO_2
5. Ferro manganese	80<1	4.5-17	Mn: 15-25 Fe: 0.3-0.5 $Na_2O + K_2O$: 8-5 SiO_2: 9-19 Al_2O_3: 3-11 CaO: 8-15 MgO: 4-6 S: 5-7 C: 1-2	a. 60-135	CO, CO_2, N_2, H_2, SO_2
6. Tin ore		2.1-3.0[d]		a. 2.1-5.7[d]	
7. Antimony ore		1.6[d]		a. 3.7[d]	
b. Open hearth					
1. Steel (no oxygen lance)	50<1	0.1-3.5	Fe_2O_3: 85-90 Also SiO_2, Al_2O_3, CaO, MnO, and S	a. 25-100	CO_2: 8-9 O_2: 8-9 N_2: balance SO_2, SO_3 and NO_x in small amounts

TABLE 4 (continued)

Source	Particulate[b]			Carrier gas[c]	
	Particle size	Outlet grain loading	Chemical composition	Carrier-gas flow rate	Chemical composition
2. Steel (oxygen lance)	69<10	0.2-7.0	Similar to no oxygen lance	a. 45-200	Similar to no oxygen lance
c. Basic oxygen (steel)	85-95<1	2-10	Fe_2O_3: 90 FeO: 1.5 Also SiO_2, Al_2O_3, and MgO	a. 35-250[d]	CO_2, CO, N^2, O_2
d. Electric-arc					
1. Steel (no oxygen lance)	84<10	0.1-2.2	Fe_2O_3: 19-44 FeO: 4-10 Cr_2O_3: 0-12 SiO_2: 2-9 Al_2O_3: 1-13 CaO: 5-22 MgO: 2-15 ZnO: 0-44 MnO: 3-12 Also CuO, NiO, PbO, and C	a. 10-100	CO_2, CO, O_2, N_2
2. Steel (oxygen lance)		1-10	Similar to no oxygen lance	a. 10-100	CO_2, CO, O_2, N_2
3. Ferroalloy	0.01-4	0.2-30	SiO_2, FeO MgO, CaO	a. 11-60	CO_2, CO, H_2, N_2, CH_4

			MnO, Al_2O_3 and K_2O		
4. Iron Foundry					
e. Reverberatory					
1. Copper		1-5	Cu, Zn, and S compounds	a. 25-60 b. 71	O_2: 5-6 CO_2: 10-17 N_2: 72-76 CO: 0-0.2 SO_2: 1-2
2. Lead		0.14-4.4		a. 1-3 b. 3.5-17.5	
3. Brass	0.05-0.5	1-8		a. 5.6-17.6	
4. Secondary	0.07-0.4	1-6	PbO, SnO, and ZnO	a. 3.1	
5. Secondary zinc		0.2-1	$ZnCl_2$, ZnO NH_4Cl, Al_2O_3, Oxides of Mg Sn, Ni, Si, Ca, and Na	a. 7-8 b. 400-800	Typical: CO_2: 2.4 H_2O: 4.5 N_2: 76.6 O_2: 15.8
6. Secondary aluminum	31<10	0.12-0.6		a. 1-9 b. 0.1	Typical: CO_2: 6.8 O_2: 8.6 CO: 0.02 N_2: 77.3 H_2O: 7.3 SO_2: trace

TABLE 4 (continued)

Source	Particulate[b]			Carrier gas[c]	
	Particle size	Outlet grain loading	Chemical composition	Carrier-gas flow rate	Chemical composition
VII. Crushing, grinding and milling					
a. Crushing iron ore	0.5-100	5-25			Air
b. Cement mill		22.3		a. 14.1[d]	Air
c. Limestone crusher		1.1-1.3	Limestone	a. 18.2	Air
d. Raymond mill (limestone)	Mass median:4.3 $\sigma = 1.6$		Limestone		Air
e. Hammer mill (limestone)	Mass median:6.0 $\sigma = 8.8$		Limestone		Air
f. Pulverizer (hydrated lime)	Mass median:4.9 $\sigma = 2.1$		Lime		Air
g. Jaw crusher (stone	5<5, 10<10 20<30		Stonedust		Air
VIII. Miscellaneous chemical					
a. Absorption towers					
1. Sulfuric acid	8-90<3	0.017-0.76	Acid mist	a. 5-62 b. 35-160	NO_x, SO_2, O_2, N_2, H_2O

(contact plant)					
2. Phosphoric acid (thermal)	Mass median:1.6	1.6-93	Acid mist	a. 3.4-30.2 b. 35-160	O_2, N_2, NO_x, H_2O
b. Lime hydrator	Mass median:2.6 σ = 8.5		Lime		Air, H_2O
c. Fertilizer (Superphosphate den and mixer)		0.09-0.15	H_2SiF_6, SiO_2		SiF_4, CO_2, H_2O,

[a] All data for uncontrolled sources unless otherwise noted.

[b] Particulates

1. Particle size: Data are presented as mass median (in μm) and geometric deviation (σ) where available, or as weight per cent less or greater than a specific diameter:

   $x < y$, $x < y$

   x = %, y = particle size (μm)
2. Outlet grain loading: grains/scf, unless otherwise noted.
3. Chemical composition: wt % (unless otherwise noted).

[c] Carrier gas

1. Flow rate: Data presented in two forms:
   a. thousands of standard cubic feet per minute, M scfm, unless otherwise noted
   b. thousands of standard cubic feet per ton of product processed, M scf/ton, unless otherwise noted
2. Chemical composition: vol %, unless otherwise noted.

[d] Actual cubic feet (ACF).

by number, are smaller than 0.1 μm in diameter and most of the particle surface area is in the light-scattering range from 0.1 to 1 μm in diameter. The mass distribution for the urban regions that have been studied show two peaks in that about half of the mass is in the 0.1- to 2-μm range while the rest is in the 2- to 30-μm range. Also, a substantial fraction of the mass of smaller particles may be liquid. Figure 1 presents particulate distribution data for Los Angeles smog. Data from Colorado and foreign areas (17) substantiate the twin-peak form of the mass and volume distribution.

Fine particles with aerodynamic diameters smaller than 2 μm have a number of important characteristics that distinguish their behavior from larger particles. Fine particles compared to large particles:

1. Scatter more light per unit mass;
2. Have greater penetration through gas cleaners;
3. Penetrate more deeply into the respiratory system but may be largely exhaled when they are very small
4. Remain airborne in the atmosphere for longer period;
5. Account for a greater number of particles in urban air pollution;
6. Are usually generated by thermal or chemical processes.

Fine particles typically undergo significant transformations between the time they are created and the time that they are collected in control equipment or emitted into the atmosphere. Such transformations include diffusion and coagulation, reactions with gases, growth by vapor condensation, evaporation, and adsorption or desorption of water.

Although generation mechanisms are not completely understood, it appears that most fumes and smokes from flames or other high-temperature processes originate from self-

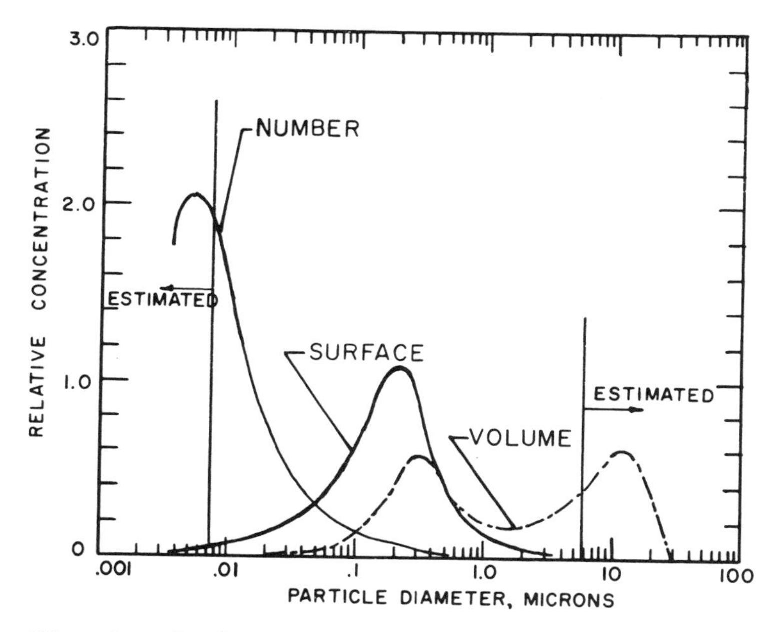

Fig. 1. Typical particulate distribution (17).

nucleation of highly supersaturated vapors just after they have left a flame front or other hot zone (12). Typically, particles formed in this way have an initial size of 0.005 to 0.05 μm and an initial number concentration of 10^8 to $10^{12}/cm^3$. At these concentrations, the aggregation rate is very high and the number concentration decreases one order of magnitude in about 1 msec. At the time of emission to the atmosphere, particles consist of hundreds of thousands (10^5 to 10^6) of 0.005- to 0.05-μm-diameter primary particle aggregates per cm^3 having a mean size of about 0.2 μm. The particle size at discharge depends on the initial particle-size concentration, temperature, retention time, and the number of primary particles that aggregate. Very small primary particle aggregates may also grow by condensation of higher vapor pressure liquids such as hydrocarbons and water (when they are present) during their transport in the ducts and stacks before emission to the atmosphere. Thus, at the point of emission into the atmosphere, fine particles may consist of aggregates of primary particles plus condensed vapors.

Primary particles are produced by:

1. Condensation. Many chemical reactions produce vapors at elevated temperatures. Heat loss by several mechanisms produces supersaturation, which results in nucleation of solid or liquid particles that are emitted from a source.

2. Chemical reactions. Particles resulting from combustion, as well as noncombustion aerosols such as sulfuric acid mist, are the indirect result of chemical reactions.

3. Comminution or powder handling. This is a prime source of large particles but, when they are reduced in size or handled, fractures of these particles produce fragments and some of these may be in the fine particle range. Soils and also sands contain fine particles which may become suspended in the atmosphere.

Secondary particles are those formed after leaving the stack. They can be produced by:

1. Condensation of vapors in the plume. The most common particles, water and hydrocarbon drops, often form at an early stage in the plume and then evaporate as the plume mixes further with the ambient air. Exceptions are sulfuric acid mist and some chemical compounds that are sufficiently hygroscopic to hold a substantial amount of water at ambient humidities or of low enough vapor pressure so that evaporation is very slow.

2. A variety of reactions, chemical and photochemical, occur outside of the stack and result in nonvolatile products that form aerosols. Sulfate-containing particles resulting from the chemical oxidation of SO_2 and combination with other compounds such as NH_3 are an example. Los Angeles-type smog is also the result of photochemical reactions of emissions (18).

3. Evaporation of droplets containing dissolved solids or nonvolatile liquids to form residue particles. Examples are sea salt particles from bubbles bursting at the ocean surface and the residues from droplets and bubbles formed in cooling towers.

There are relatively few data on size distribution and concentration of fine particles from various sources. Because most sampling and evaluation in the past has been directed at measuring mass, and since the fine particles usually do not contribute the majority of the mass, such measurements have yielded little data on fine particles.

C. TRACE ELEMENTS

Trace substances entering the environment may pose a health hazard. While the physiological effects of trace concentrations of specific substances is not known, it is a fact that significant amounts of potentially hazardous elements are emitted into the air from smelting, refining, incineration, and burning of coal and oil (19,20). Some metals that could pose health hazards in the environment and their sources and effects are given in Table 5. The concentrations are low and are not generally detected and reported in measurements such as those used to make up Table 4. The general approach of reducing and controlling emissions to meet the national ambient air standards should have the effect of indirectly reducing the potential hazard from trace substances, at least until more health information becomes available.

IV. SAMPLING AND MEASUREMENT

Particulate emission rates may be estimated from what is generally known about the source or they may be obtained

TABLE 5

Trace Metals in the Environment (19)

Element	Source	Health effects
Nickel	Diesel oil, residual oil, coal, tobacco smoke, chemicals and catalyst, steel and nonferrous alloys	Lung cancer (as carbonyl)
Beryllium	Coal, industry (new uses proposed in nuclear power industry, as rocket fuel)	Acute and chronic system poison, cancer
Boron	Coal, cleaning agents, medicinals, glass making, other industrial	Nontoxic except as boran
Germanium	Coal	Little innate toxicity
Arsenic	Coal, petroleum, detergents, pesticides, mine tailings	Hazard disputed may cause cancer
Selenium	Coal, sulfur, zinc ore	May cause dental caries, carcinogenic in rats, essential to mammals in low doses
Yitrium	Coal, petroleum	Carcinogenic in mice over long-term exposure
Mercury	Coal, electrical batteries, smelters and incinerators	Nerve damage and death
Vanadium	Petroleum (Venezuela, Iran), chemicals and catalysts, steel and nonferrous alloys	Probably no hazard at current levels
Cadmium	Coal, zinc mining, water mains and pipes, tobacco smoke	Cardiovascular disease and hypertension in humans suspected, interferes with zinc and copper metabolism
Antimony	Industry	Shortened life span in rats
Lead	Auto exhaust (from gasoline), paints (prior to about 1948)	Brain damage, convulsions, behavioral disorders, death

by direct testing of the source. As discussed in the previous section, a number of sources have been tested in the past and emission factors developed to estimate rates from production and sales data (see Tables 3 and 4).

Material balances across the process require sampling and testing of input and output streams but provide the specific information that is frequently needed. Detailed source testing and monitoring is at present only practical for large individual sources such as power stations and manufacturing plants (21). Automobiles, residential heating units, and other small mass produced equipment must generally be sampled on a statistical basis.

Gas temperature, static pressure, humidity, composition, and flow rate must be known for valid sampling. Obtaining a representative sample is a major concern. Ports for obtaining flue gas samples and flow measurements should be located in regions of stable flow. Particle samplers should be placed in vertical ducts, since particles tend to settle over long horizontal runs. Ideally, ports should be located 8 to 10 diam downstream and 3 to 5 diam upstream from bends or constrictions. If this is not possible, straightening vanes to reduce turbulence can be used. In sampling particulates, care must be taken to assure that the sample stream enters the probe at the main stream velocity, [isokinetic sampling (8,9)], so that a representative sample is obtained.

Size spectra, total particle number, and mass can be measured on a continuous, close to real-time basis using a variety of devices based on different physical principles (12):

1. Several optical particle counters are commercially available which, when properly calibrated and maintained, can be used to measure individual particles from 0.2-0.3 μm in diameter. When existing sensors are combined with

a multichannel analyzer, detailed size distributions can be obtained on a continuous, on-line basis.

2. The electrical mobility analyzer can be used in the range below a few tenths of a micron to give mass as a function of particle size with a response time of a few minutes.

3. Particle concentrations in terms of total number can be monitored continuously using condensation nuclei counters which resemble automated Wilson cloud chambers intermittently.

4. Particle mass can be monitored continuously, using a vibrating quartz crystal on which the particles are deposited by electrical precipitation or impaction.

Currently, however, there is no proven commercial instrument that can measure any of the chemical constituents of particulate pollution on a continuous, on-line basis. Commercially available instruments or techniques are of the laboratory type and are not applicable for reliable, long-term untended field operation. Such constituents as lead, sulfate ion, carcinogens, and many others are determined by collecting samples of particulate matter over periods of up to many hours and then subjecting the accumulated material to chemical analysis.

A. GAS MEASUREMENT

The gas flow rate can be found either by direct measurement in a duct, pipe or chimney, or by calculation from the quantity and composition of the materials processes. Gas flows are usually measured with differential head devices such as orifice plates, venturi meters and pitot tubes in industrial cases, and by vane anemometers for nontoxic gases at ambient conditions.

Orifice and venturi meters measure average or total flows and are usually permanently installed. Details of construction, limitations, and calibration for standard sizes are covered in handbooks and other reference material (10). Pitot tubes are more versatile and measure local velocities. To find total or average gas flow, a number of measurements must be made across the duct to obtain the velocity profile. The profile is then integrated over the area either numerically or graphically. Fluctuations in gas flow may induce errors so that several traverses are often required.

The S-type pitot tube is commonly used (see Fig. 2). It consists of two parallel tubes with openings in opposing directions at the end that is inserted into the duct. It is placed perpendicular to the stream so that one opening is pointed directly into the stream flow, allowing the measurement of static and impact pressure. Flow rate is calculated from the pressure difference.

The plugging characteristic in the standard pitot tube as shown in Fig. 2, can be overcome with the use of an electrically controlled pneumatic system designed for purging the tube when flow rate is being determined in a duct heavily loaded with liquid droplets and/or solids. A three-position manual switch actuates solenoid valves so that in the "read" position the pitot tube is connected to the manometer, in the "vent" position to the atmosphere, and in the "purge" position to a compressed air supply (9).

For most applications the velocity, u, measured by the pitot tube may be calculated from the pressure drop, Δp, using the relation

$$u = \frac{2g_c \Delta p}{\rho} \quad (1)$$

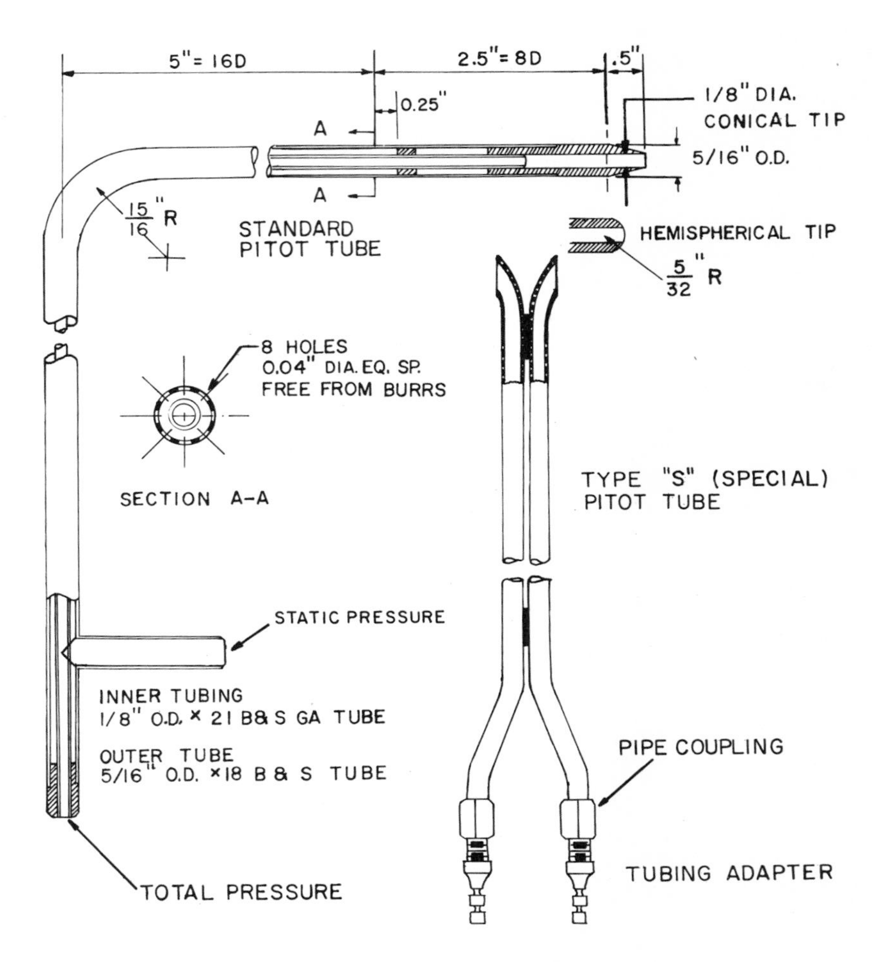

Fig. 2. Standard pitot tube (9).

Corrections for high-velocity flow and size of the impact area are available (10,11).

A knowledge of the gas temperature is essential to the thermocouples, and pyrometers are generally used for gas temperature measurement (8,10). Occasionally the gases are not at the temperature of the walls of the duct. When the gas is colder or hotter than the duct a correction must be made for radiation and conduction to the sensing element to ensure that the actual gas temperature is determined (8). In these cases there may also be a significant variation in gas temperature across the duct so that it is necessary to integrate over the temperature and velocity profiles to obtain the bulk or cup-mixing temperature.

The chemical composition of the stack gas frequently influences the choice of air pollution control equipment. Composition may be measured by chromatographic, spectroscopic, absorption, and adsorption techniques (5,8,10). Properties such as molecular weight, density, and viscosity are functions of gas composition and are important in the design and performance of control equipment.

B. SAMPLING

Methods of in-stack monitoring of pollutants usually contain the following components in the sample flow system:

1. A sampling nozzle,
2. A probe for extending the nozzle into the stack,
3. A particle collector,
4. A cooling section or condenser to remove excess moisture,
5. A gas flow measuring device,
6. A gas pump,
7. A gas temperature and pressure measuring device,
8. A flow-regulating device.

Table 6 and Fig. 3 summarize the major features of several common stack sampling trains. The differences among the systems give rise to problems in obtaining comparative results (12). For example, the line loss in the probes before the collector in the EPA and IIA trains will be greater than the loss in the in-stack collector used in the ASTM train. Because of condensation, the material collected in

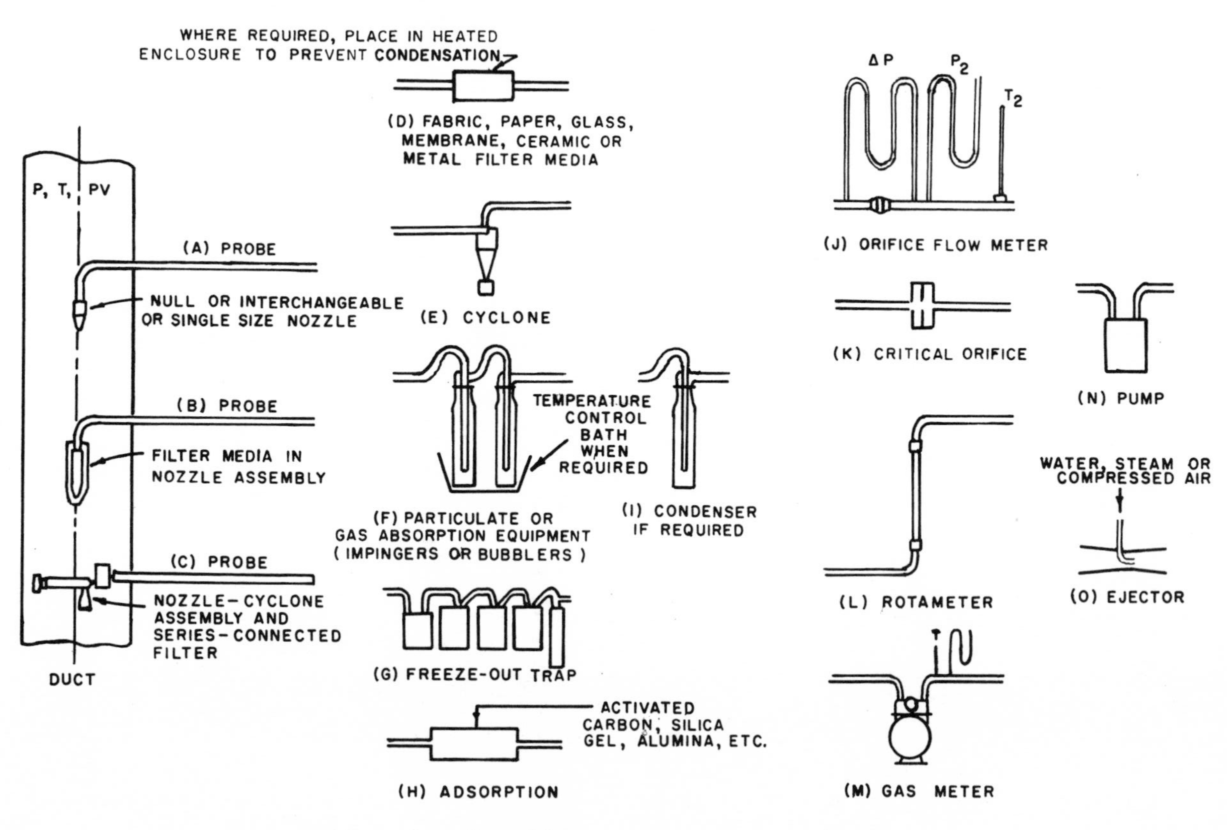

Fig. 3. Sampling system components.

TABLE 6

Major Features Of Sampling Trains

Sampling element	Environmental Protection Agency (5)	American Society of Mechanical Engineers PTC-27	American Society for Testing & Materials D2928-71	Incinerator Institute of America T-6
Nozzle	Gooseneck	Gooseneck or elbow	Gooseneck or elbow	Elbow
Probe	Glass-lined and heated	Stainless steel	Stainless steel	Stainless steel
Collector	Cyclone, glass fiber filter and impingers	Any at 99% efficiency for particles	Filter or thimble in stack	Cyclone and bag filter
Cooling section	Wet impingers	Condenser	Condenser	Condenser
Flow measurement and control	Pitot tube and nomograph	Totalizing gas meter	Totalizing gas meter or flowmeter	Null nozzle
Gas mover	Vane pump	Vane pump	Vane pump	Vacuum blower

the trains where the collector is cooled may be different in composition as well as in amount from that collected in a hot collector.

Comparative measurements made with the ASME and EPA trains on high performance fly ash electrostatic precipitators are shown in Table 7. The two systems give significantly different results; with the full EPA train indicating up to almost four times the outlet loading measured by the ASME train and the EPA "front-half" being twice the ASME for several runs (12). (The fly ash not collected is the quantity of interest in connection with air pollution.)

TABLE 7
High Performance Coal Fly Ash Electrostatic Precipitators
Measurements with Different Sampling Trains (12)

Item	Plant	Unit	Plant A	Plant B	Plant C	Plant D
Coal	Sulfur	%	1.5	2.49	2.71	3.7
	Ash	%	15.7	13.13	10.63	17.64
	Btu/lb	(As rec.)	11,800	12,566	10,349	13,837
	Elec. load	M.W.	108	519	357	540
Precipitator	Gas flow	acfm	497,100	1,515,000	1,239,000	1,630,000
	Plate area	ft^2	120,000	270,400	200,340	276,480
	A/V	ft^2/1000 cfm	248	178	162	170
	Temp.	F°	676	280	280	252
Performance[a]						
ASME train[b]	Efficiency	%	98.9	---	99.3	98.8
	Outlet	grain/scfm	0.0652	---	0.0263	0.060
	loading	lb/10^6 Btu	0.120	---	0.0597	0.124
1/2 EPA train	Efficiency	%	97.0	98.0	98.4	98.8
	Outlet	grain/scf	0.1784	0.0828	0.059	0.0615
	loading	lb/10^6 Btu	0.329	0.168	0.133	0.128
Full EPA train	Efficiency	%	96.5	96.3	97.1	97.9
	Outlet	grain/scfm	0.212	0.153	0.103	0.1025
	loading	lb/10^6 Btu	0.39	0.31	0.234	0.213

[a]Efficiency calculated from estimated ash carryover at 80% of calculated total ash. Precipitator inlet gas loading not sampled.
[b]Impinger catch excluded

Cascade impactors have been used for in situ size fractionation of stack samples (13). Satisfactory size fractionation from approximately 1 μm upwards is reported, but some dispersion of large flocs occurs. The impactor can sample the stack atmosphere directly thereby minimizing the potential detrimental effects of transport and storage. A sample large enough for chemical analysis is obtained if the chemical method is capable of great sensitivity.

Radiation interference (visible, optical, and beta) has been used for analysis of particulate loadings in emission sources. Qualitative stack effluent monitoring is often done visually using Ringelmann numbers of equivalent opacity and is part of many local codes and regulations. Optical methods based on light scattering are used for continuous monitoring of stack gases in many industries and total mass levels are frequently measured using an ASTM thimble (see Fig. 4).

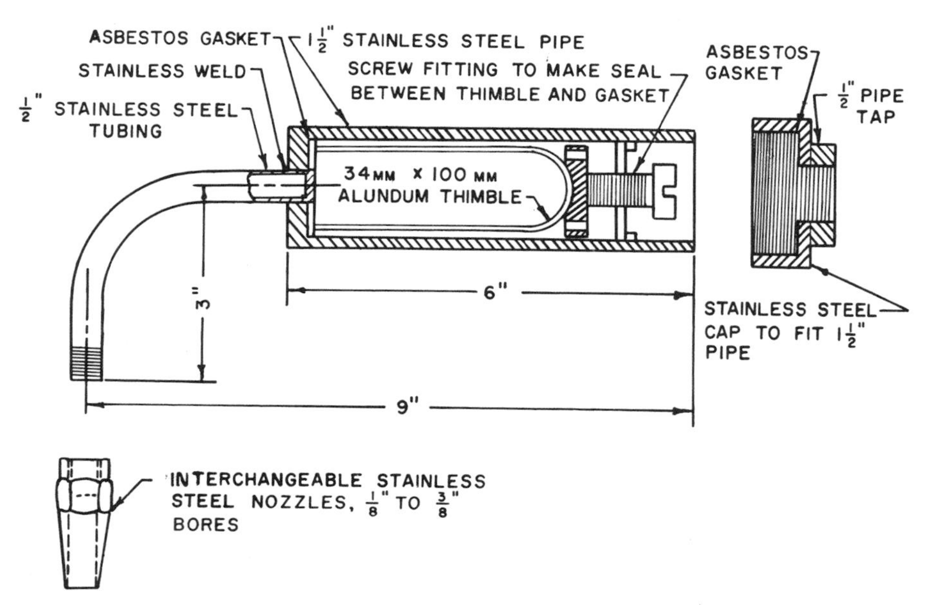

Fig. 4. Nozzle, short probe, and thimble holder (9).

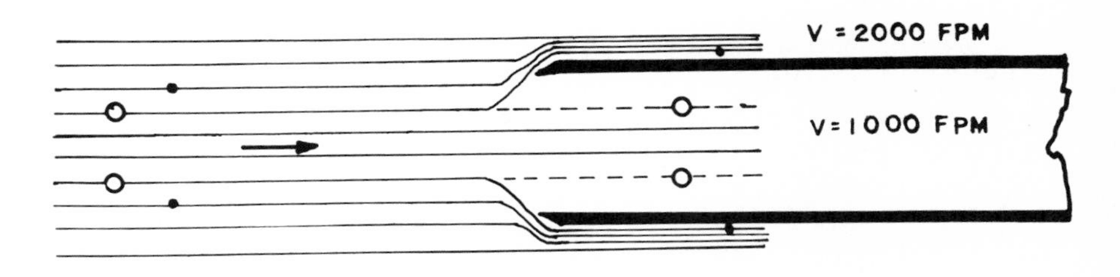

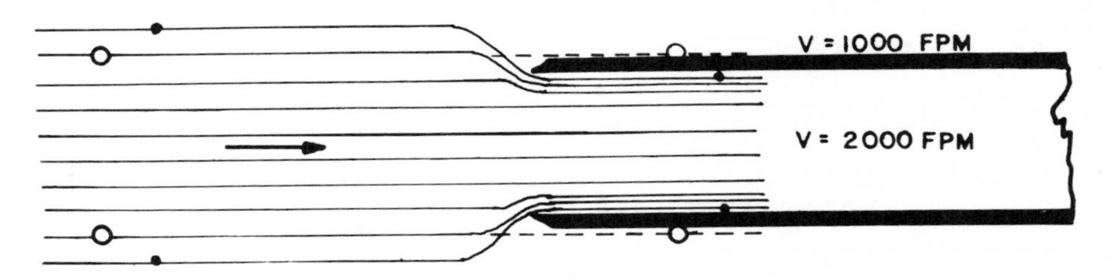

Fig. 5. Inertial effects caused by nonisokinetic sampling conditions result in sampling errors.

The fundamental requirement of a sampling system is that it collect a representative sample. The inertial effects caused by nonisokinetic sampling can be fairly well defined (see Fig. 5). Figure 6 illustrates the considerable error which can be caused by nonisokinetic sampling.

Another method of maintaining isokinetic conditions uses a series of nozzles of different diameters is shown in Fig. 4. The sampling rate is kept constant and the nozzle is selected on the basis of its inlet diameter so as to match the duct velocity. The technique is particularly advantageous when the collection device exhibits a constant pressure drop such as the impinger or the cyclone (without after-filter) and when only one pump calibration

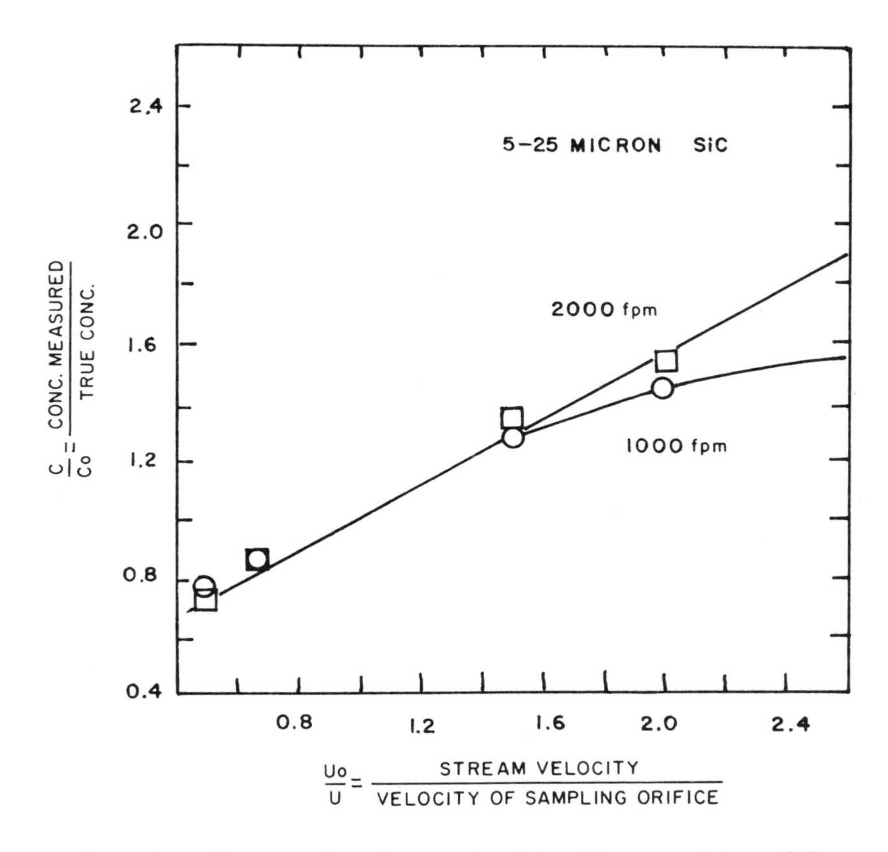

Fig. 6. Errors due to nonisokinetic sampling (9).

point, predetermined in the laboratory, is used in the field for maintaining a constant flow rate. The nozzles are changed to suit the velocity at the sampling point, nozzle size being a function of gas densities in the stack and at the flow metering device. Therefore, a means for rapid choice of nozzle diameter in the field must be provided (see Fig. 7).

A number of techniques is employed in ambient air sampling to obtain samples of particulates for direct or

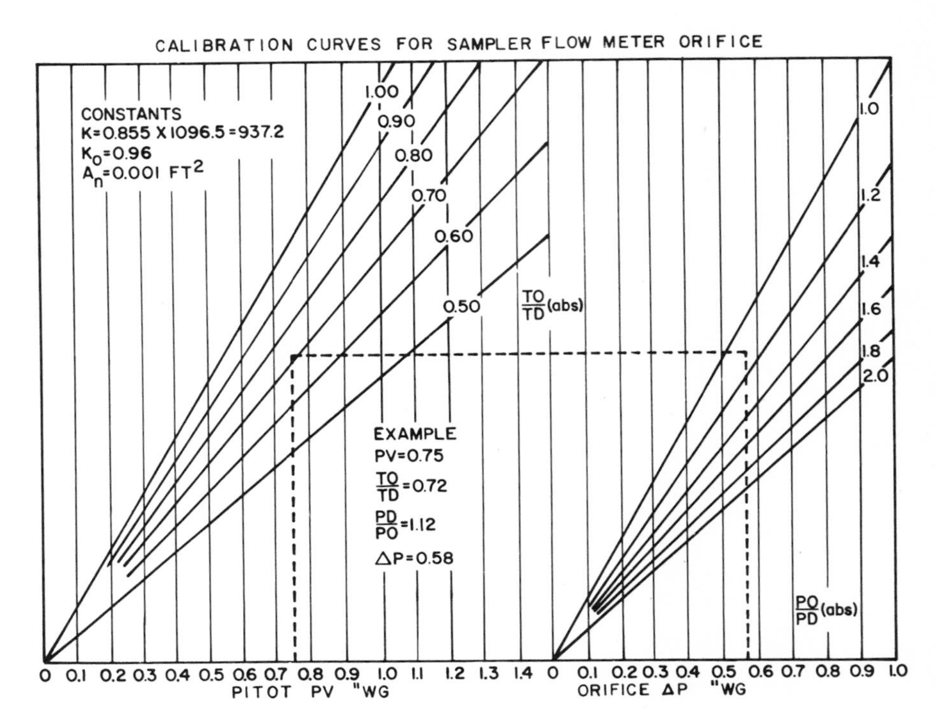

Fig. 7. Graphic aid for determining differential pressure setting in isokinetic sampling (Gas Cleaning Institute, Rye, N.Y.).

subsequent analyses and may be applicable to stack monitoring in certain cases. Among these methods are included electrostatic and thermal precipitation, and centrifugal methods, as characterized by the cascade impactor and the conical or cylindrical collector designed by Goetz. Two methods widely employed are the tape sampler and high-volume sampler. Both of these employ filter paper collectors through which air is drawn for a selected time period (1-24 hr). Estimates of the particulate concentration can be made directly on the tape sampler sections using either transmittance or reflective measurements. Transmittance measurements are usually reported in terms of Coh's (coefficient of haze) per thousand linear feet of air passing through the filter, while reflectance is reported in RUDS (reflectance unit of dirt shade) per 1000 cubic feet of air. Both of these measurements are qualitative and can usually only be used for comparative purposes.

C. PARTICLE SIZE

Information on particle size distribution in the gas stream is important in the selection of gas cleaning equipment. Particles larger than about 10 μm may be removed in inertial and cyclone separators and simple, low-energy wet scrubbers. Particles smaller than 10 μm require either high-efficiency (high-energy) wet scrubbers, fabric filters, or electrostatic precipitators.

Particle size cannot, in general, be uniquely specified by a single parameter. For irregular particles the average dimension along the three axis may be used. However, in many gas cleaning processes the hydrodynamic diameter obtained from the settling velocity has direct physical meaning, independent of particle structure, and is preferred to equivalent diameter (see Chap. 2).

Determination of particle size cannot be unique except for the special case of spherical particles. For all other cases, the results depend on the experimental method used. Sieve analysis may be used for relatively coarse particles above 44 μm, corresponding to a 325-mesh screen. Optical microscopes permit determination of particle size down to about 0.5 μm. The greater resolving power of electron microscopes extends this range down to approximately 0.05 μm. Photometric methods based on scattering or absorption of light have a lower limit of about 0.3 μm for monodisperse or uniform particle size distributions.

Inertial methods, such as sedimentation and centrifugation, are widely used. Sedimentation analysis using the settling velocity is useful for particles in the range of Stokes' diameters between 0.5 and 50 μm. Centrifugal analysis can extend the lower size limit to about 0.01 μm. Elutriation analysis separates the particles in vertically flowing fluids. Fine particles above a certain size cutoff point are carried upward with the rising fluid, and coarser particles below the cutoff point fall to the bottom of the elutriation chamber. A series of graded elutriation chambers may be used to separate particles into a series of size classes. Particles may also be separated by impaction methods in which they are deposited on a surface by impaction from an air jet. A series of graded impactors in a cascade arrangement may be used to separate the particles into size classes.

Ultrafine particles can serve as condensation nuclei for water vapor. This fact is made use of in instruments where the aerosol particles are introduced into a McLeod chamber saturated with water vapor. Sudden adiabatic expansion of the chamber causes formation of a water droplet around each aerosol particle. Light scattering techniques may be used to determine the concentration of the larger water droplets formed which relates directly to the original aerosol concentration.

Advantage can be taken of the increased diffusivity of small particles to effect a separation of sizes and an estimation of particle size. This is accomplished by the passage of the gas sample at low velocities through long, narrow channels or bundles of capillary tubes, called a diffusion battery. The smaller particles will be removed first on the walls of the channels by diffusion action and can be counted by condensation nuclei techniques or with the use of the electron microscope.

The electric mobility of a charged aerosol particle in an electric field is inversely proportional to the separation voltage applied. Also, the particle size is simply related to the electric mobility if preclassification eliminates the larger particle sizes. Further, the concentration of a monodispersed, charged aerosol is directly proportional to the separation current, i.e., the current required to collect all of the aerosol in a measured volume. These relationships have been used in electric particle counting and size distribution systems for aerosols in the 0.015 to 1 μm size range (1). The equipment consists of an unipolar sonic jet diffusion charger, which initially charges the aerosol, and a separate mobility analyzer.

The Bahco centrifugal dust classifier is used extensively for routine measurements by control device manufacturing companies. This instrument uses a form of centrifugal elutriation. Dust is fed into an air stream in the annular space between parallel rotating plates and in each stage the dust is divided into two fractions, one deposited on the periphery of the wheel and the other carried forward. By varying the air velocity a number of fractions can be collected. Instruments of this type require careful adjustment to ensure a good separation (14).

For gas-cleaning applications, the sedimentation and elutriation methods have advantages because they give results in terms of settling velocity or Stokes' diameter. But these methods usually necessitate redispersion of collected particle samples, which may be difficult, and any agglomerates present in the original aerosol cannot be reproduced in the particle size equipment. Another problem connected with sizing of industrial aerosols is the difficulty of procuring representative samples from the field because of the very large gas flows and variable conditions that characterize most industrial gas-cleaning situations.

Generally, the purpose of a particle size measurement is to discover the frequency distribution of particle size (4). The observed distribution serves as basic data from which may be derived certain representative constants, for example, the median size. Modified relative frequency distributions can also be obtained by transformation; for example, percent by weight from percent by number.

Tabular and graphical forms can be used to present the data. A table can list size versus one of many ways of expressing distribution; for example, size frequency or size cumulation. It is essential to specify which weighting process is employed, since distributions are generally radically different (e.g., number-size and weight-size distributions) (15).

Graphical methods for presenting size distributions are: (1) histograms, (2) size frequency curves, and (3) cumulative plots. Cumulative plots are used extensively and their interpretation and comparison can be enhanced by using the generally applicable log-normal distribution plot or one of its modifications (15). Figure 8 illustrate a log-normal distribution plot for particulates emitted from a pulp mill recovery furnace.

Log-normal size distributions are defined by two parameters: (1) the intercept of the cumulative curve with the 50% probability mean diameter, and (2) the polydispersity factor (geometric deviation) defined as:

$$\text{Polydispersity factor} = \frac{\text{intercept at 50\% probability}}{\text{intercept at 15.87\% probability}}$$

A completely monodisperse aerosol has a polydispersity factor of one. Figure 8 shows that 50% of the outlet fume is composed of particles smaller than 1.4 μm in diameter. The polydispersity factor is 4.2.

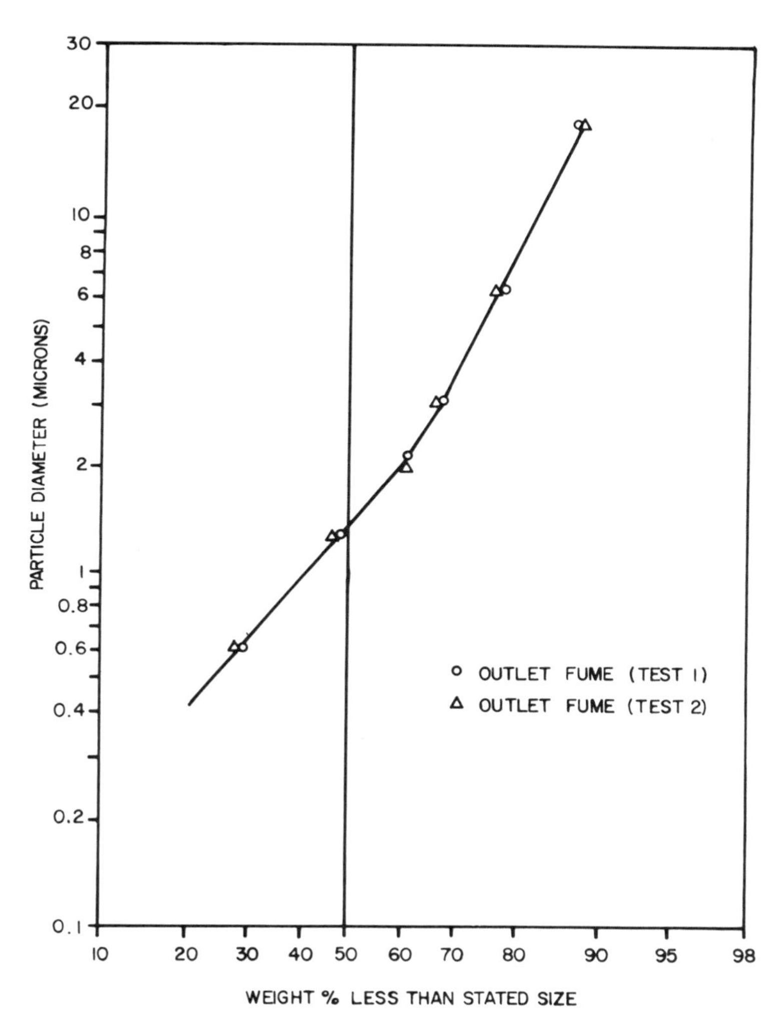

Fig. 8. Average size distributions of outlet fume kraft mill recovery furnace (4).

D. PARTICLE COMPOSITION AND SHAPE

Particle shape and surface condition influence handling characteristics, chemical reactivity, adsorption potential, and flammability limits among other particulate properties. Particle shapes of aerosols are of many types, from simple spheres to complex stars and chainlike aggregates. Fogs, mists, and some smokes are composed of spherical liquid or tarry droplets. Many fly ash particles, produced in the combustion of pulverized coal, are hollow spheres, frequently with much smaller satellite particles attached to their surfaces. Dust particles are usually irregular in shape as the result of multiple fractures that occur in crushing or grinding. Particules in many metallurgical fumes have a starlike or platelike shape; others are needlelike and tend to form agglomerated chains.

Particulate and carrier-gas chemical composition exerts an influence on choice of control and auxiliary handling equipment. Composition also influences electrical properties, toxicity, reactivity, wettability, and most other particle properties. For the size and composition of large particles microscopic investigation is usually used. McCrone et al.(16) have published a photomicrographic atlas of dust components which permits recognition of up to 90% of the particles above 10 μm in a typical urban sample. If more specific information is required, a number of colorimetric, coulometric, and spectrometric techniques are available. Two particularly sensitive methods are atomic absorption spectroscopy for determination of most metals in the microgram to nanogram range, and the ring oven technique for most ions in the same quantitative range.

In atomic absorption spectroscopy the principle employed depends on the fact that atoms in the ground state absorb radiation at discrete wavelengths, characteristic for each element, and are raised to an excited state. On returning to the ground state, the excited atoms emit radiation. The amount of energy absorbed and reradiated is a function of the number of atoms present. The equipment used for this purpose is shown in Fig. 9. Hollow cathode or vapor discharge lamps, with emitters made of the element to be determined, serve as radiation sources. The source radiation traverses the sample to be measured, which is atomized in the flame, and passes through the entrance slit of a monochromator used to separate the resonance line from others present. The intensity of the resonance line is measured by a conventional photomultiplier tube and amplifier. The chopper serves as an ac modulator for the source, allowing elimination of any dc signal contributed by emission from the flame.

The ring oven technique is a special refinement of spot test procedures that are carried out on filter paper using extremely sensitive and specific organic reagents. The operation consists of placing a piece of filter paper on the heated surface of the ring oven and introducing soluble sample material at the exact center of the paper. The addition of a few microdrops of solvent washes the sample through the pores of the paper, transporting the

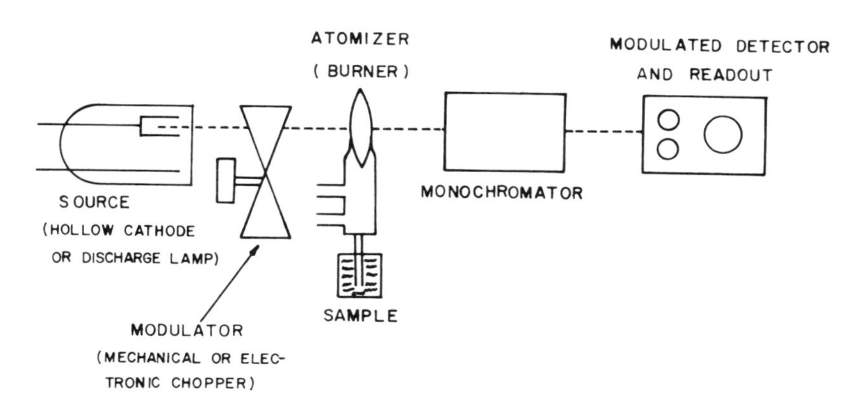

Fig. 9. Schematic diagram of atomic absorption photometer.

sample solutes to the area os the heated ring surface. The carrier solvent evaporates as the solution approaches the heated zone, depositing the solute (sample) as a sharply defined ring. The filter paper is removed from the ring oven and appropriate tests are run on sectors of the sample ring. Most of the reagents used in these tests are organic and form intensely colored insoluble products with the sample sectors of the ring.

E. PARTICLE PROPERTIES

Handling characteristics, electrical properties, wettability, toxicity, and flammability are some of the properties of particles that must be considered in designing and operating gas cleaning systems. The most important are electrical charge and conductivity. The resistivity of collected dust layers is important in electrostatic precipitator operation and is discussed in Chap. 4. Electrostatic properties of particles and fabrics play an important role in fabric filtration as discussed in Chap. 3. While of secondary importance, wettability and solubility of particles are factors in scrubber design(see Chap. 5.

Particle handling and flow properties influence dust separation equipment and auxiliary devices for all dust collectors. Powder flow properties result from a combination of factors including transmission of external and body forces through the system, particle size and shape distribution, ruggedness and resilience of the particle, cohesion, adhesion of particles to surfaces, and absorbed films, especially of water.

Particle shaking properties are important for the design of auxiliary devices for dust collectors and for the further treatment of separated dust. These properties include:

1. Shakedown weight--specific weight of highest packing density,
2. Angle of repose--angle at which piled dust begins to slide,
3. Slide angle--angle at which dust begins to slide on a base.

Generally, shakedown weight exceeds the free bulk weight by a factor of 1.2 to 1.4; angle of repose lies between 25° and 55° and the sliding angle between 35° and 65°. Both angles are influenced by particle size, moisture content of dust, particle shape, and cohesion and adhesion forces.

The abrasive behavior of dust characterizes its mechanical effect upon a surface with which it comes into contact. In gas cleaning, the inner walls of pipe conduits are subject to the highest abrasion. This dust property cannot be expressed by merely giving its hardness, since hardness expressed as resistance to the penetration of foreign bodies cannot be determined for dust particles. Furthermore, abrasive behavior is also affected by the shape and size of particles and specific weight.

REFERENCES

1. U. S. Department of Health, Education, and Welfare, Public Health Service Consumer Protection and Environmental Health Service, National Air Pollution Control Administration, "Air Quality Criteria for Particulate Matter," NAPCA Publication AP-49, Washington, National Air Pollution Control Administration. 1971.
2. "Control Techniques for Particulate Air Pollutants," NAPCA Publication AP-51, Washington, National Air Pollution Control Administration, January, 1969.
3. Shannon, L. J. et al. Emission and Effluent Characteristics of Stationary Particulate Pollution Sources, Midwest Research Institute, Kansas City, Missouri; Paper presented at the Second International Clean Air Congress, Washington, D. C. December 6-11, 1970 (MRI-1016).

4. Vandergrift, A. E., and L. J. Shannon, Handbook of Emissions, Effluents, and Control Practices for Stationary Particulate Pollution Sources, Midwest Research Institute, Kansas City, Missouri (MRI Project No. 3326-C). 1970.

5. "National Primary and Secondary Ambient Air Quality Standards," Federal Register, 36(84), April 30, 1971.

6. Federal Register, 36(67), April 7, 1971, August 14 and 17, 1971; Ibid, 36 (247), December 23, 1971.

7. Public Law 91-604, "Clean Air Amendments of 1970."

8. Strauss, W., Industrial Gas Cleaning, Pergamon , 1966, pp. 34, 47.

9. Stern, J. A., ed., Instrumentation for Air Pollution Control, Symp. Trans., Connecticut Valley Section, Inst. Soc. Am., 1968.

10. Perry, J. H., ed., Chemical Engineers Handbook, 3rd Ed., McGraw-Hill, New York, 1950, p. 396, 1272.

11. Goldman, I. B., and J. M. Marchello, A.I.Ch.E. J., 10, 5, 775, 1964.

12. Abatement of Particulate Emissions from Stationary Sources, National Academy of Engineering--National Research Council, COPAC-5, June, 1972.

13. Dorsey, J. A., and J. O. Burckle, The Characteristics of Particulate Emissions and Their Effect on Process Monitors,.....presented at the Annual Meeting of the 63rd A.I.Ch.E. Chicago, Illinois, December, 1970.

14. Herdan, G., Small Particle Statistics, 2nd ed. (rev.), Butterworths, London, 1960.

15. Irani, R. R., and C. F. Callis, Particle Size: Measurement, Interpretation, and Application, Wiley, New York, 1963.

16. J. L. McCrone, D. E. Draftz, L. P. Ronald, and L. T. Delly, The Particle Atlas, Science Publishers, Ann Arbor, Michigan, 1967.

17. Whitby, K. T., and B. Y. H. Lui,"Aerosol Measurements in Los Angeles Smog, Progress Report," Partical Technical Laboratory Report 141, University of Minnesota, 1970.

18. Robinson, E., and R. C. Robbins, "Sources, Abundance, and Fate of Gaseous Atmospheric Pollutants," SRI Project Pr-6755, American Petroleum Institute, New York, February, 1968.

19. "Trace Metals: Unknown and Unseen Pollution Threat," Chem. Eng. News, July 19, 1971.

20. "Smelter Gases Yield Mercury," Chem. Eng. September 6, 1971.

21. Dorsey, J. A., and J. O. Burckle, "Particulate Emissions and Process Monitors," Chem. Engr. Prog., 67(8), 92-96, 1971.

Chapter 2

COLLECTION AND MECHANICAL SEPARATION

John J. Kelly

Chemical Engineering Department
University College
Dublin, Ireland

Joseph M. Marchello

Division of Mathematical and
Physical Sciences and Engineering
University of Maryland
College Park, Maryland

I. ENCLOSURES AND EXHAUST SYSTEMS

In many instances sources of pollutants are in open areas, and dusts and other contaminants must be captured. Hoods and enclosures are used for this purpose. Often the captured laden gases from several sources are conveyed through a system of ducts to a single control device. Sufficient air must be provided by fans or blowers to capture and transport the contaminants.

The Bernoulli equation may be applied to the exhaust systems to determine operating conditions and design requirements. In exhaust systems the elevation term may usually be omitted. Also, pressure and temperature changes are frequently a small percentage of the absolute pressure and temperature so that the gas may be assumed to be incompressible. The integral relation for changes between points 1 and 2 becomes:

$$\frac{(P_2 - P_1)}{\bar{\rho}} + \frac{(U_2^2 - U_1^2)}{2g_c} + \frac{2fL\bar{U}^2}{g_c D} + W_s = 0 \tag{1}$$

By convention work, W_s, done by the gas is positive. Thus work done by a fan or blower on the gas would have a negative sign. Except for flow immediately downstream of an expansion section, the pressure decreases with distance, L, so that $P_1 > P_2$.

In a section of straight duct where there is no fan or blower, $W_s = 0$. For gas velocities below 0.3 of the velocity of sound, the kinetic energy change is small and Eq. (1) has the form

$$\frac{\Delta P}{L} = \frac{2f\bar{U}^2\bar{\rho}}{g_c D} \tag{2}$$

which represents the pressure drop due to friction per unit length of duct. This loss of energy, along with other losses discussed below, must be supplied by the exhaust fan at either end of the section of duct work or by the system entering pressure.

The losses in an exhaust system are due to inertial changes, orifice or flow area changes, and friction. Inertial losses arise from the energy required to accelerate the gas and from energy losses in elbows and branches of the duct.

Acceleration requirements may be obtained from

$$\frac{\Delta P}{\rho} = \Delta h = \frac{U_1^2 - U_0^2}{2g_c} \tag{3}$$

or for ambient air, from Table 1, when U_0 is small. Losses due to elbows and branches are generally expressed in equivalent length of straight duct (see Table 2).

Contraction and expansion of the duct flow area generally result in small losses (2). For taper angles up to

TABLE 1

Acceleration Head for Air

Velocity (fpm)	Head (in. water)
500	0.016
1000	0.062
1500	0.140
2000	0.249
2500	0.389
3000	0.561
3500	0.764
4000	0.998
4500	1.262
5000	1.558

TABLE 2

Air Flow Losses Due to Elbows and Branch Entries (2)

Duct diameter (in.)	Losses in equivalent feet of duct					
	Elbows[a]				Branch entry	
	90% Bend		45% Bend		45% Angle	30% Angle
	$\frac{R}{D} = 1$	$\frac{R}{D} = 2$	$\frac{R}{D} = 1$	$\frac{R}{D} = 2$		
4	7	4	4	2	5	3
6	11	6	6	3	7	5
8	14	8	8	4	11	7
10	20	11	10	6	14	9
12	25	14	12	7	18	11
16	36	20	17	10	25	15
20	46	26	23	13	32	20
24	59	33	30	16	40	25
28	71	40	35	20	47	30
36	92	52	46	26	54	35
48	130	73	64	36	60	40

[a] $\frac{R}{D}$ = Turning radius/duct diameter.

45° they may be approximated by using an equal length of duct. However, losses from orifices and at the entrance between the hood and duct can be appreciable, and vary from 1.8 times the acceleration head (Table 1) for sharp edge transitions to nearly zero for well rounded entries.

Pressure drop in straight ducts may be calculated from correlations of the friction factor, f, in terms of Reynolds number and roughness of the duct wall. For air flowing in ducts with moderate roughness Fig. 1 may be used (1). Figure 2 covers the full range of flows for different wall roughness factors, ε/D, and for any gas or liq-

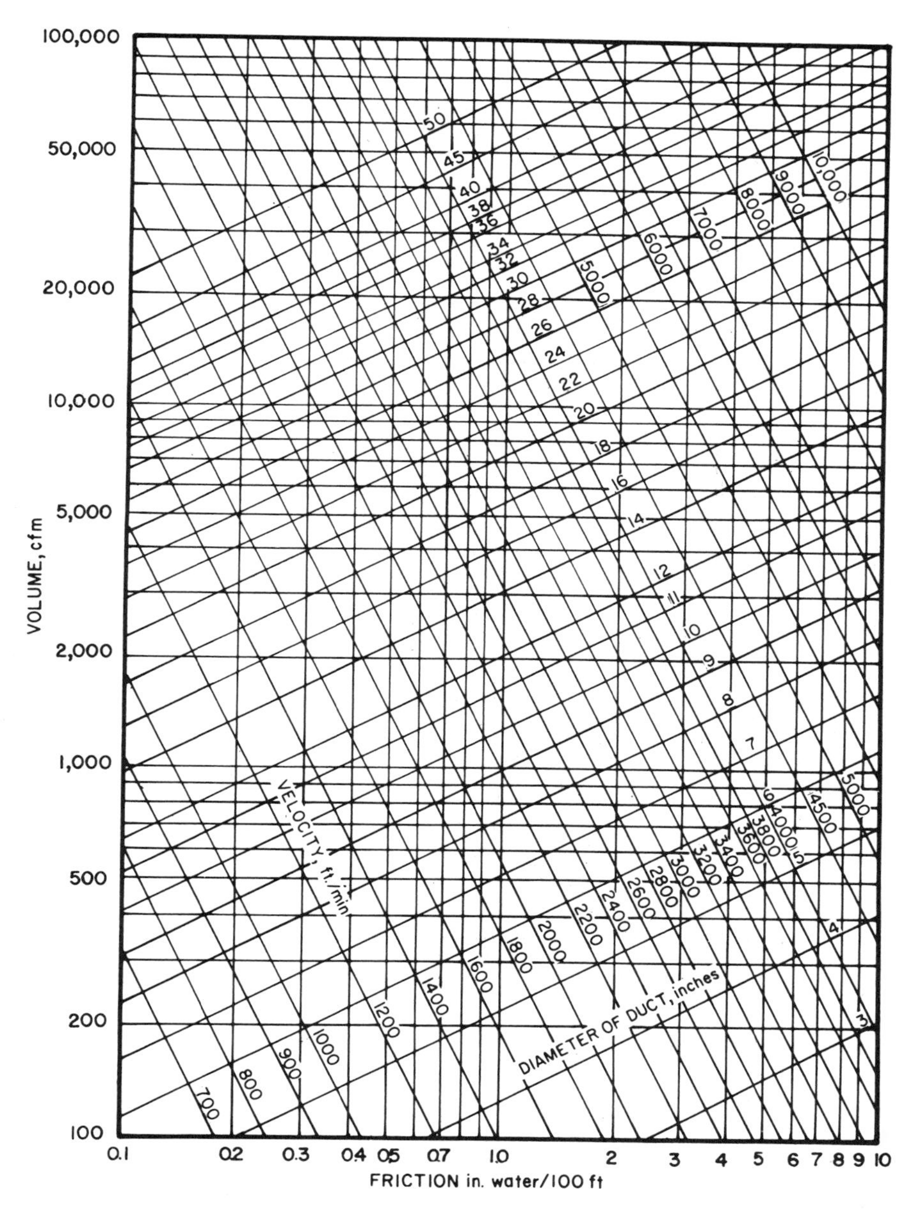

FIG. 1. Pressure loss for straight circular ducts (1).

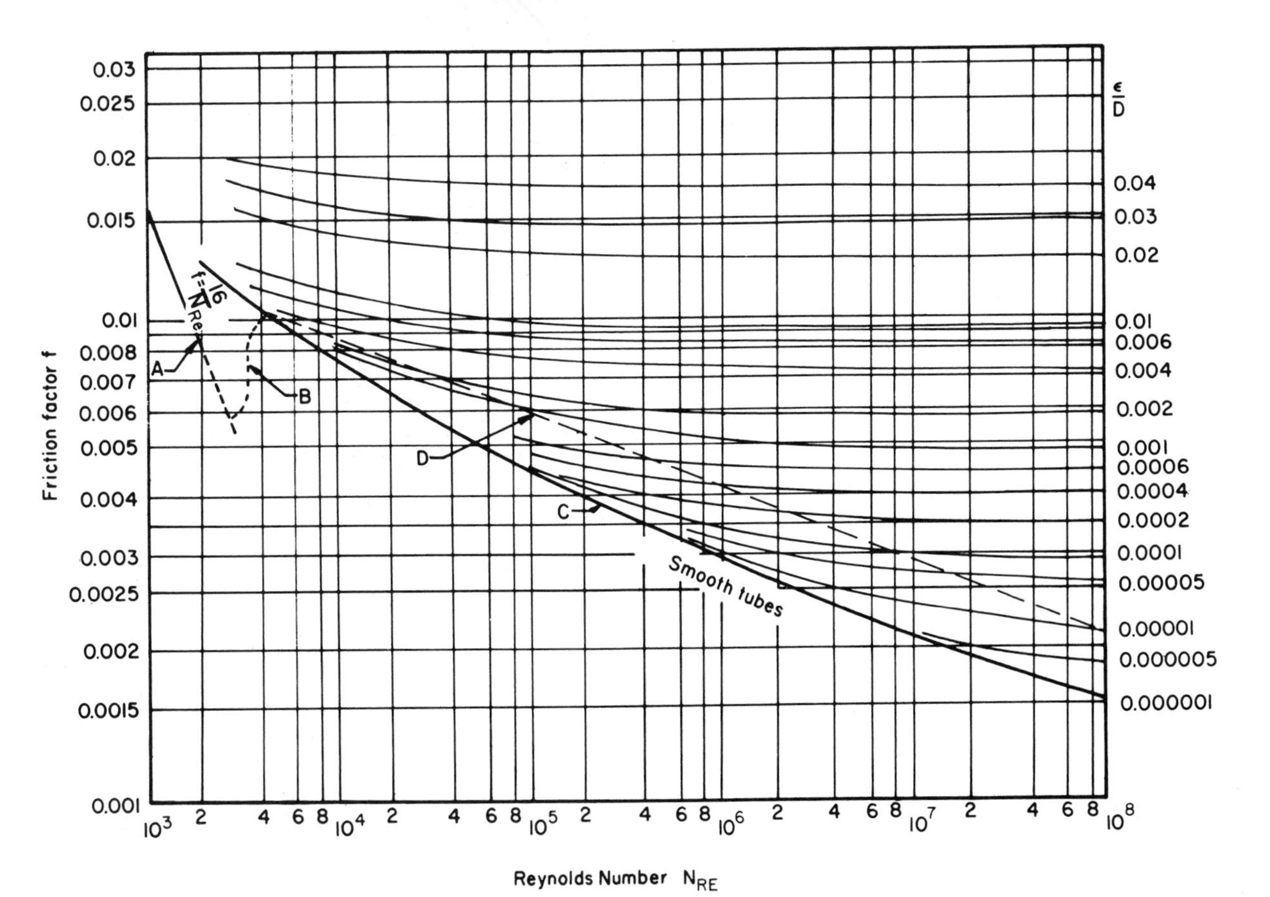

FIG. 2. Fanning friction factors for straight ducts (4).

uid. The viscosity, μ, and density, ρ, must be known in order to calculate the Reynolds number, $DV\rho/\mu$, where D is the duct diameter and V the linear velocity.

The required volumetric flow for exhaust systems may be determined from the requirements of the hoods and enclosures described below. The conveying velocities in ducts is dependent upon the nature of the contaminant (see Table 3). Duct size should be such as to provide linear velocities (1) that fluidize the dust and transport it to the collection equipment (see Table 4).

Hoods are used on equipment to capture air contaminants and heat for subsequent handling or removal. There are three general types of hoods: enclosures, receiving, and exterior. Enclosures surround the source of emissions but may have one face partially or fully open for access. Receiving hoods are located to collect emissions that are ejected toward them such as the dust particles from a

TABLE 3

Recommended Minimum Dust Velocities (1)

Contaminant	Duct velocity (fpm)
Gases, vapors, smoke, fumes, and very light dusts (zinc and aluminum oxide fumes, wood, flour, and lint).	2000
Dry, medium density dust (buffing lint, sawdust, grain, rubber, and plastic dust)	3000
Average industrial dust (sandblast, grinding, wood shaving, and cement dust)	4000
Heavy dusts (lead and foundry shakeout dusts, metal turning)	5000

TABLE 4

Reentrainment Velocities of Materials (5)

Material	Density (g/cm^3)	Median size (μm)	Pick-up velocity (ft/sec)
Aluminum chips	2.72	335	14.2
Asbestos	2.20	261	17.0
Nonferrous foundry dust	3.02	117	18.8
Lead oxide	8.26	14.7	25.0
Limestone	2.78	71	21.0
Starch	1.27	64	5.8
Steel shot	6.85	96	15.2
Wood chips	1.18	1370	13.0

grinder. Exterior hoods capture contaminants released outside and at some distance from the hood.

Flow patterns into hoods and ducts follow streamlines that depend strongly on the specific geometric shape and dimensions in each situation (1). Consideration must be given to the air flow streamlines to ensure the elimination of stagnant regions and to improve hood efficiency (see Fig. 3). Exterior hoods and exhaust systems must have adequate capacity to capture the emissions from the exterior sources. Generally one endeavors to establish a null point or null curve that locates or bounds the region that must be swept by the exhaust system to ensure capture of the contaminants.

In nearly all situations, including closed rooms, air drafts occur due to thermal and pressure changes. In industrial buildings draft velocities are usually 100-300fpm and can increase several-fold for short periods. Such drafts may prevent capture of air contaminants and must be

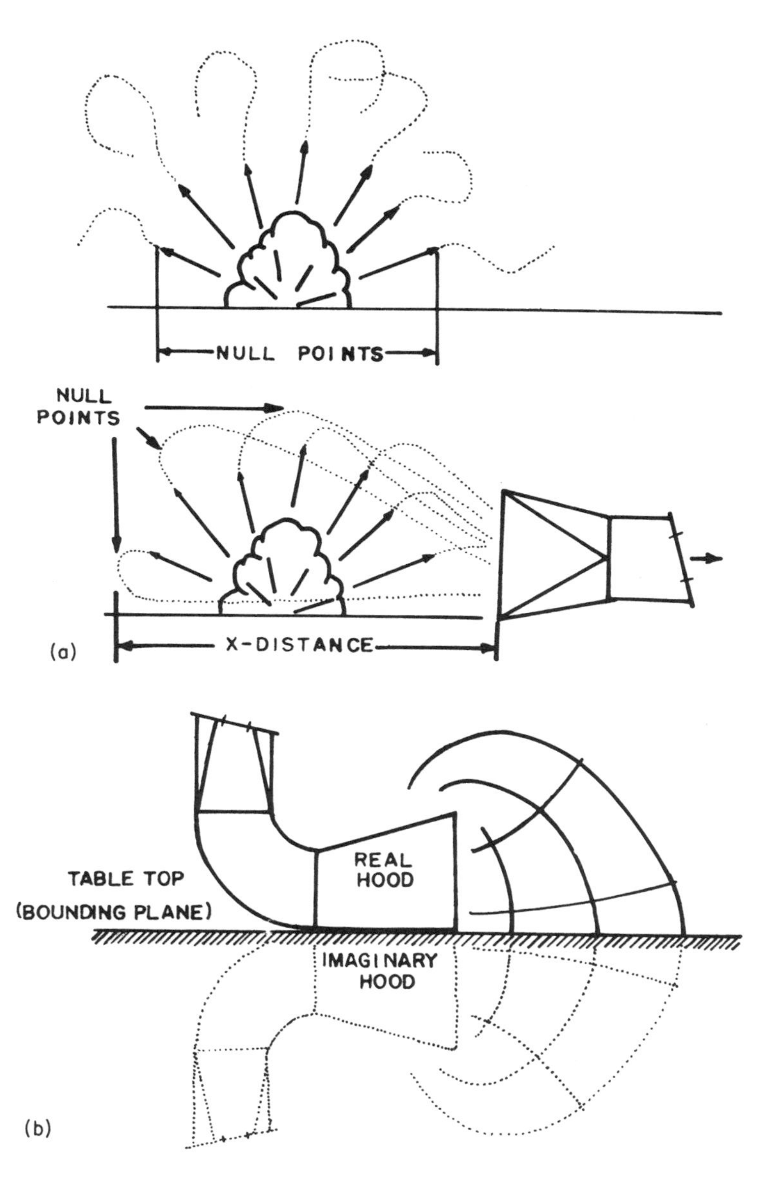

FIG. 3. (a) Dimensions used to design high-canopy hoods for hot sources. (b) Rectangular hood bounded by a plane surface.

allowed for in hood design by increasing the null region and the ventilation rate of the hood.

Recommended ventilation rates have been established for control of a variety of sources (1). For enclosures, ventilation rates are generally set between 50 and 200 fpm with 100-400 fpm intake velocities for all openings. This includes operations such as abrasive blast rooms, bagging machines, grinders, paint spray booths, and welding. When sources of heat are present in an enclosure, the total air rate must also include the thermal updraft discussed below.

Open surface tanks and other sources may often be controlled by a canopy or by slot hoods(see Fig. 4). Slot hoods are commonly employed on tanks except when the contents are very hot and give rise to strong thermal updrafts. Generally slots are provided along two parallel sides. Design of the slots and exhaust system is usually based on achieving an intake flow to the slots of 200 to 300 cfm per foot of length. The dust slot width is sized to give a linear velocity in the slot of about 4000 fpm. Baffles and increases in ventilation rate may be required to minimize air drafts in given situations.

Hot processes that release a significant amount of heat to the surroundings create a thermal updraft that frequently must be captured in a canopy hood or enclosure above the source (see Fig. 4). Sutton (1) developed an empirical equation for the spread of a rising column of hot air:

$$D_c = 0.5 \; [(2D_s)^{1.138} + y]^{0.88} \tag{4}$$

where D_c is the diameter in feet of the warm air column at a distance y feet above the hot source. The diameter of the hot source is D_s, also expressed in feet. For rectangular sources use D_s and D_c as the widths of the source

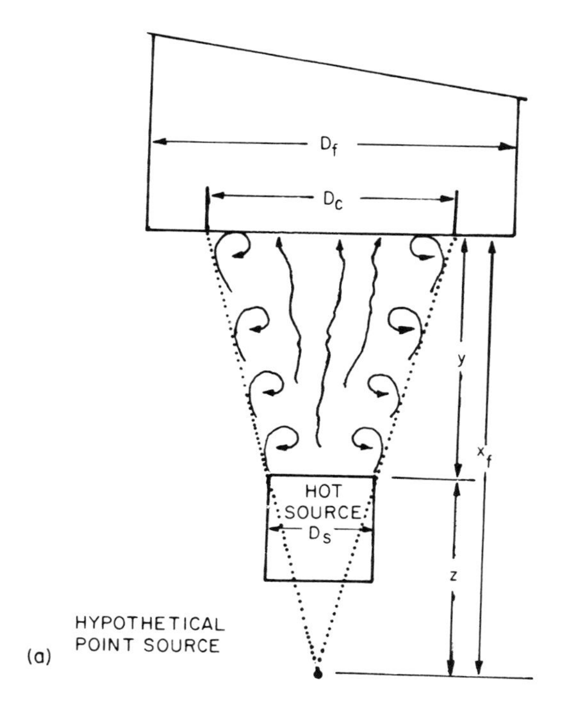

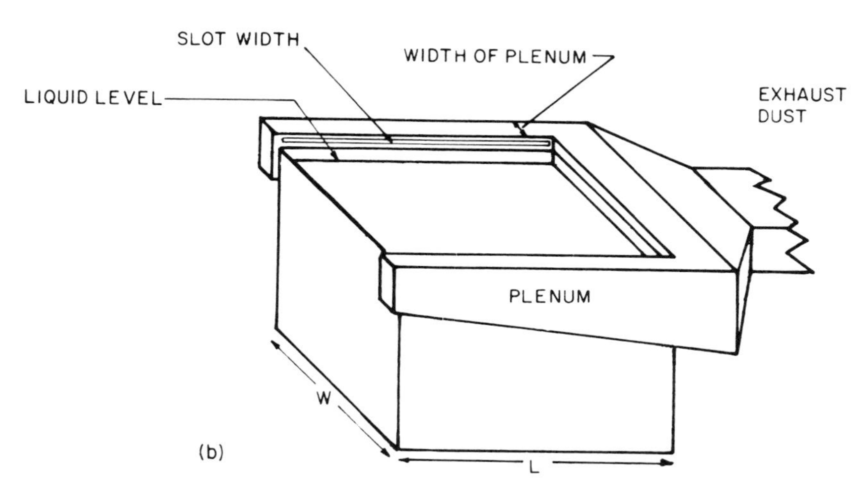

FIG. 4. (a) High canopy hood for a hot source. (b) Slot hood for control of emissions from open-surface tanks.

and air column. The length of the column is taken as the length of the source plus $(D_c - D_s)$.

The velocity of the thermal updraft may be estimated from the rate of heat transfer to the air and the equation of motion with the bouyant force included. The following empirical equation for the velocity of the thermal updraft, U_f, based on the rate of heat transfer from a hot, flat, horizontal surface may be used for most applications (1,2).

$$U_f = \frac{8(A_s)^{1/3}(\Delta T)^{5/12}}{[(2D_s)^{1.138} + y]^{1/4}} \tag{5}$$

where U_f is in fpm, A_s is the area of the hot sources in ft^2, ΔT is the temperature difference between the hot source and the ambient air in °F, and D_s and y are in feet as in Eq. (4).

The canopy hood must be somewhat larger than the thermal column to assure complete capture. One recommendation (2) is that 0.8 y be added to the updraft diameter or width and to the length. Velocity, U_r, into the face area around the thermal column should be between 100 and 200 fpm. The volumetric flow rate, Q_T, into the hood may be calculated from

$$Q_T = U_f A_c + U_r(A_f - A_c) \tag{6}$$

where A_c is the flow area of the updraft, U_f, at the hood face, A_f is the total area of the hood face, and U_r is the required velocity through the remaining area of the hood.

Hoods and ductwork may be constructed from a variety of materials. High-temperature and corrosive gases require special materials of construction and special attention to eliminate leaks. Hood depth should be sufficient to ensure capture of the contaminant. The transition between the hood and the duct should be at an included angle of 60° or less. Exhaust system layout should be such as to minimize transport distances and power requirements.

For complex exhaust systems static pressure balances must be made for each branch to ensure adequate flow. Frequently, adjustable gates are placed at branches so that flow rates can be adjusted after installation to give the desired branch flow rates. Usually the pressure drop for the branch-of-greatest-resistance is calculated in detail. This, together with the total volumetric flow of the system establishes the power requirements of the system. Flow in the other branches is then adjusted using the branch gates.

Fan and blower requirements are based on the total volumetric rate of gas to be handled, the pressure drop ducts, control equipment, and stack. Manufacturers and vendors usually provide information on the performance of their equipment such as design, speed, size, and efficiency for given applications. The ideal horsepower requirements, W_s, may be determined from

$$W_S = (0.000157) Q_T h \tag{7}$$

where Q_T is in cfm, h is the head or pressure drop in in. of water gauge, and W_S is in horsepower. Efficiencies vary with design, speed, gas density, and applications. They generally range between 40 and 70%(see Fig. 5). For preliminary calculations an efficiency of 50% is often assumed, so that the actual motor horsepower is about twice the ideal value given by Eq.(7).

Fans are generally of two types: radial flow or centrifugal; and axial flow or propeller type. They deliver

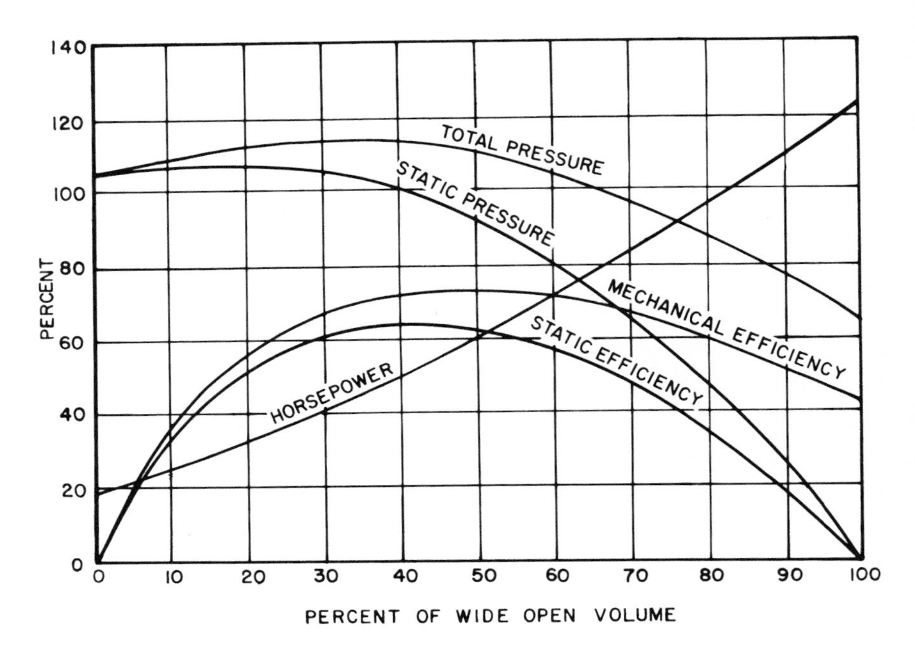

FIG. 5(a). Typical fan characteristic curves.

air with pressure increases up to about 10 in. of water. Blowers may range up to 5 psi or 140 in. of water gauge. Compressors are often used in processes and can raise the gas to very high pressures. They are normally not installed specifically for air pollution control. In complex exhaust systems and for large flows it is often desirable, because of size limitations or reliability requirements to employ more than one fan or blower in the system.

In some cases the laden gases must be cooled before they enter the control device. Methods of gas cooling include: dilution with ambient air; quenching with water; natural convection and radiation from duct work; and use of forced convection heat exchangers. Cooling by dilution and quenching involve adding sufficient air or water to lower the temperature of the final gas mixture to the de-

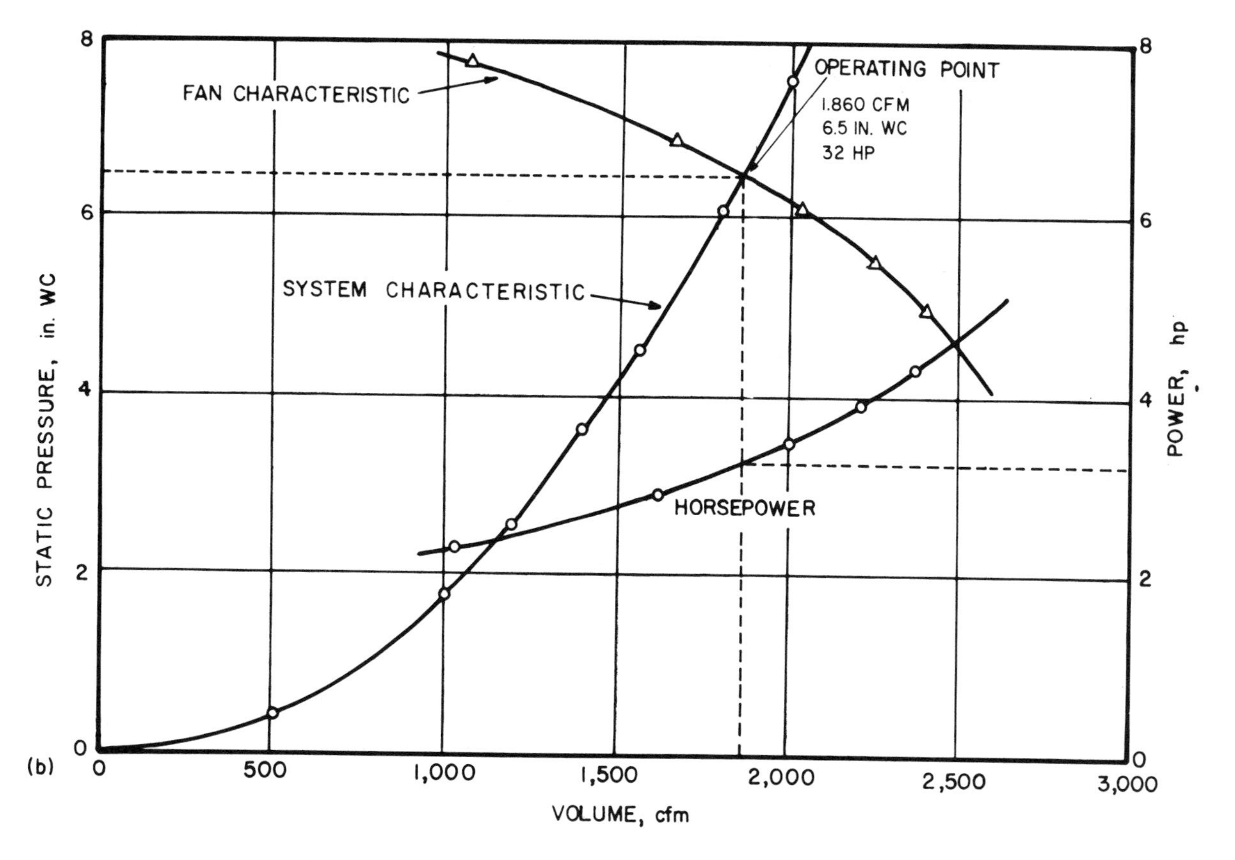

FIG. 5(b). Characteristic curves of an exhaust system.

sired level. The required amounts may be calculated from an energy balance using the sensible heat capacities of the gases and the heat of vaporization of water. Heat losses from ducts and removal by heat exchangers are determined from energy balances and heat transfer rate equations (3,4)

II. HOPPERS AND VALVES

Dry particle collection equipment must be fitted with valves and collection hoppers to remove the collected dust. Hopper sides are usually at an angle of 60° to the horizontal. When the collected dust does not flow easily, a vibrator or steeper walls can be used. The hopper size may be made such as to store dust for an extended period, or to provide an amount of dust that seals the valve from the gas so that gas leakage is minimized. It should be large enough so that the hopper valve has to be opened only occasionally because collector operation is often upset during dust removal.

Several types of valves are used as shown in Fig. 6. They may be manually or mechanically operated. In large applications they may be preset to operate automatically at certain intervals. The counter-weighted valve opens when the collected dust has sufficient head to overcome the weight. They cannot be used when the fan is upstream of the collector so that the static pressure is above atmospheric. They are also restricted to free flowing dusts. Double flap valves are often used for abrasive materials. The flaps are opened alternately by a cam mechanism to stop air from entering or gas from leaving the collector. Rotary airlocks provide a continuous discharge of material and require less space than other valves.

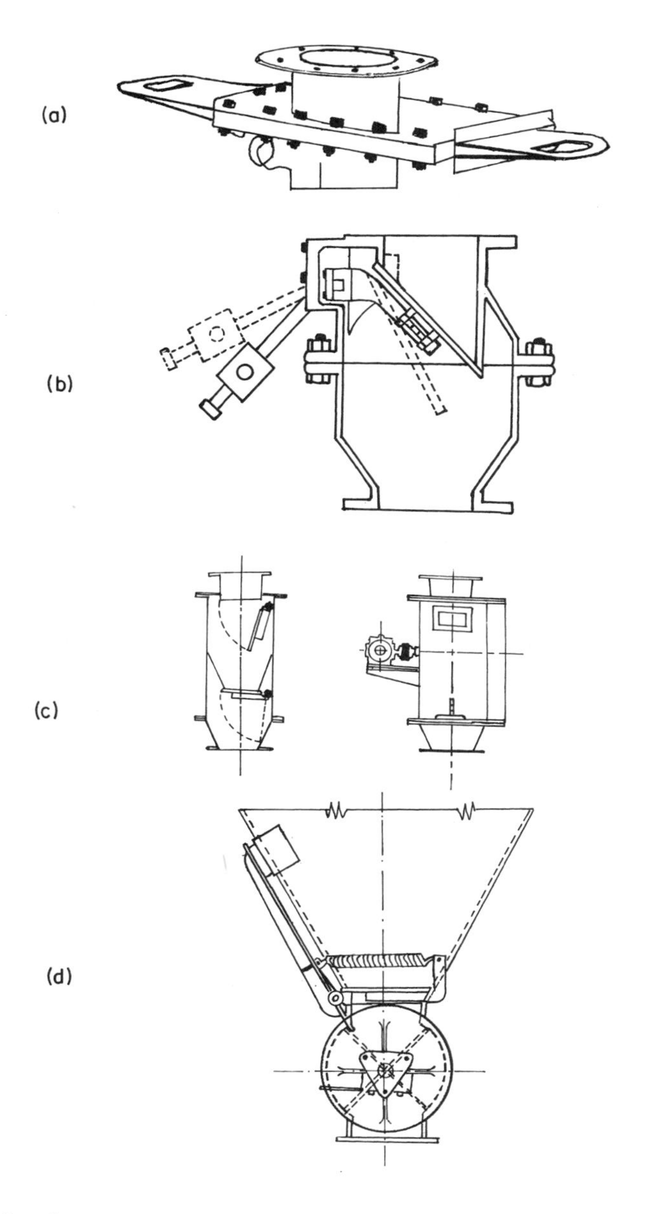

FIG. 6. Hopper valves: (a) push-pull slide, (b) counter-weighted flap, (c) double flap, and (d) rotary air locks (1,5).

Hopper valves discharge the collected dust into bags, bins, dust conveyors, or closed cars to transport the material to disposal, recycling, or other use.

Example 1--A galvanizing plant is considering two hot-dip kettles. The molten zinc in the kettles is covered with a layer of molten flux containing ammonium chloride. During operation a fine fume of flux rises from the kettles with the thermal column of hot air. The flux surface is at 800°F. The kettles are 6 ft wide and 15 ft long. They are at opposite sides of the building and 60 ft apart. An overhead crane delivers material from the cleaning and prefluxing tanks to the kettles.

Control of the fume would require use of a heated, precoated baghouse to be located on the roof of the building 40 ft above one of the kettles. Pressure drop in the baghouse is 3 in. of water. Overhead canopy hoods are to be used to capture the fume. They must be 6 ft above the kettle surface to permit installation and use of small cranes to work the kettles.

a. Estimate the hood face areas, duct diameter, and specify bends and branches.
b. Estimate the pressure loss of the system and the fan horsepower.

Solution

a. From Eq. (4), the air column width is

$$D_c = 0.5\ ([(2)(6)]^{1.138} + 6)^{0.88} = 7.7\text{ ft}$$

The air column length is

$$L + (D_c - D_s) = 15 + (7.7-6) = 16.7\text{ ft}$$

To assure capture of the column it is recommended that 0.8 y = 0.8(6) = 4.8 ft be added. Rounding off the totals, a hood face 12 ft wide and 22 ft long should be adequate.

From Eq. (5) the updraft velocity is

$$U_f = \frac{(8)[(6)(15)]^{1/3}\ (800-70)^{5/12}}{([(2)(6)]^{1.138} + 6)^{1/4}} = 255 \text{ fpm}$$

The area of the air column is $A_c = (7.7)(16.7) = 128\ \text{ft}^2$, and the hood face area is $A_f = (12)(22) = 264\ \text{ft}^2$. Assuming that $U_r = 150$ fpm, Eq. (6) gives

$$Q_T = (255)(128) + 150\ (264-128) = 53{,}000 \text{ cfm}$$

For a duct velocity of 5000 fpm (Table 3), the diameter would be

$$D = \left(\frac{4Q}{\pi U}\right)^{1/2} = \left|\frac{(4)(53{,}000)}{(3.14)(5{,}000)}\right|^{1/2} = 3.7 \text{ ft}$$

Make the duct 4 ft in diameter and the duct velocity will be 4200 fpm.

Summary: The duct diameter is 4 ft. The total air rate is 106,000 cfm. Each hood face is 12 ft by 22 ft and 6 ft above the kettle.

b. The air outside the thermal column must be accelerated from 0 to 4200 fpm and the air in the column must be accelerated from 255 to 4200 fpm. From Eq.(3) or from Table 1, neglecting the column velocity, the pressure drop needed to accelerate the entering air is 1.1 in. of water gauge.

The length of the hood side for a 60° angle is L = 12/[2 sin (60°)] = 7 ft. This is the equivalent length of straight duct for the hood duct contraction. To clear the traveling crane, let the duct for the furthest kettle rise 24 ft above the hood.

It then makes a 90° (R/D = 2) bend and runs 60 ft to a 45° branch entry with the other duct. From Table 2, the eqi-valent lengths of duct are 73 ft for the bend and 36 ft for the branch. The large duct entry to the fan and baghouse will be 6 ft in diameter. The fan is immedia-ately in front of the baghouse and at the end of a 45° entry merging the two 4-ft ducts into the 6-ft fan entry. The fan entry equivalent length is 75 ft.

Summary:

Equivalent length ft	
7	hood duct
24	vertical duct
73	90° bend
60	horizontal duct
36	45° branch entry
75	45° fan entry
275	

The total equivalent length of duct for the longest run is 275 ft. From Fig. 1, the pressure drop in the 4-ft duct hand-ling 53,000 cfm at 4200 fpm will be 0.70 in. of water per 100 ft or 1.9 in. for 275 ft. The total pressure loss that the fan must supply is

$$\Delta P_T = 1.1 + 1.9 + 3.0 = 6.0 \text{ in. water}$$

From Eq. (7) the ideal fan horsepower is

$$W_s = (0.000157)(106{,}000)(6.0) = 100$$

III. SETTLING AND MOMENTUM SEPARATORS

Particles may be collected by gravity settling and by inertia when the gas changes direction of flow. In settling chambers the gas stream is slowed down to allow particles to settle out. For most installations, size restrictions limit settling chambers to the collection of particles with aerodynamic diameters greater than about 70 μm. Momentum separators are more compact and are able to collect particles down to 20 μm in many cases.

A falling particle will accelerate until the frictional drag of the air balances the gravitational acceleration, after which it will continue to fall at a constant velocity given by the general Stokes law expression

$$U_t = \sqrt{\frac{4gd_p(\rho_p - \rho_a)}{3\rho_a C}} \tag{8}$$

where C is the drag coefficient which is related to Reynolds number, $N_{Re} = d_p\rho_a U_t/\mu_a$. ρ_a is the density and μ_a the viscosity of air ($\rho_a = 0.0013$ g/cm^3 and $\mu_a = 0.00018$ g/cm-sec at 20°C and 1 atm).

The general drag coefficient curve for spherical particles may be represented by three relationships

$$C = 24\ N_{Re}^{-1.0} \qquad \text{for} \qquad N_{Re} < 2$$

$$C = 18.5\ N_{Re}^{-0.6} \qquad \text{for} \qquad 2 < N_{Re} < 500$$

$$C = 0.44 \qquad \text{for} \qquad 500 < N_{Re}$$

The changes in the drag coefficient for $N_{Re} > 2$ are due to fluid displacement effects. On the other hand for $N_{Re} < 0.0001$ which corresponds to particle sizes near or below the mean free path of the air molecules, the air resistance becomes discontinuous and the Cunningham correction, K_M, must be multiplied by the value of U_t from Eq.(8) to obtain the actual terminal settling velocity (4,5).

For horizontal settling chambers (Fig. 7) the residence time of the gas in the chamber, t_r, is

$$t_r = \frac{V}{Q} = \frac{L}{U} \tag{9}$$

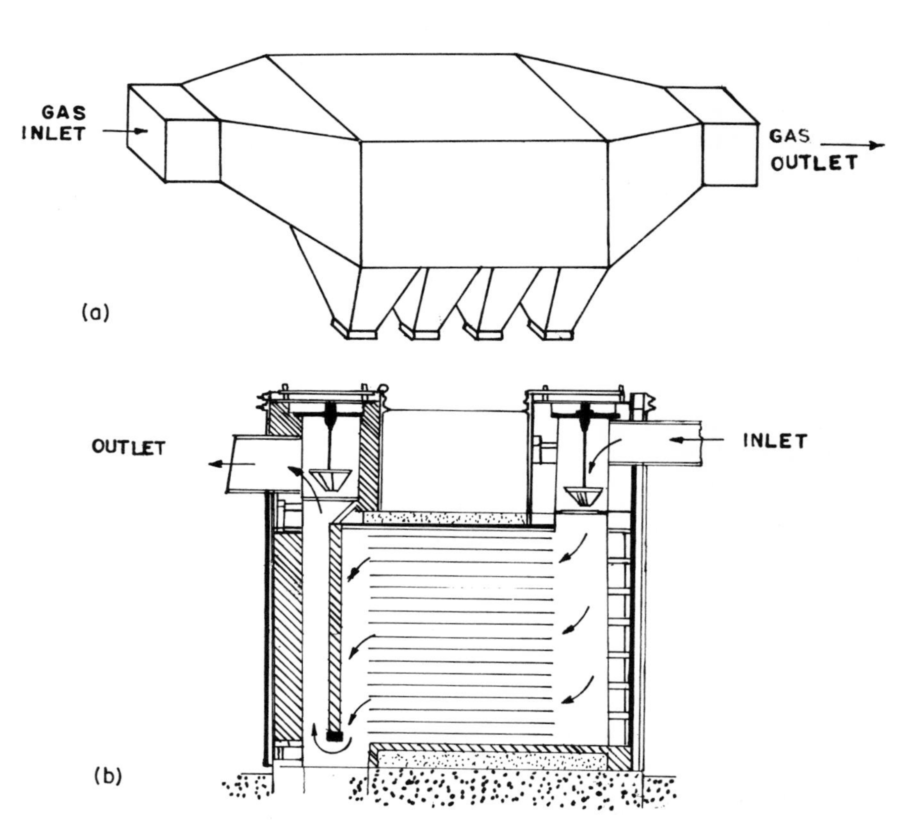

FIG. 7. (a) Horizontal settling chamber (5). (b) Multi-tray settling chamber (4,5).

where V is the chamber volume, Q is the volumetric flow rate, L is the chamber length, and U is the average linear gas velocity in the chamber. The time required for a particle in the size range d_p to settle a distance h is given by

$$t_{s,p} = \frac{h}{U_{t,p}} \tag{10}$$

where $U_{t,p}$ is the settling velocity of a particle with diameter d_p given by Eq. (8).

The particles must be the at the height of the chamber, h, or less so that the collection efficiency, n_p, to a first approximation is

$$n_p = \frac{t_r}{t_{s,p}} = \frac{LU_{t,p}}{hU} \tag{11}$$

This is the collection efficiency for particles of size d_p. When Eq. (11) gives $n_p > 1$ use $n_p = 1$.

Factors which cause deviations from Eq. (11) include the time for particles to reach the terminal settling velocity, hinderance of settling at high particle concentrations, and reentrainment. In general a chamber velocity below 10 fps is satisfactory to avoid pickup or reentrainment (Table 4).

There is a variety of designs for settling chambers, depending upon the application. Many are installed on buildings, such as penthouse and cupola arrestors, and in exhaust systems prior to transport to control equipment that remove smaller particles and gases. Frequently they are used to recover material for recycle to a process.

Improvement in settling chamber efficency can be obtained by decreasing the height a particle has to fall, for example through the use of trays (see Fig. 7).

The major disadvantage of settling chambers is the large space required. Some advantages are: simple construction; low pressure losses; no pressure and temperature limitations except for those of the construction materials; dry disposal; and minimum abrasion.

Baffle chambers use changes in the direction of gas flow to impinge the particles on a surface (Fig. 8). While the space required is less, the pressure drop is increased. Momentum or impingement chambers may have venturi-shaped flow areas, special metal collector curtains and shutters. In the louvred collector, the dust laden gas enters the large end of a cone section containing many louvred slots. The gas changes direction and flows through these slots. The dust is not reflected and passes out the small end of the cone with a portion of the gas. This smaller volume of gas enters a baffled settling chamber and perhaps a

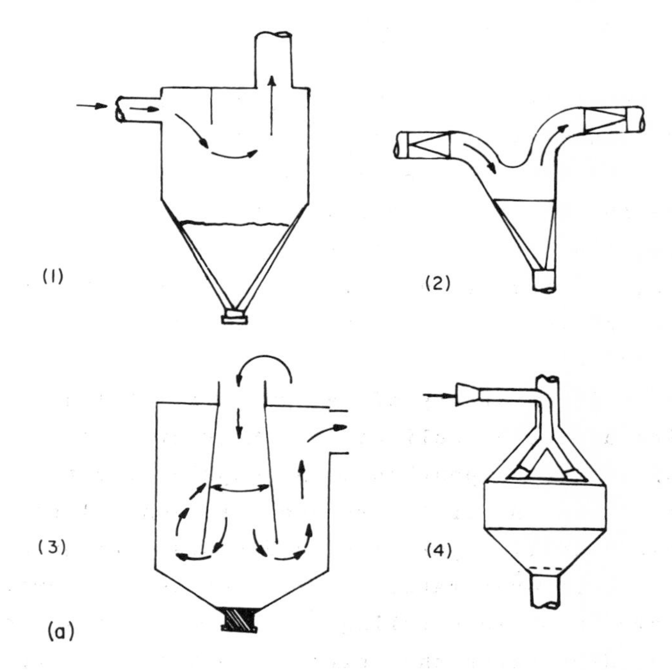

FIG. 8(a). Baffle chambers: (1) simple baffle type, (2) rounded trap, (3) downward entry, and (4) accelerated settling chamber (5).

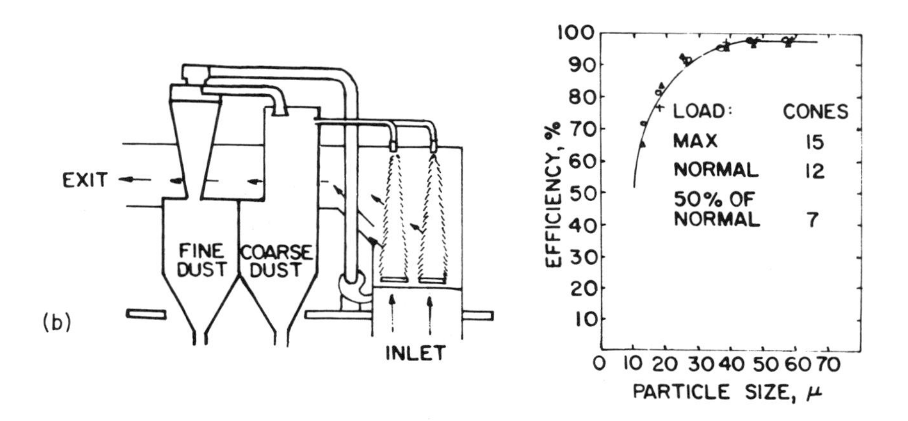

FIG. 8 (b). Louvre cone collector with baffle chamber and cycle, and efficiency curve (5,6).

cyclone to remove fine dust. This gas is returned to the cone and finally leaves as cleaned gas.

IV. CYCLONES

Cyclonic collectors are generally of two types: the large-diameter lower-efficiency cyclones; and the smaller-diameter, multitube high-efficiency units. The larger cyclones have lower efficiencies especially on particles less than about 30 μm. However, they have low initial cost and usually operate at pressure drops of 1 to 3 in. of water. The multitube cyclones are capable of efficiencies exceeding 90% on particles greater than 10 μm , but the cost is higher and pressure drop is usually 3 to 5 in. of water. They are also more susceptible to plugging and erosion. Cyclones are often used as a part of the process when the gas stream is heavily laden with the product, such as in spray driers, coal driers, alfalfa dehydrators, and milling operations. They are widely used in grain elevators, sawmills, asphalt plants, and detergent manufacture.

The centrifugal force on particles in a spinning gas stream can be many times greater than gravity. The vortex motion can be applied to the gas stream in several ways: with a fixed impeller, with a rotating turbine, or by admitting the gas tangentially in an annular space (see Fig. 9). In the conventional reverse-flow cyclone the gas enters the annular upper section and spirals down to the apex of the conic section, where it reverses direction and flows up out the inner cylinder (5,6).

Theoretical treatment of cyclone performance is incomplete, primarily due to the complex flow patterns of the gas and particles. Generally collection efficiency increases with energy expended. The pressure drop, Δp, may be approximated by (1)

$$\Delta p = \frac{13 \rho_g U_i^2 A_d}{2 g_c D_e^2} \tag{12}$$

where U_i is the gas inlet velocity, A_d is the inlet duct area, and D_e is the diameter of the cyclone exit duct.

Efficiency of particle removal by cyclones depends upon specific design of the unit, particle density and

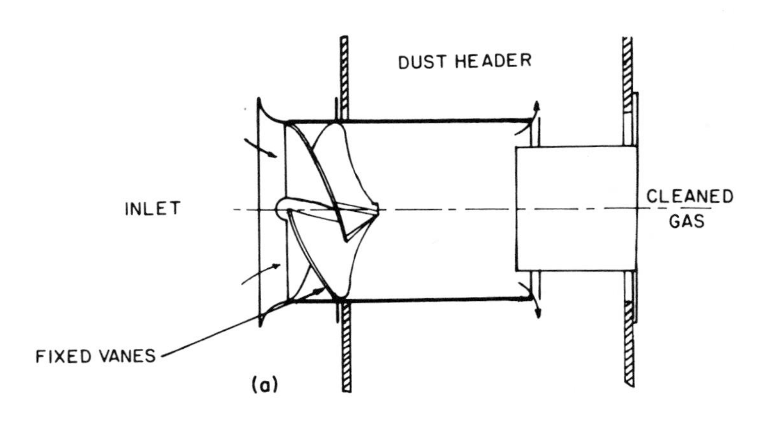

FIG. 9(a). Fixed impeller cyclone (5).

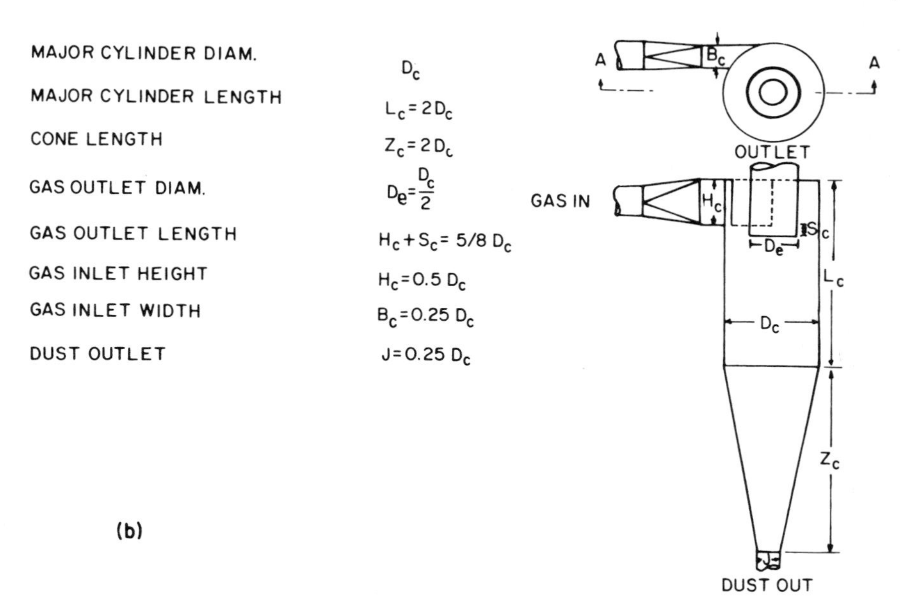

FIG. 9(b). Conventional cyclone with dimension ratios (1).

size, and gas flow. Lapple (1) correlated collection efficiency in terms of the cut size, D_{pc}, which is the diameter of those particles collected with 50% efficiency. Collection efficiency for particles larger than D_{pc} will be greater than 50% while for smaller particles the efficiency will be less. The particle cut size may be calculated from

$$D_{pc} = \sqrt{\frac{9\mu B_c}{2\pi N_e U_i\ (\rho_p - \rho_g)}} \tag{13}$$

where B_c is the cyclone inlet width, N_e is the effective number of turns the gas makes in the cyclone (generally between 1 and 10), and μ is the gas viscosity. To a first approximation $N_e = L_c/H_c$ (see Fig. 9). The efficiency for each particle size range may then be determined using Fig. 10.

Many types of commercial cyclones have been developed by manufacturers and performance information for specific

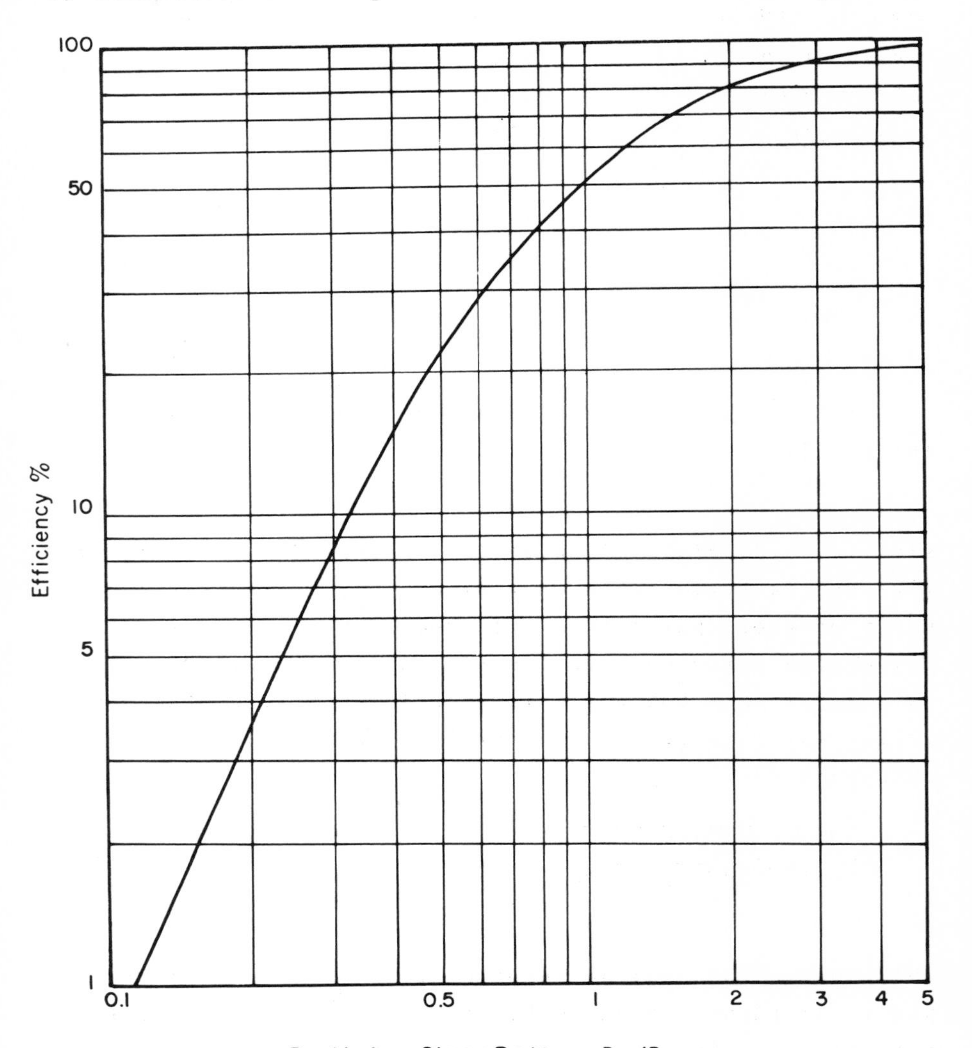

FIG. 10. Cyclone efficiency.

applications is generally available (1,5,6). Frequently straight concentrators are used in which the concentrated dust is drawn off with a portion of the gas. A secondary collector, which may be a conventional cyclone, is then used to remove the dust. Multicellular units are frequently used as preliminary cleaners (see Fig. 11). Multiple cylcones may have only marginally higher efficiencies than one large cyclone. However, they usually require less space. When liquid or wet particles are collected the design must provide for drainage of the collected liquid layer. Usually the conical bottom section is made hemispherical and a false bottom with a slot around the periphery is used to avoid reentrainment of liquid.

Example 2--A flour mill uses a two-stage reverse-flow cyclone unit to remove dust from the air leaving the milling room. The stages are identical in size and have $B_c = 20$ in., $N_e =$ 8 turns, $U_i = 70$ ft/sec, and $\rho_p = 1.0\ g/cm^3$. The air is at 20°C and contains 2 grains/ft^3 entering the first cyclone. Calculate the efficiency for each grade size, the overall efficiency, and the pressure drop of the unit.

Grade size (μm)	% in grade range
100 to 20	50
20 to 10	20
10 to 2	20
2 to 0	10

Solution

$$\rho_p = (1)(62.4) = 62.4\ lb/ft^3 \text{ and } \rho_p >> \rho_g$$

$$\mu = 0.00018\ g/cm\text{-}sec = (1.2 \times 10^{-5}\ lb/ft\text{-}sec)$$

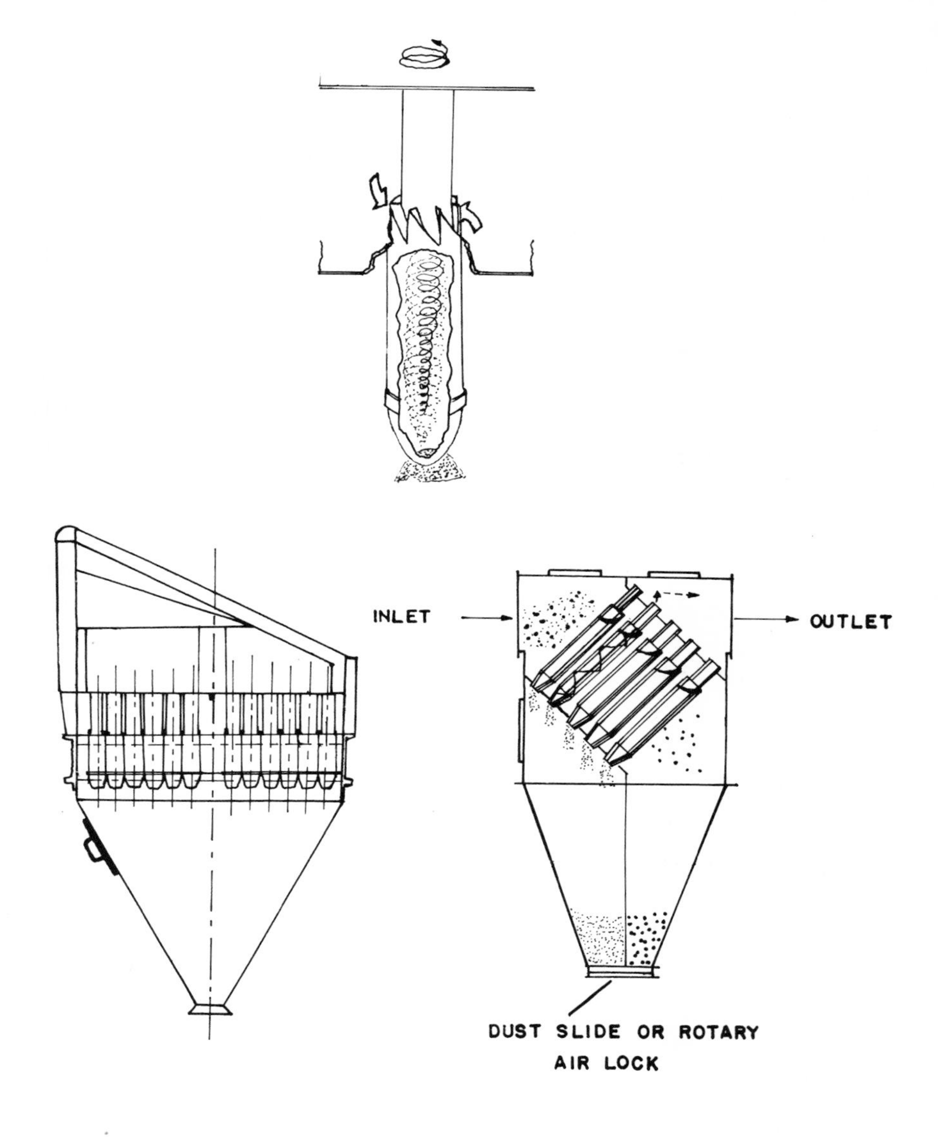

FIG. 11. Multiple conventional cyclone designs.

From Eq. (13)

$$D_{pc} = \sqrt{\frac{(9)(1.2 \times 10^{-5})(20)}{(2)(3.14)(8)(70)(12)(62.4)}}$$

$$D_{pc} = 2.9 \times 10^{-5} \text{ ft} = 8.8 \ \mu m$$

From Fig. 10, make up the table

Grade size μm	Average D_p μm	D_p/D_{pc}	% Efficiency
100 to 20	60	6.8	99
20 to 10	15	1.7	73
10 to 2	6	0.68	33
2 to 0	1	0.114	10

These efficiencies will be the same for each stage and the following table can be constructed:

Grade size μm	First Stage		Second Stage		Both Stages
	% in	% removed	% in	% removed	% removed
100 to 20	50	49.5	0.5	0.495	49.995
20 to 10	20	14.6	5.4	3.950	18.55
10 to 2	20	6.6	13.4	4.420	11.02
2 to 0	10	0.1	9.9	0.099	0.199
	100	70.8	29.2	8.964	79.764

The overall efficiency of the first stage is 70.8% and for the second stage it is 8.964/29.2 or 30.7%. Check the general relation for collectors in series (6)

$$\eta = \eta 1 + \eta 2 \ (1-\eta 1)$$

$$\eta = 70.8 + (30.7)(100-70.8) = 79.76$$

which agrees with the value in the table. The grain loading of the exit gas is (2)(1-0.796) = 0.408 grains/ft^3 of gas.

From Eq. (12) and Fig. 9, with $B_c = D_c/4 = 20$ in., $H_c = D_c/2 = 40$ in., $A_d = B_cH_c = 800$ in.^{2}, $D_e = D_c/2 = 40$ in., and $\rho_g = 0.075$ lb ft^3, substituting

$$\Delta p = \frac{(13)(0.075)(70)^2(800)}{(2)(32.2)(40)^2(144)}$$

Then

$$\Delta p = 0.257 \text{ psi} = 7.21 \text{ in. of water for each stage, or 14.4 in. for the unit.}$$

REFERENCES

1. Danielson, J.A., ed., "Air Pollution Engineering Manual," Dept. of Health, Education and Welfare, U.S.P.H.S. Publ. No. 999-AP-40, 1967 pp. 25-99.

2. *Industrial Ventilation*, 7th ed., Am. Conf. of Govern. Industrial Hygienists, Comm. on Industrial Hygiene, Lansing, Michigan, 1962.

3. Bennett, C.O. and J.E. Myers, *Momentum, Heat, and Mass Transfer*, McGraw-Hill, New York, 1962, pp. 239-395.

4. Perry, John H., ed., *Chemical Engineers' Handbook*, 4th ed., McGraw-Hill, New York, 1963, Section 9.

5. Strauss, W., *Industrial Gas Cleaning*, Pergamon, 1966, pp. 144-211.

6. Stern, A.C., ed., *Air Pollution*, Vol. III, Academic, New York, 1968, pp. 360-400.

Chapter 3

FABRIC FILTERS

Paul G. Gorman
A. Eugene Vandegrift
and Larry J. Shannon

Midwest Research Institute
Kansas City, Missouri

I. SUMMARY

This chapter discusses the theory and characteristics of fabric filters and presents design and cost information for filter systems in industrial applications. Most of the filters installed to control particulate emission from industrial operations provide high collection efficiency, in excess of 99%, at pressure drops of 2-6 in. of water. However, the design air-to-cloth ratio may vary from 1 cfm/ft^2 up to 30 cfm/ft^2 depending on the specific application and the characteristics of the particulate matter.

Several types of natural and synthetic fabrics are used in industrial filters, including cotton, Dacron, and fiberglass. The fabric may be cleaned by mechanical shaking or, as in many newer installations, by various methods of reverse air flow.

The purchase cost of fabric filters varies from $0.50/cfm up to $2.00/cfm, while the total installed cost ranges from $1.50/cfm up to $6.00/cfm. The typical annual cost for the fabric filter systems, based on 15-year depreciation, is about $1.05/cfm.

Fabric filter systems are used in several industries. some of the largest users are in cement manufacture, iron and steel production, and nonferrous metal smelting operations. Filters also have potential application for coal-fired power plants and incinerators.

II. INTRODUCTION

A simple and very effective method of removing particulate matter from a gas stream is to filter it through cloth or some other porous material. One of the earliest applications of this technique was to cover the mouth and nose of miners with cloth masks to prevent inhalation of dust. This same technique was later extended to the cleaning of exhaust gases from metallurgical operations in order to recover valuable constituents and to prevent injury to the surrounding inhabitants and livestock. Single bags were used at first; shortly thereafter units consisting of several bags were adopted. The method of cleaning was by manual shaking. Large fabric filter installations were applied to smelting operations as early as 1890. Mechanical cleaning methods were applied in the early 1900s and the Hersey reverse jet filter was developed in the 1940s. This was followed by other reverse air cleaning techniques.

Present-day industrial fabric filters, often referred to as "baghouses," usually consist of several tubular or envelope-shaped elements of woven or felted fabrics. These fabrics may be cotton, wool, fiberglass, or a variety of synthetic materials. When operating properly, they have collection efficiencies in excess of 99.9% with pressure drops ranging from 2 to 6 in. of water. Cleaning may be accomplished by mechanical shaking or several variations of reverse air flow to dislodge the collected dust layer. Operating temperatures vary from ambient up to 550°F.

The principal advantages of fabric filter systems in such installations are (1):

- Particle collection efficiency is very high and can be maintained at high levels.
- Efficiency and pressure drop are relatively unaffected by large changes in inlet dust loadings for continuously cleaned filters.
- Filter outlet air may be recirculated within the plant in many cases.
- The collected material is recovered dry for subsequent processing or disposal.
- There are no problems of liquid waste disposal, water pollution, or liquid freezing.
- Corrosion and rusting of components is usually not a problem.
- There is no hazard of high voltage, thus simplifying maintenance and repair and permitting collection of flammable dusts.
- Use of selected fibrous or granular filter-aids permits the high-efficiency collection of submicron smokes and gaseous contaminants.
- Filter collectors are available in a large number of configurations, resulting in a range of dimensions and inlet and outlet flange locations to suit installation requirements.

Some limitations in the use of fabric filters include:

- Fabric life may be shortened in the presence of acid or alkaline particles or gas constituents and at elevated temperatures.
- Temperatures much in excess of 500°F require special refractory mineral or metallic fabrics that are still in the development stage.
- Hygroscopic materials, condensation of moisture, or tarry adhesive components may cause crusty caking or plugging of the fabric, or may require special additives.
- Certain dusts may require fabric treatments to reduce seeping of the dust, or in some cases, to assist in the removal of the collected dust.
- Concentrations of some dusts in the collector(~50 g/m^3) may represent a fire or explosion hazard if a spark is admitted by accident. Fabrics can burn if readily oxidizable dust is being collected.
- Replacement of fabric may require respiratory protection for maintenance personnel.

A considerable amount of work has been done in investigating the theory of fabric filtration. Some of the important factors regarding this work are discussed in the following section. However, most of this work has been directed toward particle collection on bare fibers, which represents only the initial phase of filtration, that is before any dust layer has accumulated on the filters. The effects of deposited material on the subsequent performance of the filter are considerably more important, but have not been studied as extensively, making the correlation of theory with operating practice very difficult.

The actual operation of fabric filters and their characteristics are described in Sec. IV of this chapter, which deals with several important design parameters, including configuration, air-to-cloth ratio, cleaning methods, fabric selection, and precooling requirements. These parameters relate to the important evaluation of the economics of fabric filters. This evaluation is presented in Sec. V. A description of some of the common industrial applications of fabric filters and special aspects of these applications is given in Sec. VI.

III. THEORY

The collection of particles from a gas stream by means of a fabric filter is the result of several collection mechanisms. These mechanisms include impaction, diffusion, electorstatic forces, and gravity, all of which may contribute to the collection of particles as the gases pass through the accumulated layer of dust on the fabric. To visualize the process, imagine that the dust layer accumulated on a filter is not really a layer at all. It is probably more like a maze of passageways through which the gas must pass before it reaches the downstream side of the dust and fabric. A particle entering any pathway through the maze will be blocked if any opening in the path is smaller than the particle. Also, the many changes in direction will promote impaction and the numerous voids will enhance capture by diffusion. Electrostatic and adhesive forces may also be involved. The cumulative effect of all of this produces the high over-all collection characteristic of fabric filtration. Table 1 presents a simple method of judging the importance of each mechanism, based only on particle size as developed by The American Petroleum Institute (2). Consideration of the factors influencing each of these mechanisms is presented in the following sections.

TABLE 1

Particle Size Range of Importance
For Different Collection Mechanisms (2)

Mechanism	D_p particle size range of importance (μ)
Impaction	> 1
Diffusion	< 0.01 to 0.5
Electrostatic	0.01 to 5.0
Gravity	> 1

A. COLLECTION MECHANISMS

1. Inertial Impaction

Inertial impaction occurs when a particle approaches an obstacle that is part of the filter medium or part of the already deposited cake. If the particle is of such a size that the gas flow around the obstacle does not deflec the particle sufficiently, its inertia will bring it into contact with the obstacle. Collection efficiency, due to impaction, has been studied as a function of an impaction parameter (I) defined as the following (3):

$$I = \frac{C_s \rho_p D_p^2 U_s}{9 \mu_f D_o} \quad (1)$$

Equation (1) shows that the impaction parameter is directly proportional to the particle density (ρ_p), the undisturbed stream velocity (U_s), and the square of the particle diameter (D_p). D_o is the diameter of the collecting obstacle, μ_f is the fluid viscosity, while C_s is the Cunningham-Millikan aerosol particle slip correction factor.

The parameter (I) has a value of about 1.0 for filtration of a 10-μ particle at a velocity of 2 ft/min (3). The associated collection efficiency would be on the order of 30% for a single collection sphere. However, in an operating filter, the velocity through the collected layer is not the same as the undistrubed stream velocity and there are many potential collection spheres in the varied particle path through this layer. Collection by impaction becomes more significant for particles greater than 1 μ and becomes a major factor for larger particles.

2. Diffusion

The migration of particles to a collection fiber or deposited dust particle has been shown to be a function of the Peclet number (P_e) defined as (3):

$$P_e = D_o U_s / D_{se} \tag{2}$$

where D_o is the diameter of the collecting object, U_s is the undistrubed stream velocity, and D_{se} is the particle diffusion coefficient. Collection by diffusion is only significant for particles less than approximately 1 μ in diameter. The efficiency increases as velocity decreases because of the increased time available for random Brownian or thermal motion to bring a particle into contact with a collecting surface.

3. Electrostatic Forces

When particles in a gas stream or in the collection object, or both, have an electrostatic charge, this will also influence the collection efficiency. Particles conveyed in a gas stream may acquire an electrostatic charge, but it has been difficult to measure or assess the effect of such charges. Some experimental work has been done in this area with qualitative results indicating substantial improvement in efficiency due to the effect of electrostatic charge (4). Although such phenomena may improve collection efficiency, they may also affect fabric cleanability. Most designs of filter systems do not consider electrostatic forces, although this may be inherent in selection of a lower or higher air-to-cloth ratio for a particular application.

4. Gravitational Settling

When a dirty gas stream enters a fabric collector, the gas velocity is decreased and gravity settling may bring about removal of particles without actual collection by the filtering surfaces. Collection by settling is a function of the terminal settling velocity, which is given in Table 2 (5).

TABLE 2

Terminal Velocities of Rigid Spheres of Unit Density in Air at 760 mm Hg Pressure and 20°C (5)

Diameter (μ)	Terminal velocity (cm/sec)
0.1	8.71×10^{-5}
0.2	2.27×10^{-4}
0.4	6.85×10^{-4}
1.0	3.40×10^{-3}
2	1.29×10^{-2}
4	5.00×10^{-2}
10	3.03×10^{-1}
20	1.20
40	4.71
100	24.7

Effects of gravity on submicron aerosol filtration through fiberglass, sand, and lead shot (500μ) were studied by Thomas and Yoder (6) with typical results, as shown in Fig. 1. It is apparent that penetration was reduced (greater efficiency) for downflow, even for particles of less than 1 μ as used in these studies. [Experimental verification of the effects of collection by diffusional and impaction mechanisms at different particle sizes may be observed in these data, maximum penetration or minimum efficiency occurring in the vicinity of 0.35μ (5)].

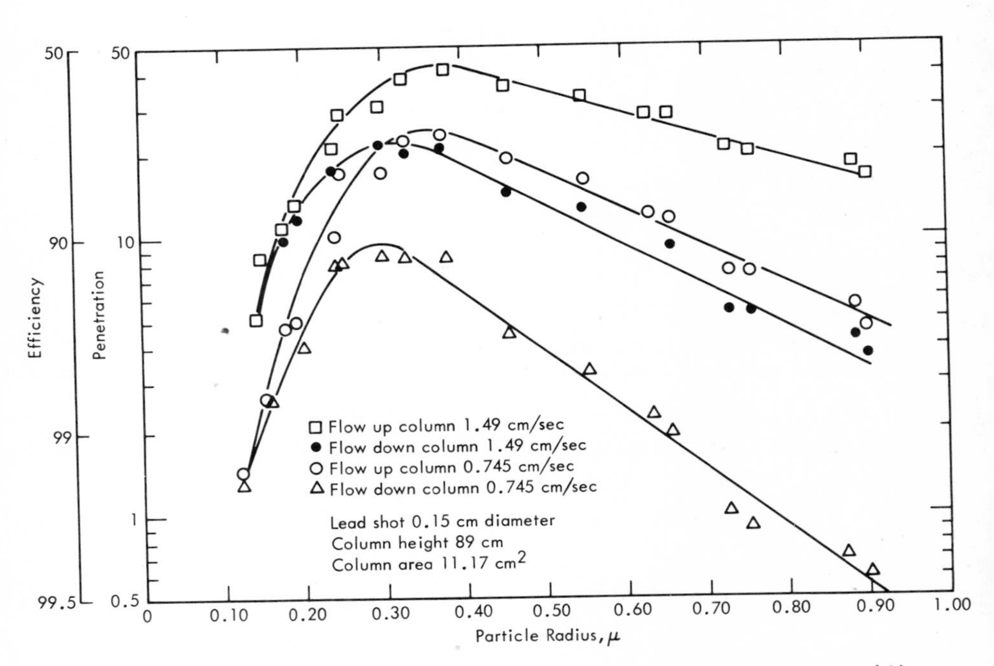

FIG. 1. Filteration of aerosols through lead shot (6).

B. PRESSURE DROP THROUGH FABRIC FILTERS

A factor of major importance in a fabric filter is the pressure drop caused by the combined resistances of the fabric and the accumulated dust layer. The resistance of the fabric alone is affected by the type of cloth and the weave. This resistance varies directly with the air flow. The permeability of various fabrics is usually specified by the manufacturer and expressed as the air flow rate (cfm) through 1 sq ft of fabric when the pressure differential is 0.5 in. of water. At normal filtering velocities, the resistance of the clean cloth is usually less than 10% of the total resistance, so it is relatively insignificant in comparison with the resistance of the dust layer. The openings between the fibers are, for the most part, many times larger than the size of the particles that are collected. Thus a new filter bag has a lower collection efficiency initially and low pressure drop. Soon after the bag is in service, the first coating of particles is formed. This bridges and partially plugs up the openings so that the normally high efficiency of filtration is attained. Even after the first and subsequent cleaning cycles, the efficiency remains high because the accumulated dust is not entirely removed and that which remains serves as a precoating for the next cycle of operation.

The pressure drop through the accumulated dust layer may be considered as that of flow through a packed bed. In this case, the pressure drop may be expressed by Darcy's equation as:

$$\frac{\Delta P}{L} = \frac{\mu_f U_s}{K} \tag{3}$$

where L is bed depth, μ_f is gas viscosity (see Fig. 2), K is permeability of the bed, and U_s is gas velocity (cfm/ft^2).

The permeability (K) is determined by the properties of the dust, including particle size and shape. Laboratory and field measurements of permeability of different dust cakes show wide variation, and it is difficult to predict permeability on the basis of particle properties. One of the most critical properties accounting for this variation

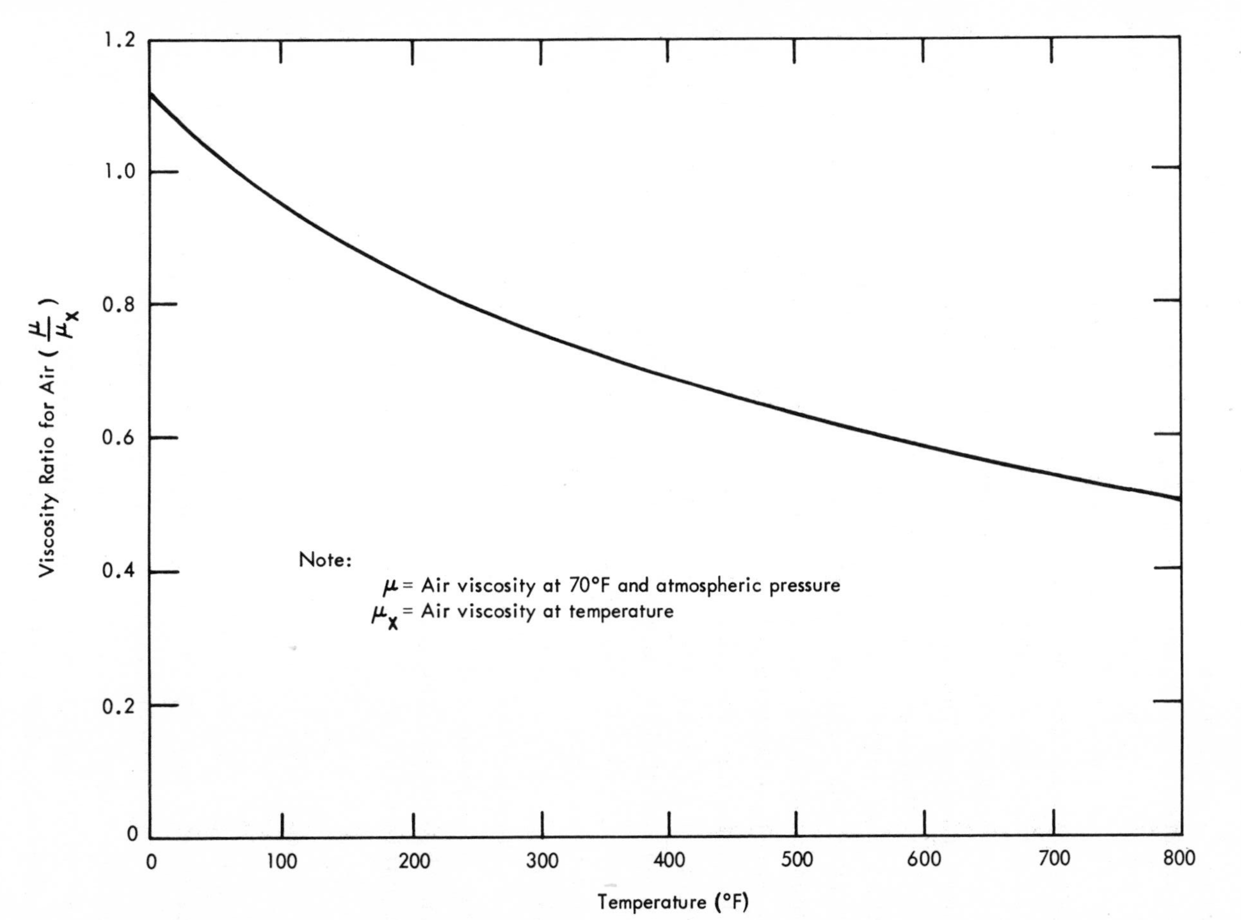

FIG. 2. Air viscosity ratio as a function of gas temperature (7).

is the wide range of porosities in fabric filter dust cakes. Other pressure drop relationships have been attempted, but because of dependence on porosity, they generally correlate the pressure drop only to within a factor of 2 at best (8).

The pressure drop through the dust layer has been found to be directly proportional to the thickness of the accummulated dust deposit. It was also found that resistance increases with decreasing particle size. Equation (3) indicates that the pressure drop would also be directly proportional to filtering velocity. However, test results on filtration of fly ash by Borgwardt et al. indicate that the resistance of the dust layer increases with the 1.5 power of the filtering velocity (9).

These tests on fly ash also showed that frequency of cleaning is not an important factor in minimizing pressure drop and "...suggests that the shorter cycles which are customarily used may greatly overclean bags producing excessive wear with little or no real benefit" (9). The explanation for this observation may be the residual drag of the fabric and dust that remains after each cleaning cycle, as shown in Fig. 3 from this study. Residual drag is a major part of the total pressure drop and it might be expected that increased frequency of cleaning would provide decreasing benefits in removing the residual dust embedded in the fabric.

Although a considerable amount of study has been devoted to filtration theory, it has been difficult to relate the collection efficiency and pressure drop to the actual operation of industrial fabric filters. Therefore most filter designs are based on previous experience and general industry practice. The characteristics of operating filters and their design considerations are discussed in the following sections.

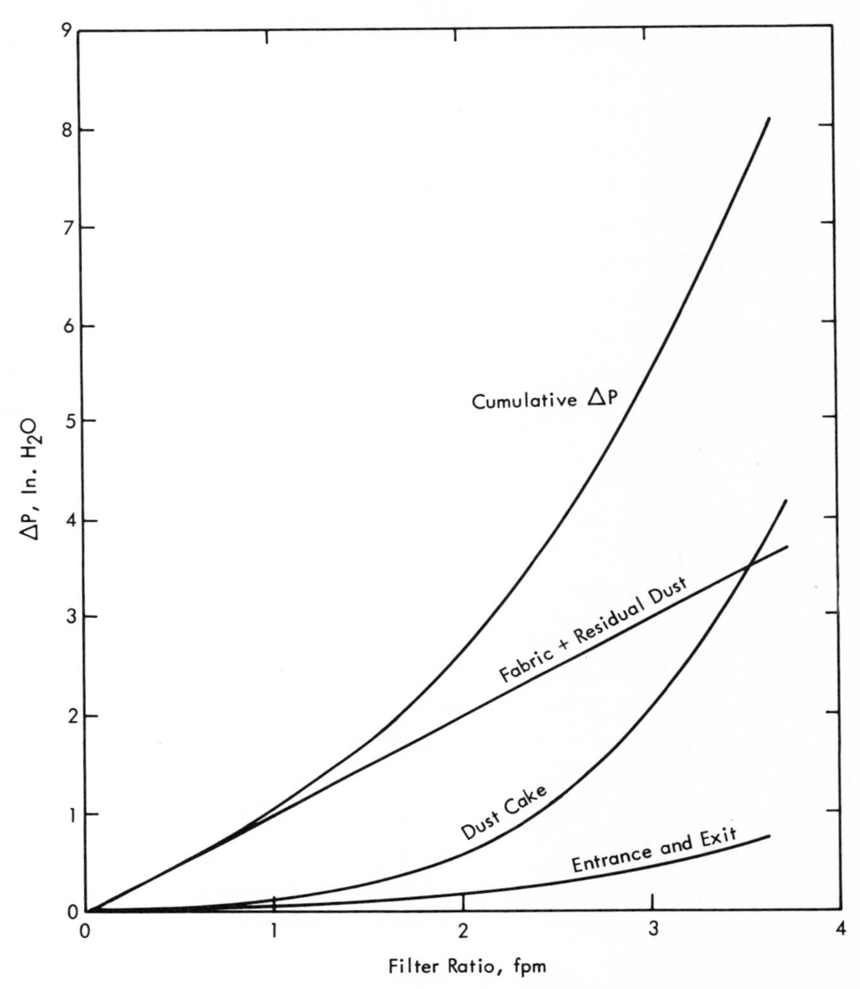

FIG. 3. Contribution of filter components to pressure drop (9)

IV. CHARACTERISTICS OF FABRIC FILTERS AND DESIGN FACTORS

A. CHARACTERISTICS OF FABRIC FILTERS

Almost all filters are one of two types, either envelope or cylindrical (bag). The cylindrical type may use outside filtering or inside filtering, as depicted in Fig. 4 (10). One of the advantages of inside filtering is that the filter can be entered for inspection during operation if the gas is not toxic and if operating temperature permits. The size of the filter (surface area of fabric) is

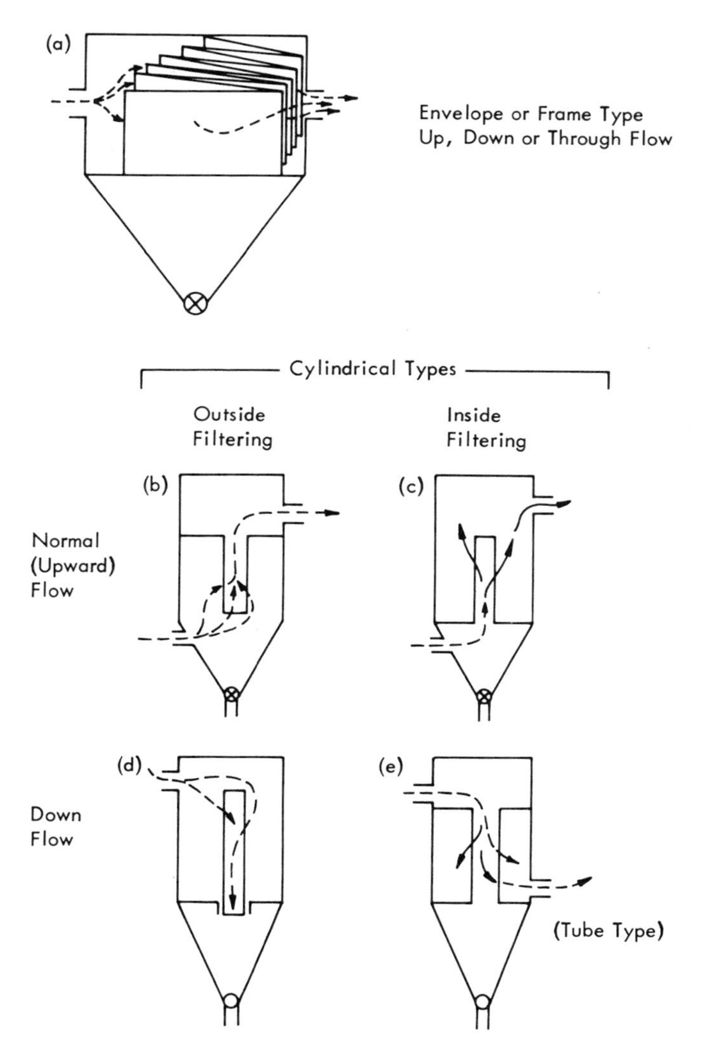

FIG. 4. Configurations of fabric filters (10).

very small in some cases (<10 ft^2) while some are as large as 1 x 10^6 ft^2. The larger units are often compartmented. Pressure drop in fabric collectors is 2-4 in. of H_2O, although the fan necessary for this type of collector is usually designed to provide about twice this amount to overcome ducting and entrance losses. The conveying velocities in the ductwork are normally 2000-3000 ft/min for economic reasons and to help prevent sedimentation [see Tables 3 and 4 (11)].

Normal inlet dust loadings for fabric filters range from 0.1 to 10.0 grains/ft^3 and mainly effect the required frequency of cleaning. Outlet dust loadings are usually less than 0.01 grain/ft^3, although in some cases it may be as high as 0.18 grain/ft^3 (12). Higher penetration of dust is commonly caused by bag wear. The actual collection efficiency of filters in good operating condition generally exceeds 99.99%. Even with high inlet loadings, efficiency remains above 99.9% and is primarily dependent on cleaning mechanisms and cleaning cycles.

B. AIR-TO-CLOTH RATIO

One of the major factors in the design and operation of a fabric filter is the air-to-cloth ratio used. This is the ratio of the quantity of gas entering the filter (cfm) to the surface area of the fabric (ft^2). It is therefore expressed in units of cfm/ft^2, or is sometimes referred to as the filtering velocity (ft/min). The air-to-cloth ratio may be as low as 1 cfm/ft^2 up to as high as 30 cfm/ft^2. Many filters are designed for 10-15 cfm/ft^2. In general, a lower air-to-cloth ratio will be used for gases containing smaller particles or particles which may otherwise be "difficult" to collect. A higher air-to-cloth ratio will be used for gas streams containing larger particles or those that are easier to collect.

TABLE 3

Conveying Velocities For Dust Collecting (11)

Material	Velocity (ft/min)	
	From	To
Wood flour sander dust	1500	2000
Sawdust, light, dry	2000	3000
Sawdust, heavy, wet or green	3000	4000
Shavings, light, dry	2000	3000
Shavings, heavy, wet or green	3000	4000
Wood blocks, edgings, heavy, wet or green	3500	4500
Hog waste	3500	4500
Grinding dust	3000	4000
Foundry dust, tumbling barrel, shake-out	3500	4500
Sand blast dust	3500	4000
Buffing lint, dry	2500	3000
Buffing lint, sticky	3000	4000
Metal turnings	4000	5000
Lead dust	4000	5000
Cotton	2500	3000
Cotton lint, flyings	1500	2000
Wood	3000	4000
Jute lint, flyings	2800	3000
Jute picker stock, shredded bagging	3000	3300
Jute dust shaker waste	3100	3400
Jute butts (conveying)	3600	4500
Grain	2500	4000
Ground feed, $\frac{1}{2}$ in. screen	4500	5000
Grain dust	2000	3000
Coffee beans	3000	3500

TABLE 3 (cont.)

Material	Velocity (ft/min)	
	From	To
Shoe dust	3000	4000
Rubber dust	2000	2500
Bakelite molding powder	3000	3500
Bakelite molding powder dust	2000	2500
Granite dust and surfacer chips	3000	4000

TABLE 4

Velocities For Low-Pressure Pneumatic Conveying (11)

Material	Velocity (ft/min)	
	From	To
Wood flour	4000	6000
Sawdust	4000	6000
Hog waste	4500	6500
Pulp chips	4500	7000
Tanbark, dry	4500	7000
Tanbark, leached, damp	5500	7500
Cork, ground	3500	5500
Metal turnings	5000	7000
Cotton	4000	6000
Wool	4500	6000
Jute	4500	6000
Hemp	4500	6000
Rags	4500	6500
Cotton seed	4000	6000
Flour	3500	6000

TABLE 4 (cont.)

Material	Velocity (ft/min)	
	From	To
Oats	4500	6000
Barley	5000	6500
Corn	5000	7000
Wheat	5000	7000
Rye	5000	7000
Coffee beans, stoned	3000	3500
Coffee beans, unstoned	3500	4000
Sugar	5000	6000
Salt	5500	7500
Coal, powdered	4500	6000
Ashes, clinkers, ground	6000	8500
Lime	5000	7000
Cement, portland	6000	9000
Sand	6000	9000

The selection of the air-to-cloth ratio for a fabric filter installation is primarily made on the basis of normal industry practice or the recommendation of the filter manufacturer. Field performance data, including air-to-cloth ratios, were compiled in a recent study funded by EPA and are shown in Table 5. However, many factors must be considered in selection of the appropriate air-to-cloth ratio and design of the filter system. Without prior experience, it is best to consult with the manufacturer of these collectors, some of whom are listed in Ref. 14.

One of the more important factors that enters into the determination of the air-to-cloth ratio is the method of cleaning the fabric. Most manufacturers can provide filters using a variety of cleaning methods; these methods are discussed in the next section.

TABLE 5

Field Performance Data for Selected Fabric Filter Installations (13)

Category, dust name, process	Efficiency (%)	Dust: av. particle size (μ)	Concentration (grains/ft³)	K-value	Gas: inlet temperature (°F)	Corrosive?	Continuous?	Steady?
Combustion								
Coal ash-boiler		20	1.9	8.5	270	--	N	Y
Coal ash-boiler	99.5					--	Y	Y
Coal ash-boiler				(7)[d]		--	Y	Y
Coal ash-lab		10	3.0	(8)	Amb	N	N	--
Coal ash-lab	99.9	16	Var.		Amb	N	Y	Y
Oil ash-boiler	99.9				300	Y	Y	Y
Oil ash-boiler			0.013	(4.7)	285	Y	Y	Y
Municipal incin.	99.8	Var.	0.46	25	(500)	--	Y	N
Municipal incin.	96.5	Var.	0.33	180	480	Y	N	N
Refuse incin.	99.6		0.25		410	--	N	N
Food and Feed								
Flour mill	99.9	10	14	4.3	110	N	Y	Y
Inorganic Chemicals								
Lime kiln			7.5	9	500	N	Y	Y
Lime kiln		5	8.6		(450)			
Dolomite (MgO) millin		40	7.0	(670)	150	N	Y	N
Alumina handling		100		0.2	Amb	N	N	Y
Hypochlorite mfgr.		50	2.3	15	Amb	--	Y	--
Chromium salts								
Organic Chemicals								
Carbon black mfgr.	99.5		15	56	425	--	Y	Y
Carbon black mfgr.	99.99		25	38	375	--	Y	Y
PVA handling	99+	3	0.05	(25)	Amb	Y	Y	N
Misc. resin handling		5	3	8.0	Amb	N	Y	N
Nonmetallic Minerals								
Cement kiln-dry	99.8	10	10		510	--	--	--
Cement kiln-dry	99.6	10	2.8		510	--	Y	Y
Cement kiln-wet	99.9	10	12		320	--	--	--
Cement kiln-wet	99.5				600	--	Y	Y
Cement kiln-wet					550	--	Y	Y
Cement kiln-dry					650	--	Y	Y
Cement kiln-wet	99+	(15)	0.5	12	525	--	Y	Y
Cement milling			(5)	70	325	--	Y	--
Cement loading		10			Amb	N	N	Y
Cement bagging					Amb	N	N	N
Cement--general		10			Var.	--	--	--
Stucco classifying		5			220	N	N	Y
Stucco conveying		40			200	--	N	Y
Stucco board trim		Var.	0.7	9	140	N	N	N
Glass frit smelter			0.3		185	--	--	--
Glass mfgr. handling		100			Amb	--	Y	N
Abrasive mfgr.					Amb	--	--	--
Asphalt rock proc.	99.9		27		400	--	--	--
Asphalt rock proc.		5	1.6	(1.7)	150	N	--	--
Asphalt rock proc.		(20)			Amb	N	N	N

TABLE 5 (cont.)

Gas Flow 1000 acfm	Air-to-cloth ratio(cfm/ft^2)	Pressure difference (in. of H_2O)	Cloth: cloth name[a]	Cloth life (Yr)	Filter Type[b]	Cleaning Mechanism[c]	Costs: ($/cfm) installed	Operating	Maintenance	Total annual
20	2.3	5.6	Gl	--	I.D	RFC				
1.0	(2.5)	4	Gl	(2)		RFC,V				
10	(3.0)	(2.5)	Gl		I.U	RF				
0.032	4.0	(3)	Var.		I.D	Sh				
0.50	8	(5)	Wo		O.U	RP				
2.5	7.7	3.2	Gl		I.U,D	RFC				
820	6.5	5.7	Gl	1.0	I.U	RFC	2.34			0.46
0.375	5	4	Gl	> 1	O.U	RF				
14	2.5	3.6	Gl	> 2	I.U	RFC	2.01			
5.3	10)	3.7								
8.4	8.6	4.5	Wo	1.5	I.D.	RJ	2.36	0.20	0.07	0.39
140	2.3	5	Gl	1.2	I.U	RFC,FP	7.94	0.12	0.26	
	2.1		Gl	2			1.80	(0.06)	(0.14)	
3.3	0.7	3	Da	1	I.U	RF		0.07	0.43	
0.11	1.9	1.8	Da	0.25	O.U	RP	50		10.00	
10	3.3	3	Dy	0.6	I.U	Sh		0.07	0.39	
26	2.4	5	Wo		I.D	RJ				
43	1.6	7.5	Gl	1	I.U	RF		0.18	0.33	
18	1.1	5.8	Gl	2	I.D.	RF,Sh				
14	10	(5)	Wo	0.33	I.D.	RF		0.12	0.22	
4	2.7	3	Co	1	I.U	Sh		0.07	0.24	
300			Gl							
300	2.0		Gl	2.0	I.U	RFC	3.33			
146			Gl							
140	2.0	(7)	Gl	3.0		RFC	1.16	0.14	0.17	
95	2.1	3	Gl	(2)	I.U	RFC				
		2.5	Gl	(2)	I.U	RFC				
343	1.5	4	Gl	1	I.U	RFC				
10	1.8	2.5	Co	1	I.U	Sh			0.36	
2.2	1.8		Co	3	Env.	RF			0.21	
8.3	2.5	2.5	Co	0.75	Env.	RF	1.30	0.08	0.27	
(0.03)	3.0	3	Var.	Var.	I.U	Sh				
6	7.5	2.7	Da	2.5	O.U	RP		0.11		
10			Co	3	Env.	RF				
7	3.4		Co	3	I.U	Sh			0.36	
9.3	2.2		Da	(3)						
0.17	2.3	3	Co	0.33	O.U	Sh		0.53	0.94	
Var.	2.5	3	Co	1		Sh				
41	7.5		Ny	(2)	O.U	RP		0.15	(0.17)	
30	9.0	3.5	Wo		I.D	RJ				
10			Co	(5)	I.U	Sh		0.45	0.17	0.86

TABLE 5 (cont.)

Category, dust name, process	Efficiency (%)	Dust: av. particle size (μ)	Concentration (grains/ft^3)	K-value	Gas: inlet temperature (°F)	Corrosive?	Continuous?	Steady?
Iron and Steel								
Foundry cupola			0.70	120	450	--	N	N
Foundry cupola		20	(2)		275	--	N	N
Electric furnace	99	1	0.33	715	225	N	Y	N
Electric F.O.L.			2.3		275	N	Y	N
Electric F.O.L.	75	0.5	0.18		140	--	N	N
Electric furnace	97	(1)	2.7		400	--	Y	N
Electric furnace		1	0.76	45	110	--	N	N
Electric F.O.L.			0.5	(50)	150			
Electric furnace	98+		0.5	(49)	140			
Electric F.O.L. shop	99		0.09	(100)	175			
Electric F.O.L. shop		10	0.54	65	195	--	Y	N
B.O.F. - (Pompey)	99.9	(0.5)	2.2		215	--	--	N
B.O.F. - reladling		75	0.5	(120)		--	N	N
B.O.F. - reladling		80	0.5	233	200	--	N	N
Open hearth	99.5	(5)	0.8	35	430	--	N	N
Sinter drop off		Var.	1.9	12	285	--	--	--
Shake out, shot blast			7	3	Amb	N	N	N
Perlite expansion		Var.			225	--	--	--
Nonferrous Metals								
Zinc-conc. roasting	99+	8	4	50	450	Y	Y	--
Zinc-sintering		(50)	7.8	7	200	--	Y	--
Zinc-fuming kiln					400	Y	Y	--
Zinc-muffle furnace					180	--	--	--
Zinc-galvanizing		2	(0.2)	180	Amb	N	--	N
Copper-primary smelt			1.7	65	450	--	--	--
Copper-"Flammofen"	99.8		1		400	--	--	--
Copper-secondary smelt.			0.75		250	--	--	--
Copper-secondary smelt.		(1)	8	40	600	--	Y	N
Copper-converter	99+	(100)	1.8	(15)	350	Y	N	N
Copper-blast furnace	99+	(1)	1.2	18	375	--	Y	--
Lead blast furnace	99.8	(1)	3		225	--	--	--
Lead slag fuming	99.7	(1)	4		250	--	--	--
Lead blast furnace	99+	(1)	2.3	57	275	--	--	--
Aluminum spray disposal		(2)	100	0.7	(400)	--	--	--
Uranium processes	99.99	(1)	14			--	--	--
Beryllium sinter		(5)	0.0008		250	--	--	--
Miscellaneous								
Fibrous grinding		Var.	0.7	5	Amb	N	N	N
Motor room	94				Amb	N	Y	Y
Ultrafiltration, Beryllium Plant					Amb	N	--	--

[a] Cloth Name: Gl = glass; Wo = wool; Co = cotton; Nom = Nomex; Var = various fabrics reported; Da = Dacron; Ny = Nylon; Polyp = polypropylene; Dy = Dyne; Or = Orlon.

[b] Filter Type: I = dust collected on the inside of the filter element; O = collected on the outside; U = upward flow through the filter house; D = downward flow.

TABLE 5 (cont.)

Gas flow 1000 acfm	Air-to-cloth ratio(cmf/ft^2)	Pressure difference (in. of H_2O)	Cloth: cloth name[a]	Cloth life (yr)	Filter type[b]	Cleaning Mechanism[c]	Costs: ($/cfm) installed	Operating	Maintenance	Total annual
120	2.1	7.5	Gl	0.3	I.U	Sh	(7.30)		0.09	
			Or			Sh				
60	1.43	6	Da	(1.5)	I.U	RFC	> 3.90			
340	2.5		Da		I.U	RFC		0.26	0.19	
110	4	4	Var.	1.5		RF,RP	2.62			(1.84)
80	1.5		Gl	2.5	I.U	RF				
115	2.8	1.9	Or	2.5	I.U	Sh				
170	1.9	2.4	Gl		I.U	RFC			0.20	
26	1.9	3.5	Or		I.-	Sh				
700	(3)	(6)	Or		I.U	Sh	2.86		0.52	
525	1.95	5.8	Gl	4	I.U	RFC	4.85			
							7.3			
46	1.8	(8.5)	(Da)	> 4		RF,Sh				
1	4.5	(6)	Co,Da			Sh	(4.0)			
100	2.5	8.2	Da	(1)	I.U	Sh	(5.20)			
(100)	2.0	3.3	Gl	(2)	I.U	RFC,Son				
240	2.3	5.0	Gl	2	I.U	RFC				
12	5.0	(6)	Co	1.6	I.D	RJ				
	3.0		Da			Sh				
0.8	3.0	7.5	Gl	(1.5)		RFC			1.65	
90	2.3	3.5	Or	(1.5)	I.U	RFC,Sh			1.14	
100	3.0		Nom	2		Sh			1.38	
30	1.8		Or			Sh			1.15	
36	2.7	4	Co	7.5		Sh				
200	0.9	9	Gl	2	I.U	Sh,RFC				
4	(2.7)	6.5	Gl	2	I.U	RFC				
300	1.5		Or		I.U	Sh	3.60			
3.6	0.6	3	Gl	1	I.U	Sh				
150	1.3	7	Nom	3.5	I.U	Sh				
190	1.25	9	Gl	0.67	I.U	Sh	5.20			
260	2.4	6	Or	(1.5)			5.70		1.48	
195	3.5	6	Or	(1.5)			11.00		0.24	
25	1	2	Da	1.75	I.U	RF,Sh				
0.002	3	1.7	Gl			RF				
60	17	7	Wo	3 wk		RJ				0.35
40	20		Polyp	2	I.D	RP				
10	6		Co	0.1		Sh			0.45	
48	5.9		Co	9	I.U	Sh				
162	6.0		Or			Sh				

[c] Cleaning Mechanism: RFC = reverse flow accompanied by flexural collapse; V = vibration; RF = reverse flow with little or no flexure; Sh = shake; RP = reverse pulse; RJ = reverse jet ring; Son = sonic cleaning.

[d] ( ) = estimate.

C. CLEANING METHODS FOR FABRIC FILTERS

Various cleaning methods are used to remove collected dust from the filter bags to maintain a nominal pressure drop of 2-6 in. H_2O. Since the pressure drop would continue to increase as the thickness of the accumulated dust layer increases, a method must be provided to remove the dust layer. Manual or mechanical shaking methods or air flow reversal is generally used to force the collected dust off the cloth.

Many mechanical shaking methods have been used to clean fabric filters. High-frequency agitation, with little cloth travel, can be very effective. Since it is easier to vibrate or rap a compact group of taut filter elements, this cleaning method is usually employed on envelope filters. It is especially effective with deposits of medium to large particles adhering rather loosely. In this case, high filtering velocities can be used and higher pressure drops can be tolerated without danger of blinding the cloth.

Reverse jet cleaning is another method that has been in use for quite some time. It employs a carriage attached to a doughnut-shaped ring surrounding the outside of each bag. This ring contains a slot around the inner circumference next to the bag. Air is supplied to each ring, which blows through the bag to remove the dust from the inside of the bag as the ring travels up and down the bag. The air is usually supplied by a blower at pressures of 5-20 in. of H_2O. The volume required is about 5% of the primary filtering flow.

One of the newer cleaning methods uses an intermittent pulse jet of high-pressure air (100 psi) directed down into the bag. This pulse of clean air forced through the bag removes the collected dust. In some designs, lower-pressure air is used, but this may require a greater quantity of cleaning air. Felted fabrics are often used in conjunction with the pulse jet cleaning method. Higher air-to-cloth ratios (10-30 cfm/ft^2) are usually attainable with felted fabrics so their cost may be justified. It has been reported that the maintenance cost for this type of filter is lower than for those using mechanical shaking. Efficient use of the compressed air is important, since the power requirement for the compressor can equal that for the primary fan. A comparison of cleaning methods is shown in Table 6 (15).

The fabric to be used in a specific filter application often influences the cleaning method that may be used. Selection of fabrics is considered in the following section.

D. SELECTION OF FABRICS

The selection of fabric material is usually determined on the basis of the operating temperature and the resistance of the fabric to abrasion and corrosiveness of the gases. The characteristics of various fabrics in this regard are shown in Table 7 (16), and the abrasiveness of specific dusts and other dust properties are given in Tables 8 (17) and 9 (18).

As is shown in Table 7, fabrics may be constructed of a variety of cloth materials including cotton, wool, fiberglass, and manmade fibers. Many fabric weaves are also used, or the fabric may be felted, whereby the identity of the separate yarns tend to be replaced by a more uniform mat. The felted fabrics are almost always cleaned by reverse jet or pulse-jet methods. Fabric characteristics may also be altered by further treatment for specific

TABLE 6

Comparisons of Cleaning Methods[a] (15)

Cleaning method	Uniformity of cleaning	Bag attrition	Equipment ruggedness	Type fabric
Shake	Av	Av	Av	Woven
Rev. flow, no flex	Good	Low	Good	Woven
Rev. flow, collapse	Av	High	Good	Woven
Pulse-compartment	Good	Low	Good	Felt, woven
Pulse-bags	Av	Av	Good	Felt, woven
Reverse jet	V. Good	Av-high	Low	Felt, woven
Vibration, rapping	Good	Av	Low	Woven
Sonic assist	Av	Low	Low	Woven
Manual flexing	Good	High	---	Felt, woven

[a]These value judgments do not permit comparison of performance aspects, only of methods.

[b]Fabric limited.

TABLE 6 (continued)

Filter velocity	Apparatus cost	Power cost	Dust loading	Maximum temperature[b]	Submicron efficiency
Av	Av	Low	Av	High	Good
Av	Av	Med. low	Av	High	Good
Av	Av	Med. low	Av	High	Good
High	High	Med.	High	Med.	High
High	High	High	V. high	Med.	High
V. high	High	High	High	Med.	V. high
Av	Av	Med. low	Av	Med.	Good
Av	Av	Med.	--	High	Good
Av	Low	--	Low	Med.	Good

TABLE 7

Filter Fabric Characteristics (16)

Fiber	Operating exposure (°F) Long	Short	Supports combustion	Air permeability[a] (cfm/ft^2)
Cotton	180	225	Yes	10-20
Wool	200	250	No	20-60
Nylon[d]	200	250	Yes	15-30
Orlon	240	275	Yes	20-45
Dacron[d]	275	325	Yes	10-60
Polypropylene	200	250	Yes	7-30
Nomex[d]	425	500	No	25-54
Fiberglass	550	600	Yes	10-70
Teflon[d]	450	500	No	15-65

[a] cfm/ft^2 at 0.5 in. w.g.

[b] P = poor, F = fair, G = good, E = excellent.

[c] Cost rank, 1 = lowest cost, 9 = highest cost.

[d] Du Pont registered trademark.

TABLE 7 (cont.)

Composition	Abrasion[b]	Mineral acids[b]	Organic acids[b]	Alkali[b]	Cost[c] rank
Cellulose	G	P	G	G	1
Protein	G	F	F	P	7
Polyamide	E	P	F	G	2
Polyacrylonitrile	G	G	G	F	3
Polyester	E	G	G	G	4
Olefin	E	E	E	E	6
Polyamide	E	F	E	G	8
Glass	P-F	E	E	P	5
Polyfluoroethylene	F	E	E	E	9

TABLE 8

Apparent Density and Abrasiveness For Some Industrial Dusts Collected in Fabric Filters (17)

Dust	SW	BW	Code[a]
Alfalfa meal	34	17	N
Alum	103	55	N
Alumina	250	60	VA
Aluminum	165	160	M
Amonium chloride	94[b]	52	M
Antimony	414	417	VA
Asbestos, shred	153	23	M
Ashes, hard coal	31	35	VA
Ashes, soft coal	43	43	VA
Asphalt, crushed	87	45	VA
Ammonium sulfate	113	45	M
Bagasse	20	8	M
Bakente, powdered	100[b]	40	N
Baking powder	80	41	N
Bauxite, crushed	158	80	VA
Beans, meal, etc.	82[b]	41	N
Bentonite	110[b]	51	VA
Bicarbonate of soda	137	41	N
Bonemeal	75	55	M
Bones, ground, minus 1/8 in	100[b]	50	M
Boneblack	65	23	M
Bonechar	80	40	M
Borax, powdered	109	53	VA
Bran	35	15	N
Brass	530	165	M
Brewers grain, spent, dry	65	28	N
Brick	118	135	A
Calcium carbide	137	80	A
Calcium carbonate	169	147	A
Carbon, amorph., graph.	260	130	M

TABLE 8 (cont.)

Dust	SW	BW	Code[a]
Carbon black, channel	15	5	M
Carbon black, furnace	15	5	M
Carborundum	250	195	VA
Casein	80	36	M
Cast iron	450	200	VA
Caustic soda	88	40[b]	M
Cellulose	94	80[b]	M
Cement, portland	100	80	VA
Cement, clinker	131	78	VA
Chalk, minus 100 mesh	143	73	A
Charcoal	25	21	N
Cinders, coal	46	43	A
Clay, dry	85[b]	63[b]	A
Coal, bituminous	87[b]	50	A
Coal, antracite	100[b]	55	A
Coca, powdered	70[b]	35	N
Coconut, shredded	45[b]	22	N
Coffee	48	25	N
Coke, bituminous	83	30	A
Coke, petroleum	110	40	A
Copra (dried coconut)	45	22	N
Cork, fine ground	30	15	M
Corn, cracked, etc.	70	50	N
Cornmeal	80	40	N
Cullet (broken glass)	140[b]	100	A
Dicalcium phosphate	144[b]	43	M
Dolomite	181	100	A
Ebonite, crushed	72	59	N
Egg powder	35[b]	16	N
Epsom salts	162[b]	45	M
Feldspar	160	70	A

TABLE 8 (Cont.)

Dust	SW	BW	Code
Ferrous sulphate	118	60	A
Fish meat	80[b]	40[b]	N
Flour	50	35	N
Flue dust, dry	235	117	M
Fluorspar	200	82	A
Fly ash	85	40	VA
Fullers earth	95[b]	47	A
Gelatin, granulated	65[b]	32	N
Glass batch	162[b]	95	A
Glue, ground	80	40	M
Gluten meal	80[b]	40	N
Grains, distillery, dry	70[b]	30	N
Graphite	132	40	A
Gypsum	145	75	A
Ilmenite ore	312	140	VA
Iron oxide	330	100[b]	VA
Lead	710	710	A
Lead arsenate	400	72[b]	A
Lead oxide	567	180[b]	A
Lignite	85	50[b]	A
Lime, ground	87	60	VA
Lime, hydrated	81	40	A
Limestone	163	85	VA
Litharge	560	180[b]	A
Magnesite	187	187[b]	VA
Magnesium	109	100[b]	VA
Magnesium chloride	138	33	A
Malt, dry	44	26	N
Manganese sulfate	125	70	A
Maple, hard	4	43	N
Marble	168	96	A

TABLE 8 (Cont.)

Dust	SW	BW	Code
Marl	120	80	A
Mica, ground	175	15	M
Milk, dried, powdered	70	35	N
Monel metal	554	550	M
Muriate of potash	160	77	M
Naphthalene flakes	71	45	N
Oak	47	15	N
Oxatic acid crystals	104	60	N
Phosphate rock	160	80	VA
Phosphate sand	190	95	VA
Porcelain	150	75	M
Quartz	165	100	VA
Resin	67	35	M
Rubber, ground	72	23	N
Rubber, hard	74	59	N
Rubber, soft	69	55	N
Salt, rock	136	45	A
Salt, dry, coarse	138	50	A
Salt, dry, pulverized	140[b]	75	A
Saltpeter	132	80	N
Sand	150	100	VA
Sandstone	144	95	VA
Sawdust	35	12	N
Shale, crushed	175	87	A
Slag, furnace, granulated	132	62	VA
Slate	172	85	A
Soap, chips, flakes	30	15	N
Soap powder	50	25	N
Soapstone talc	175	62	M
Soda ash, light	74	35	M
Soda ash, heavy	134	65	M

TABLE 8 (Concluded)

Dust	SW	BW	Code
Sodium nitrate	134	70	A
Sodium phosphate	94	45	A
Soybeans, meal	90	45	N
Starch	96	40	N
Steel	487	100	A
Steel chips, crushed	487	60	A
Sugar	105	53	N
Sulfhur	126	60	N
Talc	169	60	M
Tanbark, ground	110	55	M
Tin	457	459	A
Titanium	280	100	VA
Tobacco	50	25	N
Vermiculite ore	160	80	A
White lead	120	74	A
Zinc oxide	360	35	A

[a]Code: VA = very abrasive; A = abrasive; M = mildly abrasive; N = less abrasive; SW = specific wt. lb/ft^3; BW = bulk wt. lb/ft^3.

[b] Estimated.

purposes, such as to decrease adhesion or improve wearability. Silicones are often used on fiberglass to reduce the fiber abrasion.

E. FILTER HOUSING

The configuration and design of the filter housing depends on the fabric surface area required and on the temperature, moisture content, and corrosiveness of the gases. When the baghouse is designed so that the dirty gas enters the inside of the bags under positive pressure, it may not be necessary to provide any housing, except for weather protection. If a method is used which requires a housing, it is usually insulated for one of two reasons: (a) if the temperature is above 160°F or (b) in order to minimize condensation. The most common construction material for the housing is steel.

Many smaller collectors are assembled as a unit at the factory, while larger units may be shipped disassembled or as compartments that are joined in the field. The largest unit that can usually be shipped by rail is approximately 10 ft x 40 ft x 12 ft high. Care must be taken in assembly to ensure tight seals, since loose seals have been a major complaint from fabric filter users (19).

The diameter of the filter bags also influences the size of the baghouse. For example, 1750 ft^2 of filtering area can be provided in about 80 ft^2 of floor area by using 6 in.-diameter by 10 ft-long bags. If 12 in.-diameter bags were used instead, they would need to be about 14 ft long to provide the same filtering area in the same floor space, though 12 in.-diameter bags can easily be obtained 20 ft long or more when there is adequate head room. This (12 in. x 20 ft) would result in a baghouse having about 2500 ft^2 of filtering area in the same floor space (80 ft^2)(20). The length/diameter ratio affects the stability of the

TABLE 9

Particle Size Distribution, Bulk Density, Porosity, and Flowability of Some Typical Powders (18)

Solid[a]	Av. particle size (μ)	Mesh size: Powder (%)	Mesh size: Granule (%)	Bulk: Loose
Aluminum				
Granule	840	0.0	100.0	99
Powder	< 30	100.0	0.0	48
Coal				
Granule	3000	5.0	95.0	43
Granule powder	< 74	80.0	20.0	28
Limestone				
Granule	2000	1.0	99.0	85
Granule powder	< 30	80.0	20.0	55
Powder	< 30	100.0	0.0	42
Salt				
Granule	250	0.0	100.0	74
Powder granule	70	25.0	75.0	63
Granule and powder	30	65.0	35.0	46
Silica				
Granule	80	0.0	100.0	97
Granule and powder	60	60.0	40.0	51
Powder	< 74	100.0	0.0	51
Powder	5	100.0	0.0	27
Powder	0.01	100.0	0.0	2.1
Powder (porous)	3.5	100.0	0.0	1.9
Soda Ash				
Granule	250	1.0	99.0	63
Powder granule	100	20.0	80.0	36
Granule and powder	74	60.0	40.0	33
Sugar				
Granule	150	0.5	99.5	50
Granule powder	74	72.0	28.0	29
Powder	< 74	100.0	0.0	23
Sulfur				
Powder granule	220	8.5	91.5	70
Powder and granule	74	52.0	48.0	35
Powder (lumped)	< 74	100.0	0.0	36

[a]Defines: particles > 200 mesh (74 μm) as granular, free-flowing units for average weight material of average M.W. (inorganic) of 25-40 lb/ft^3 ⟶ : particles < 200 mesh as "powders," nonflowing. The lighter a material, the coarser is its powder; a heavier material will have fine powdered entities.

TABLE 9 (cont.)

density Packed	density Working	True density (gm/cm^3)	Loose powder porosity[b] (void fraction)	Flowability[c]	Floodable (or can be fluidized)
114	101		0.40	E	No
74	57	2.65	0.71	P	Very
54	46		0.53	G	No
36	30	1.45	0.69	P	Very
105	89		0.48	E	No
81	62	2.62	0.66	P	Yes
66	51		0.74	P	Yes
86	75		0.46	G	No
79	67	2.22	0.54	Pa	No
62	50		0.67	P	Yes
108	98		0.41	E	No
79	61		0.69	P	Yes
73	58	2.65	0.69	P	Yes
44	34		0.84	VP	Yes
2.7	2.4		0.9872	VVP	Yes (dusty)
2.7	2.0		0.9885	VVP	Yes (dusty)
76	65		0.53	E	No
47	39	2.15	0.73	Pa	No
51	37		0.75	P	Yes
57	51		0.53	G	No
43	33	1.69	0.72	P	Yes
36	28		0.78	VP	Yes
86	73		0.45	Pa	No
46	38	2.02	0.72	P	Possible
50	40		0.71	VP	No

[b] $\varepsilon = 1 -$ bulk density/true density

[c] E - excellent, G - good, Pa - passable, P - poor, VP - very poor, VVP - very very poor.

vertical bags, so care must be taken to ensure that bags do not rub together during operation or cleaning. The bag length-to-diameter ratio may vary between 5 and 40, but it more commonly varies between 10 and 25.

Consideration must be given during design to allow adequate space below the filter bags for the collecting hopper. The hoppers are commonly designed with 45° or 60° sloping sides to provide adequate sliding and they can be treated on the inside to minimize adhesion of the dust. The collected dust in the hopper can be removed by screw conveyors, rotary valves, trip gates, air slides, and other methods.

F. PRECOOLING OF GASES

Because most filter fabrics are limited to temperatures below 550°F, it is often necessary to precool the gases to be filtered. Cooling can represent a considerable portion of the cost of the installation. It is also necessary to evaluate the methods of precooling, prior to design of the filter, in order to determine the volume of gas (cfm) to be filtered. The three precooling methods available are:

1. Radiation and convection cooling using large U-tube coolers or heat exchangers.
2. Spray cooling with water.
3. Addition of outside air.

Radiation or convection cooling with large U-tube coolers exposed to the outside air may take up considerable space and be expensive to install. The heat transfer rate for these units is about 1.0 Btu/hr ft^2°F. Therefore to cool 100,000 cfm of air from 600°F to 300°F would require a U-tube cooler having a surface of approximately 100,000 ft^2. A unit of this size may represent a relatively large investment.

Spray cooling of the gases is accomplished by spraying water into the hot gases where it vaporizes, thereby cooling the gas stream. This usually requires the lowest capital investment, but precautions must be taken to ensure that the cooled gas stream temperature is maintained at least at 75°F above the dew point to prevent condensation and plugging of the bags. The quantity of water vaporized must also be computed and added to the quantity of gas in order to determine to total gas volume to be filtered.

Cooling by air addition is a good method of reducing the temperature of the gas stream, but it can require large volumes of air, which correspondingly increases the size of the baghouse and fan. Therefore the cost of using this method often becomes prohibitive.

If the temperature of the gas stream is not too high (above 550°F) the use of the more expensive high-temperature fabrics, such as fiberglass, may be economically justified as an alternative to precooling. Whatever method is selected, the total volume of gas to be filtered must be determined to complete the design and economic evaluation of the filter system. The economics of fabric filters is discussed in the following section and in Chap. 6.

V. FABRIC FILTER ECONOMICS

The over-all cost of a fabric filter collection system, or any other means of dust removal, is composed of the initial cost of the basic device and of other associated equipment (such as ducting, fans, etc.), plus operating and maintenance costs. These costs may vary considerably, depending on the specific installation. The recent study conducted by GCA for EPA included surveys of equipment manufacturers and users and summarized cost data as a guide for estimating the cost of fabric filter systems.

Typical cost information resulting from this study is shown in Fig. 5 (21) and Table 10 (22). Figure 5 indicates several important aspects of the economics of a fabric filter system:

1. The cost of the collector is only a few per cent of the total cost; therefore the collector cost taken alone is a poor criterion to use in selecting the system.

2. Fabric replacement (labor plus material) represents about one fifth of the total annual cost; therefore more serviceable fabric may lead to substantial savings. Also, fabric replacement cost is about four times greater than collector cost, so the collector should be very carefully designed to promote fabric life.

3. Labor costs are nearly one-third of the total annual cost and are twice as mush as the initial costs.

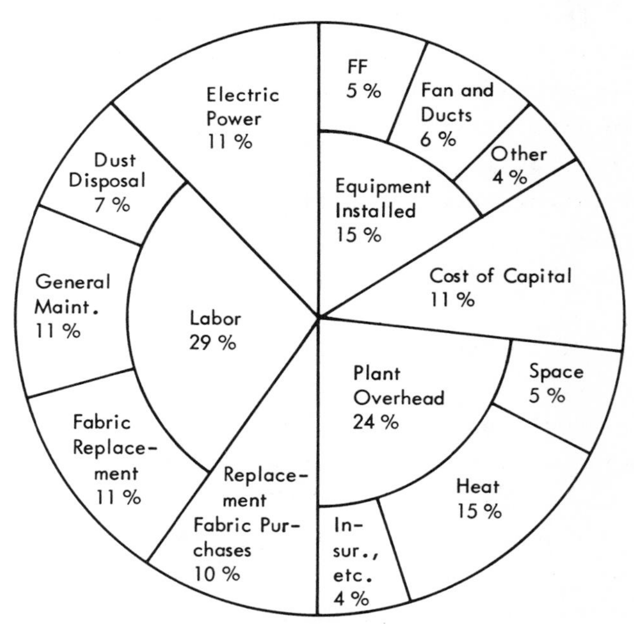

FIG. 5. Fabric filter annual cost distribution (21).

TABLE 10

Typical Costs Of Fabric Filtration (22)

I. Installed cost - $2.38/cfm		
Planning and design	$0.10	
FOB baghouse	0.80	
Freight	0.05	
Fan and motor	0.25	
Ducting	0.65	
Disposal equipment	0.10	
Instrumentation	0.05	
Foundation and installation		
Labor	0.28	
Start-up	0.10	
Total	$2.38	
II. Annual cost - $0.77/cfm/yr		
Electric power	$0.12	
Labor	0.30	
Plant overhead	0.25	space: 0.055; heat: 0.15; insurance, etc.: 0.045
Cloth purchases	0.10	
Total	$0.77	
III. Total cost of operation - $1.05/cfm/yr		
Annual cost	$0.77	
15-Year amortization of the installed cost	0.16	
Interest on the unamortized portion of installed cost, at 10%	0.12	
Total	$1.05	

TABLE 11

User Survey Cost Data (23)

Dust	Process	Total cost installation 1969 basis ($000)	Size (ft^2)	Cost ($\$/ft^2$)	Size (kcfm)	Cost ($/cfm)
Fly ash	Oil comb.	2,190	147,500	14.9	820.0	2.7
Fly ash	Coal comb.		3,400		10.0	
Ash	Mun. incin.		75		0.3	
Ash	Mun. incin.	29	5,520	5.25	13.6	2.14
Flour	Milling	19.8	976	20.3	8.4	2.4
Dolomite	Kiln	11.5	4,590	2.5	3.3	3.5
Hypochlorite	Transport	9.4	3,020	3.11	10.0	0.94
Aluminum hydrate	Transport	5.15	58	89.0	0.1	51.0
Carbon black	Mfgr.		33,175		42.5	
Carbon black	Mfgr.		15,900		17.9	
Resin, fiber	Transport		2,190		4.0	
PVA	Transport	20	1,340	15.0	14.0	1.43
Cement clinker	Cooler	1,440	102,000	14.0	275.0	5.2
Cement	Mill	20.4	5,440	3.75	10.0	2.0
Cement	Bagging	10.8	3,370	3.2	8.3	1.3
Cement	Kiln		194,400		300.0	
Cement	Transport		1,500		1.8	
Lime	Kiln	1,140	65,800	17.3	140.0	8.1
$CaSO_4$	Dryer		800		6.0	
Wallboard	Trim Saw		2,040		7.0	
Stone	Crushing		2,000		(6.0)	
Abrasive	Transport					
Raw materials	Glass mgfr.		72		0.17	
	Bagging	5.3	390	13.5	0.75	7.0
Fe, Zn oxides	El. fnce. shop	2,530	290,000	8.7	645.0	3.9
Fe, Zn oxides	Elec. furnace	512	59,600	8.5	80.0	6.4
Fe_2O_3	Elec. furnace	176	19,600	9.0	46.0	3.8
Fe_2O_3	Elec. furnace	(20.4)[a]	13,000	(1.57)	(32.0)	(0.64)
Fe_2O_3	Elec. furnace	147	44,200	3.3	115.0	1.28
Fe_2O_3	Elec. furnace	233	42,000	5.5	60.0	3.9
Fe_2O_3	Cupola		2,000		25.0	
Fe_2O_3	Cupola	880	63,600	13.8	120.0	7.3
Kish	Pouring	412	40,000	10.3	100.0	4.1
Fe, coke	Sinter line	235	98,500	2.4	241.0	1.0
Sand, FeO	Casting clean.	35.2	2,390	14.7	12.0	2.9
Fe, FeO	Grinding					
Atmos.	Motor room vent.		4,640		48.0	
ZnO	Brass smelting	134	6,400	20.9	3.6	37.0
Be, BeO	Sinter, machine		2,080		40.0	
Pb, PbO	Smelt.	291	25,600	11.4	25.0	11.6
ZnO, PbCl	Blast furnace	119	153,600	0.78	190.0	0.63
Misc. oxides	Cu refining	910	116,000	7.8	150.0	6.1
Filler, fiber	Plastics molding		1,660		10.0	

[a]( ) = estimate.

TABLE 11 (cont.)

Air-to-cloth ratio	Annual maint. (cloth and all labor) $	Annual maint. (cloth and all labor) ($/cfm)	Total annual use (hr)	T≥200°F	Fabric life	Fabric purch. $/bag	Fabric purch. $/yr	Replacement labor (% of annual maintenance)
6.0	57,500	0.070	8,000	Yes	2 Yr	50.00	30,000	(22.0)
3.0				Yes				
4.5				Yes	10.5 Mo	20.00		
2.5			1,100	Yes	> 2 Yr			
8.6	650	0.078	8,000	No	1.5 Yr	(50.00)	(500)	6.0
(0.7)	1,200	0.36	8,000	No	1 Yr	2.80	907	20.0
3.3	4,450	0.445	8,000	No	7 Mo	6.05	3,350	9.0
1.9	1,000	10.0	1,100	No	3 Mo	25.00		(100.0)
1.6	22,100	0.52	8,000	Yes	1 Yr	9.65	20,000	9.5
1.1	5,850	0.33	8,000	Yes	2 Yr			(32.0)
2.7	1,120	0.28	8,000	No	1 Yr	(2.80)	864	5.5
10.4	3,800	0.27	8,000	No	4 Mo	40.00	3,000	11.0
2.5			8,000	Yes				
1.8	3,100	0.31	8,000	Yes	1 Yr		1,700	13.0
2.5	1,500	0.18	1,350	No	9 Mo		(500)	33.0
1.5			8,000	Yes	1 Yr			
	275	0.13	570	No	3 Yr		60	36.0
2.3	46,200	0.33	8,000	Yes	1.2 Yr		36,300	10.0
7.5	200	0.033	5,700	Yes	2.5 Yr			
3.4	1,300	0.185	5,700	No	3 Yr	6.00	(75)	2.0
	(1,200)	(0.20)	1,900	No	4 Yr			
2.5			6,000	No	1 Yr			
2.3	(50)	0.29	8,000	No	4 Mo			
1.9			8,000	No				
2.2	142,800	0.078	8,000	No	5 Yr	25.00	14,250	(5.0)
1.3	(7,900)+	(0.10)+	8,000	Yes	2.5 Yr	(25.00)	(7,900)	
3.4	(5,000)+	(0.11)+	200	Yes	3 Yr	35.00	3,300	33.0
2.3	1,370	0.043	8,000	Yes	3 Yr	(3.00)	720	18.0
3.0	9,250	0.08	2,850	No	2.5 Yr	5.00	4,500	14.0
1.4	9,300	0.16	8,000	Yes	>1.5 Yr	30.00		
12.5				Yes				
2.1	135,000	1.125	4,300	Yes	4 Mo	10.00	43,800	28.0
2.5			800	No	>8 Mo			
2.4	(18,000)+	(0.075)	8,000	Yes	2 Yr	31.00	18,000	
5.0	3,650	0.304	2,300	No	1.6 Yr	13.50	650	35.0
	1,225		5,200	No	13 Yr		275	12.0
10.5			8,000	Yes	>9 Yr			
0.6	4,400	1.23	8,000	Yes	<1 Yr	2.80	1,900	23.0
20.0				Yes	2 Yr			
1.0	4,900	0.20	8,000	Yes	1.75 Yr	13.70	2,500	29.0
1.2	210,000	1.11	8,000	Yes	8 Mo	27.00	105,000	14.0
1.3	80,400	0.535	5,000	Yes	3.5 Yr	27.00	22,500	40.0
6.0	(4,300)	0.43	3,800	No	1 Mo	1.00	(2,760)	30.0

Therefore the equipment should be designed to require a minimum of attention.

4. Direct operating costs are half the total annual cost; therefore, even after the system is bought and installed, substantial savings may result from careful use.

Table 10 is based on a typical 15-year amortization of the installed cost. However, it may be possible to amortize this cost on the basis of a 5-year depreciation period for new pollution-control systems as per Sec. 169 of the Internal Revenue Act of 1969.

The GCA report stated that "as a rough rule of thumb, the total initial investment cost including material and labor will average about $2.50 per cfm of gas filtering capacity, with an expected range of $1 to $7 from one installation to another depending on the severity of the problems." The same report stated that, "while the scale factor from FOB collector cost up to installed-system cost is typically about three (see Table 10), this value is obviously highly dependent on the individual system." Some of the important items that help make up the cost of a fabric filter system are discussed below. Actual user cost data, by industry category, are shown in Table 11 (23).

A. INITIAL FILTER COST

The purchase cost of the basic filter equipment, not including fans and ducting, etc., will usually vary from $1 to $2 cfm for 10,000 cfm capacity, down to $0.50 to $1 cfm for 100,000 cfm capacity (24). The higher figures are more applicable to units operating at higher temperatures, while the lower figures are probably more applicable for lower-temperature units.

The cost of an air compressor for pulsed cleaned equipment is normally included with the initial fabric filter cost. These compressors typically provide 2 scfm/1000 cfm of gas filtered, at delivery pressure of 75 to 100 psig. If not provided with the filter, the compressor may add from $1000-$3000 to the purchase cost of the filter.

B. FANS

The fabric filter systems normally require no more than 10 in. design pressure. General cost of fans in this range are:

Capacity (cfm)	Cost ($/cfm)
10,000 or more	0.10-0.20
1,000	0.30-0.55

C. DUCTING COSTS

The ducting cost is dependent on the gauge of metal and diameter of the ducting as well as the length. Table 12 gives the estimated installed cost per foot for several gauges and diameters; typically it is around $10/ft(25). Table 12a gives the estimated ducting cost before installation. Labor for installation of ducting is about 85 to 100% of the material cost. A survey of nine large filter installations averaging 50,000 cfm indicated the costs of the associated ducting to be $0.60 cfm (ranging from $0.25-$1 cfm) (26).

D. INSTRUMENTATION

Most filter installations include some instrumentation to monitor gas temperatures, pressure drop, dust levels in hopper, and possibly other conditions, depending on the sophistication of the system. The cost of installed instrumentation may vary from $0.01 cfm for a large baghouse to roughly $0.10/cfm for a 1000-cfm unit.

TABLE 12

Approximate Costs for Installed Duct Systems ($/ft) (25)[a]

Metal gauge (U.S. std.)	Thickness (in.)	Duct diameter (in.) 6	12	24	48
6	0.203	--	--	108.00	216
14	0.078	--	20.75	41.50	83
16	0.062	8.25	16.50	33.00	66
18	0.050	6.60	13.20	26.40	53
20	0.0375	4.90	9.80	19.60	39
22	0.0312	4.10	8.20	16.40	31
24	0.025	3.30	6.60	13.30	--
26	0.0188	2.50	5.00	10.00	--
30	0.0125	1.65	3.30	--	--

[a]The 1969 cost basis of $1.85/lb for galvanized sheet steel, riveted and soldered, including a nominal number of fittings.

TABLE 12a

Dust Fitting Costs ($/item, not installed)[a]

Description	Gauge	Diameter (in.) 6	12	24
Straight pipe, per ft	20	2.40	3.90	6.50
	24	0.56	0.95	2.97
Same, with self-locking seal	24	1.50	2.80	--
	26	1.18	2.24	7.60
Reducer	20	$18 (12 x 6)	$36 (24 x 12)	
Flange	20	$23 (12 in.)	$35 (24 in.)	
Tee, 45 deg	20	$33 (12 x 12 x 8)	$50 (24 x 24 x 12)	

[a]The 1969 Boston area prices ($/item, not installed) in lots of 10 fittings each.

E. PLANNING AND DESIGN COSTS

Any user of a fabric filter will devote some time to planning and engineering, even on a turnkey project. Evaluation of bids and liaison with the contractor will require some portion of the project engineer's time. This may range from as little as 10 man-hr up to as much as 1000 man-hr or more. If this time is charged at a cost of $15/man-hr for 100 hr, it would add $1500 to the project cost, or on the order of $0.10/cfm of system capacity.

F. FOUNDATION AND INSTALLATION

Since the basic filter units weigh from 500 to 2000 lb/1000 cfm of capacity, they require substantial foundations and supporting members. The location of the filter would thus have considerable effect on the cost of foundations and support members, but a reasonable estimate would be that it comprises about 10 to 15% of the total installed cost.

G. OPERATING AND MAINTENANCE COSTS

Typical operating and maintenance costs for fabric filter systems are shown below (27):

Cost item	Typical cost range
Power	
Fan power for filtering	$0.10-0.25/cfm-yr
High-pressure air for cleaning	$0.00-0.25/cfm-yr
Labor	
Fabric replacement	$0.02-0.20/cfm-yr
General maintenance	$0.02-0.20/cfm-yr
Dust disposal	$0.01-0.15/cfm-yr
Plant overhead: Space, heat, lights, insurance, etc.	$0.05-0.50/cfm-yr

Not included in the above list is the cost of replacement bags for the filter. This is discussed below, along with other specifics about operating and maintenance cost factors.

1. Power Cost

The cost of pumping the gas is given by:

$$P = 0.93 \frac{CQH}{e} \tag{4}$$

where P is power cost in $/yr (assuming 8000 hr of operation/yr), C is cost of electricity ($/kw-hr), Q is volume flow rate (cfm), H is average differential fan pressure (in. of H_2O), and e is fan-motor efficiency (usually 0.60).

The efficiency (e) for a fan-electric motor unit is usually assumed to be 60%. The fan pressure (H) is typically 5-10 in. H_2O for a fabric filter system. The cost of the electricity (C) is generally about $0.012/kwh.

2. Cost of Maintenance Labor

A survey of 30 fabric filter installations reported that about 0.047 man-hr/yr/cfm were required for maintenance, equivalent to about $0.21/yr/cfm. This includes labor for bag replacement, general maintenance, and labor associated with disposal of collected dust (28).

3. Plant Overhead

The cost of plant overhead, as charged to the filter, is an accounting of such items as space occupied by the filter, heating requirements, insurance and taxes, etc. It may vary depending on company accounting methods. Such items as charges for space may be interpreted as rent of floor space, and a typical value is 50 cfm/ft^2 of floor area with a rental charge of $2.75/ft^2/yr. Thus a typical space cost is $0.055/yr/cfm. A heating charge may be assessed against the filter if it is exhausting air which originated from the heated air inside the building. If the exhausted gas originated as unwanted process heat, this charge may not be applicable.

4. Cost of Replacement Bags

Average bag life is about one year, so the maintenance cost must include charges for purchasing a new set of bags once each year. The cost of each bag is dependent on its size and the material. Relative cost of bags and actual retail costs are shown in Tables 13 (29) and 14 (30). The GCA survey of fabric users indicates that a typical bag replacement cost is $0.10/yr/cfm (cotton and glass bags were the most widely used fabrics reported in the survey).

TABLE 13

Fiber, Temperature Range, and Relative Cost (29)

Type	Typical name	Recommended temp. range: Max.	Contin.	Relative approx cost
Cotton	--	225	(160-190)	1
Rayon, acetate	--		210	1.1
Wool	--	250	(180-235)	2.75
Acrylic	Orlon	275	(200-275)	2.75
	Dynel	240	(150-180)	3.2
Vinyls	--	--	250	2.7
Polyester	Dacron'	325	(250-280)	2.8
Polyethylene, polyolefin	Polyfain		200	2.0
Saran			160	2.5
Polyamide	Nylon	250	200	2.5
	Nomex	500	425	8.0
Polypropylene	--			1.75
Poly-TFE	Teflon	500	(225-450)	30.0
Glass	--	600	(450-550)	5.5
Asbestos	--		500	3.8
Ceramic	Fibrefrax	2800	2300	~75.0
Metal	Brunsmet	--	--	~100.0

TABLE 14

Typical Filtration Fabric Costs (30)

Basic material	Fiber (cost/lb)	Woven fabric (cost/yd)	Felted fabric (cost/yd)	Selected retail bags[a]	
				Cost	Length x diameter
Cotton	0.40	0.41	--	13.60	21 ft x 9 in.
				1.50	5 ft x 5 in.
Wool	(Wide var.)	1.77		50.00[b]	14 ft x 7 in.
			3.97	35.00[b]	7 ft x 12 in.
Orlon	0.66-0.80	1.01	--	5.00	9.5 ft x 6 in.
				3.00	13.5 ft x 5 in.
		--	4.82	--	--
Dacron	1.40	1.04	--	30.00	30 ft x 10 in.
				22.00	25 ft x 11.5 in.
				13.70	22.5 ft x 12 in.
				2.80	9 ft x 6 in
		--	4.82	25.00[b]	5 ft x 5 in.
Nomex	2.50-6.00	--	--	50.00	25 ft x 11.5 in.
				27.00	20 ft x 8 in.
			11.50	--	--
Nylon	1.00	0.70	--	--	--
Fiberglass	0.60	0.98-1.68	N.A.	27.00	20 ft x 8 in.
				20.00	25 ft x 12 in.
				16.50	25 ft x 11.5 in.
				10.00	22.5 ft x 8 in.
				9.63	12 ft x 5 in.
				2.80	6 ft x 7 in.

Teflon	--	8.00 (8 oz)	--	75.00	25 ft x 11.5 in.
			36.70 (23 oz)	--	--
			29.20 (19 oz)	--	--
Fibrefrax[b]	20.00	38.00	--	--	--
Brunsmet[b]	~40.00	--	--	--	--

[a] Bag cost and fabric costs are not related, since the information is from separate sources.

[b] Felted fabrics.

VI. SPECIFIC INDUSTRY APPLICATIONS OF FABRIC FILTERS

Fabric filters have been used in many industries for years and are used to control a variety of sources or operations within these industries. These include feed and grain operations, iron and steel, cement, lime, primary and secondary nonferrous metals, ferroalloy manufacture, iron foundries, asphalt batch plants, and carbon black manufacture. These applications are discussed in the following section. Also included is short discussion of coal-fired power plants and incineration where fabric filters have not been used but which are significant sources where fabric filtration may find increasing application in the future.

A. GRAIN AND FEED

Feed and grain operations have many sources of dust emission that have been controlled in the past by cyclones and, to some extent, by baghouses. It is desirable to control these sources not only to reduce pollutant emissions, but also to decrease weight loss from the product stream and to reduce the explosion hazard associated with the highly combustible grain dust. Recent emphasis on reduction of emissions from grain handling operations has led to increased use of baghouses.

Flour mills have made extensive use of baghouses to maximize product recovery. Such mills may employ up to ten separate units, since many of the milling operations require air aspiration for proper separation of intermediate products. Increased use of pneumatic conveying has also led to increased utilization of fabric filters. It has been found that the filters are usually designed with air-to-cloth ratios of 10-15 cfm/ft^2 and the installed cost is reported to vary between $2 and $4/cfm (31).

B. IRON AND STEEL

The largest use of fabric filters in the various iron and steel processes is in the electric arc furnaces. A few have also been used in sinter plants and scarfing operations. The estimated installed and operating costs are shown in Figs. 6, 7, 8, and 9 (32). The fabric filters

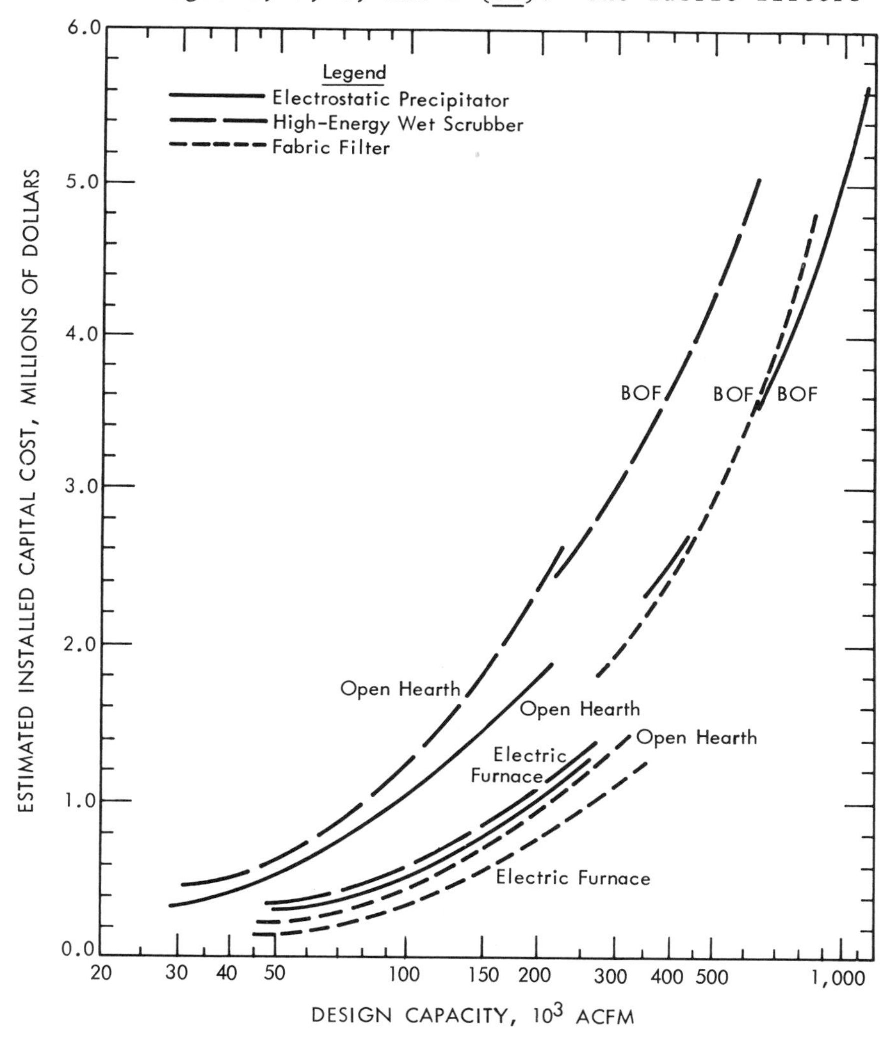

FIG. 6. Installed cost (1968) of control equipment for steel-making processes (32).

have been applied on electric arc furnaces ranging up to 150 net tons capacity and for multiple-furnace shops as well as single-furnace shops. The fiberglass bags usually used in this application are attacked by fluorides in the off gas when fluorspar is employed as a fluxing agent. Therefore other synthetic fabrics must be used in this case.

Electric furnaces are increasing in use and the characteristically small particle size emitted requires the high collection efficiency of devices such as the fabric filter. However, the high temperature of the gases from these furnaces requires the use of large volumes of tempering air, evaporative cooling, or radiation chambers ahead of the filter

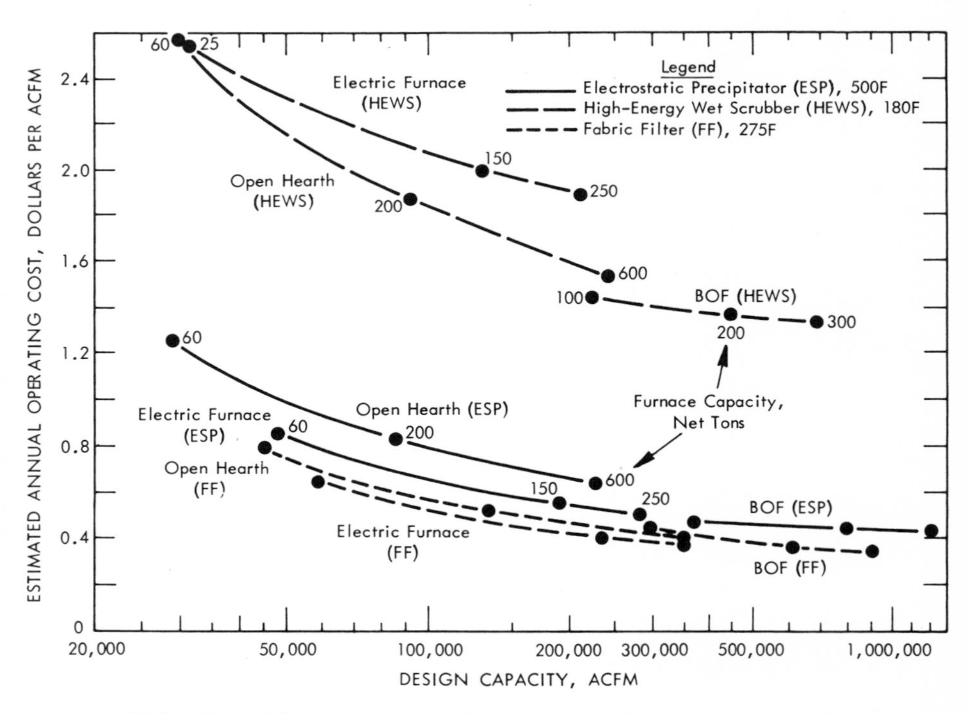

FIG. 7. Direct operating costs for control of steel-making process (32).

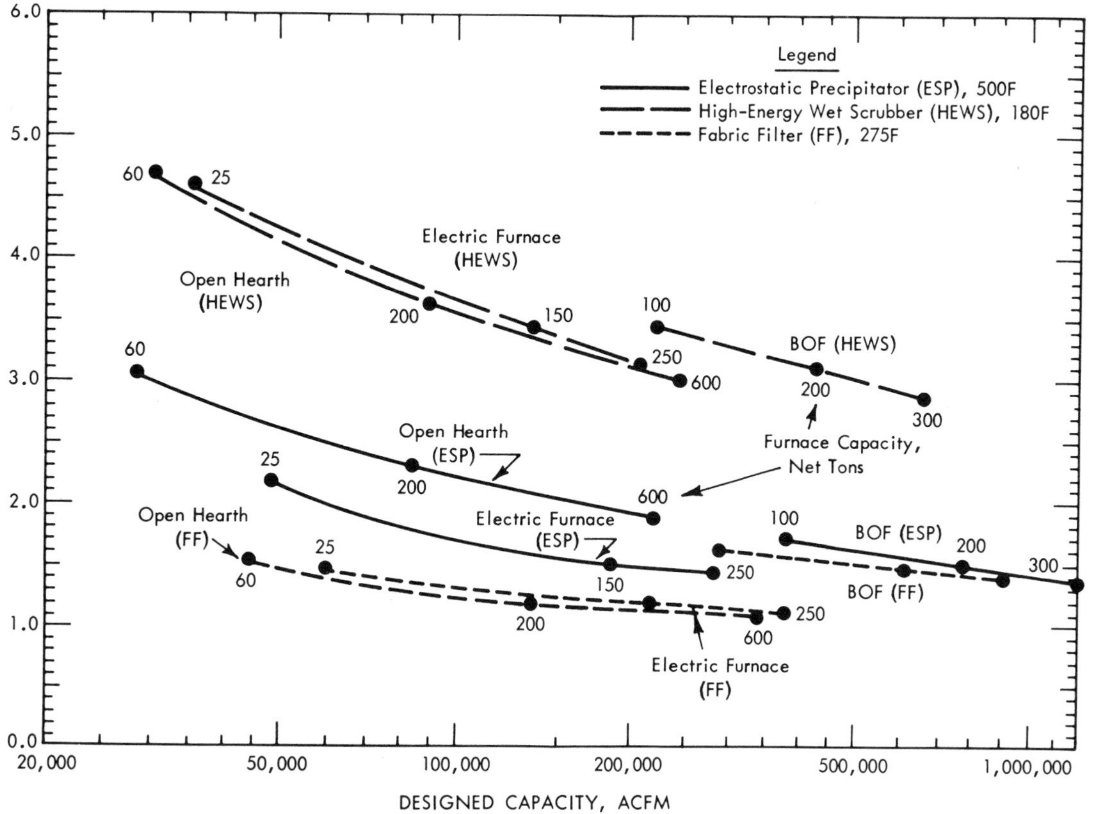

FIG. 8. Total operating costs for control of steel-making processes (32).

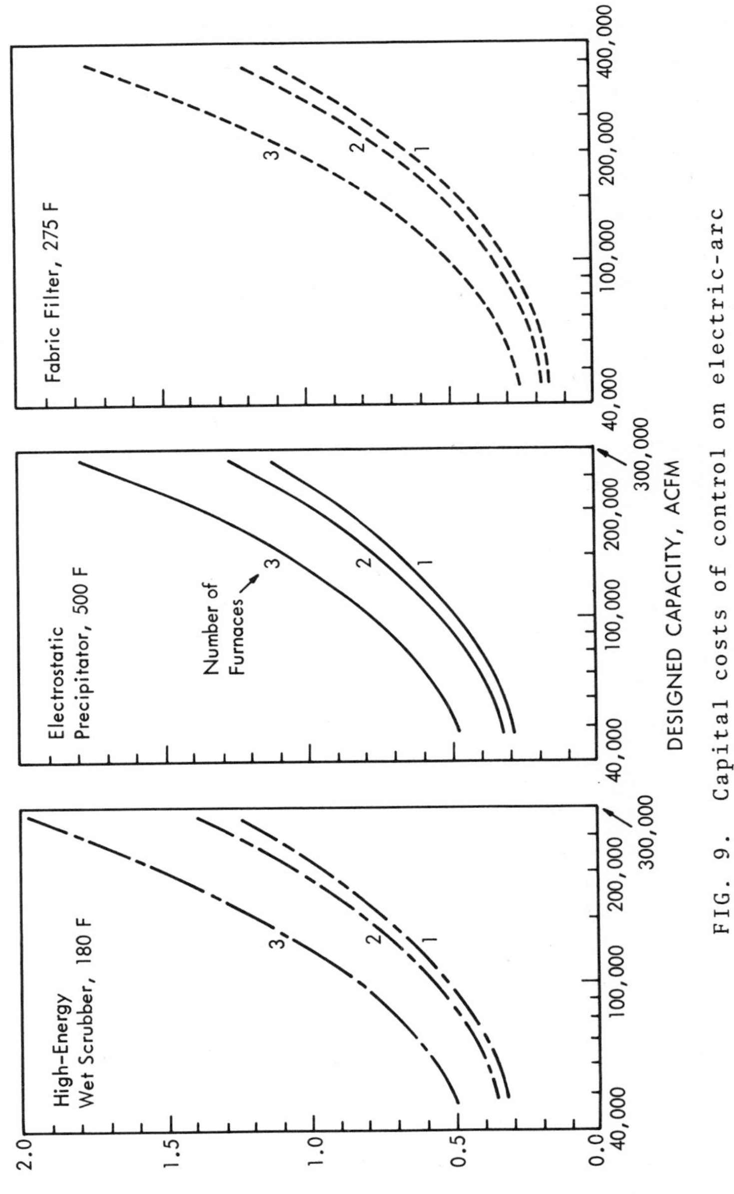

FIG. 9. Capital costs of control on electric-arc furnaces (32).

C. CEMENT MANUFACTURE

The manufacture of portland cement is another large operation that makes extensive use of fabric filters. They are used not only in the crushing and transport operations, but also on the large quantities of effluent from the kilns. The hot kiln gases usually range from 500-600°F in temperature and contain acid gases such as H_2S and SO_2 plus varying amounts of water with total gas flows from 49,000-250,000 acfm.

Cyclones and electrostatic precipitators have been widely used for control of the kiln exhaust gases, but the development of siliconized fiberglass bags has led to increased application of baghouses in this service. These have been installed on both wet-and dry-process cement plants. A 1962 survey (33) of the users of fiberglass baghouses on cement kilns showed that even then, at least 12 plants had units in this service operating at an average temperature of 500°F and at pressure drops from 3-9 in. of water. Bag life was reported to be from $2\text{-}2\frac{1}{2}$ yr. At that time, the gas-to-cloth ratio varied from 1:1 to 2.5:1. However, all of the filters were cleaned by the bag-collapse method, which utilizes dampers to cause a reverse flow of air and partial collapse of the bag. More recent installations may utilize the newer reverse air cleaning techniques, and higher air-to-cloth ratios.

The high moisture content of kiln exhaust gases requires that precautions be taken to ensure that the temperature is maintained above the dew point, especially during startup and shutdown. For this reason, the ducting and walls of the baghouse are often insulated.

It is interesting to note that even though the SO_2 content of the exhaust gases from coal-fired kilns is quite high, the dust cake on the bags removes a significant quantity of this pollutant, thereby serving a two-fold purpose.

D. LIME MANUFACTURE

The manufacture of lime requires a calcining kiln similar to that used in cement manufacture. Likewise, the control practices are similar and fiberglass baghouses are utilized. A number of installations make use of the fiberglass baghouses handling gas flows as high as 150,000 acfm at temperatures in the range of 350-550°F. Design air-to-cloth ratio has been reported to be approximately 2:1 (34). A 350 t/D lime kiln producing 100,000 acfm of exhaust gas at 550°F was equipped with a baghouse with costs reported as follows (35):

Installed cost	$150,000
Operating cost	$14,000/yr
Maintenance cost	$8,000/yr

The above installed cost is equivalent to $1.50 acfm, while another installation reports an installed cost of $1.80/acfm (34).

Fabric filters are also used for control of emissions in other lime manufacturing operations such as stone drying, grinding, and bagging. The baghouses used are sometimes preceded by cyclones for removal of larger particles before the gases enter the filter. The air-to-cloth ratios reported for the filters used in these operations show wide variations of 1:1 to 10:1.

E. PRIMARY NONFERROUS METALLURGY

Primary nonferrous metallurgy includes smelting and refining of copper, lead, zinc, and aluminum. These processes characteristically emit metallic fumes, generally submicron in size. Electrostatic precipitators are often used for control in this service. The disadvantages of the cloth filter are (a) high wear and (b) expensive bag replacement. However, a baghouse has been used downstream of a precipitator in a copper smelter for further cleaning of combined reverberatory furnace and converter gas stream (36). A German publication reports the use of a similar system for a lead smelter reverberatory furnace. (37). Baghouses have also been used for collection of fumes from lead blast furnaces (38) at an installed cost (1957) reported to be $4.35/acfm.

Baghouses have not found extensive application for aluminum reduction cells partly because of the fluoride content of fume emitted. Wet scrubbers have more often been used because of their ability to remove up to 95% of the fluorides. However, a process now in the development stage is directed toward use of a baghouse to remove particulates and fluorides from the effluent of aluminum prebake cells. Alumina powder is added to the effluent gases and is collected on the filter, along with the particulates, and absorbs up to 95% of the fluorides. The collected material is returned directly to the prebake cell (39).

F. ASPHALT

The rotary rock dryer is the largest source of particulate emission in asphalt batch plants. Cyclones followed by wet scrubbers have commonly been used for reducing the emissions from the dryer. Fabric filters are applicable to this source and are increasing in use as more stringent pollution-control codes are adopted. These filters provide excellent collection efficiency with little or no visible emissions. They may be more expensive than wet scrubbers, but the collected dry fines are sometimes usable in concrete mixes. The intermittent nature of the drying cycle requires that precautions be taken to prevent overheating and condensation. The fabric filters used in this application operate at air-to-cloth ratios ranging from 3-6 cfm/ft^2.

G. FERROALLOY MANUFACTURE

In the nonferrous metal industry, silicon is used primarily as an alloying agent for other metals. Many types of furnaces are employed in the manufacture of ferroalloys, but the majority of furnaces are electric arc, producing large quantities of CO, which may or may not be burned at the mouth of the furnace, depending on the hooding arrangement. When the gases mix with air and are burned, as in the open furnace, the quantity of effluent gas is increased, which proportionately increases the size of filter required. As in other electric furnace operations, the high temperature of the gases requires cooling through heat transfer surfaces or by air dilution. The properties of the fume may limit the air-to-cloth ratio to 2 acfm/ft^2 thereby requiring large numbers of bags. Cost for filters applied to open furnaces is reported to be $2.90-$3.60/acfm (40). It should also be possible to apply fabric filters to covered furnaces, with their high CO content, but no applications to this type of ferroalloy electric furnace are known.

H. IRON FOUNDRIES

Grey iron foundries show a wide range in furnace sizes and may employ many different types of furnaces, including cupola, electric arc, electric induction, and reverberatory air furnaces. Most foundries use the cupola furnace which is equipped with a low-pressure drop wet scrubber (i.e., "wet cap"). The emissions from the cupola furnace vary widely in composition and quantity over the operating cycle of the furnace.

Although wet scrubbers have commonly been used for control of cupola emissions, there are several fabric filter installations using synthetic filter bags preceded by appropriate cooling methods to protect the baghouse from high temperature. The installed cost of baghouses on cupolas is reported to range from $0.93-$3.50/acfm (41). Data on three fabric filter installations are shown in Table 15 (42).

I. SECONDARY NONFERROUS METALS

The secondary nonferrous smelting and refining industry includes those establishments that recover nonferrous metals and alloys, primarily from scrap. The metals recovered include copper, lead, zinc,and aluminum. The emissions from the associated furnace operations vary widely through the operating cycle from charging of scrap to pouring the melt. These furnace emissions are often controlled with baghouses. Smaller fabric filter systems are also used in conjunction with hoods over kettles, furnace doors, and other sources.

The baghouse has been generally accepted in the brass and bronze ingot industry. A recent EPA report of the air pollution aspects of this industry shows that the concentration of particulate matter escaping the baghouses ranges from 0.006-0.036 grain/scf and operating efficiencies were generally between 95 and 99.6%. Filter ratios varied from 2.0-2.7 cfm/ft^2 and pressure drops were 2-6 in. of H_2O (43). Other data for baghouses on brass and bronze furnaces are given in Tables 16 (44) and 17 (43).

TABLE 15

Cupola Dust Removal Installations with Fabric Filter (42)

Data	Installation No. 1	2	3
Furnace type	Cold blast	Hot blast	Hot blast
Melting rate (tons/hr)	5.0	33.0	17.6
Pressure drop (in. W.G.)	14	About 14, filter plus pipeline	3-6
Gas temperature (°F)			
Cupola exit	1,380	930	1,830
Filter inlet	446	390	480
Filter exit	230[b]	248	---[a]
Gas quantity before filter			
Scfm	8,400	13,000	65,000
Scfm dry	4,500	7,300	38,000
Dust content (grains/scf dry)			
After cupola	---[a]	3.52	---[a]
Before fabric filter	1.19	2.84	0.35
After fabric filter	0.0015	0.041	---[a]
Collection efficiency (%)	99.83	99.75	99.00
Dust discharge in atmosphere (lb/hr)	0.073[b]	0.530	---[a]

[a] No data available.

[b] Before clean gas measuring point after fabric cleaner, ~1200 scfm false air was drawn in.

TABLE 16

Brass-Melting Furnace and Baghouse Collector Data (44)

Furnace data			
Type of furnace	Crucible	Crucible	Low-frequency induction
Fuel used	Gas	Gas	Electric
Metal melted	Yellow brass	Red brass	Red brass
Composition of metal melted(%)			
Copper	70.6	85.9	82.9
Zinc	24.8	3.8	3.5
Tin	0.5	4.6	4.6
Lead	3.3	4.4	8.4
Other	0.8	1.3	0.6
Melting rate (lb/hr)	388	343	1600
Pouring temperature (°F)	2160	2350	2300
Slag cover thickness (in.)	1/2	1/2	3/4
Slag cover material	Glass	Glass	Charcoal
Baghouse collector data			
Volume of gases (cfm)	9500	9700	1140
Type of baghouse	Sectional tubular	Sectional tubular	Sectional tubular
Filter material	Orlon	Orlon	Orlon
Filter area (ft^2)	3836	3836	400
Filter velocity (ft/min)	2.47	2.53	2.85
Inlet fume emission rate (lb/hr)	2.55	1.08	2.2[a]
Outlet fume emission rate (lb/hr)	0.16	0.04	0.086
Collection efficiency(%)	93.7	96.2	96.0

[a] Includes pouring and charging operations.

TABLE 17

Baghouse Information Summary - and Bronze Ingot Institute (43)

Size of baghouse (cfm)	No. of bags-diameter (in.)	Bag material	Furnaces vented to the baghouse: No. and type:
2 Units, each 24,000 (design) 18,000	1944-6	Orlon and dacron	2 Rev: 2 Rot: 4 Elec:
3 Units, each 13,000 (design)	2268-6	Dacron	2 Rev: 2 Rot: 1 Cupola
10,000 (design)	100-15	Glass	1 Radiator sweater:
27,500 (design) 18,000 (actual)	324-8	Orlon and dacron	4 Rev:
30,000 (design) 29,000 (actual)	528-8	Orlon and dacron	3 Rot:
	320-5	Orlon	---[b]
	900-8	Orlon	3 Rev: 1 Rot:
Square filter type 26,000 (design)	400-10	Orlon	Cupola 2 Rev: 5 Rot:
Multiple-bag type	15,000-6	Orlon	---[b]
Custom design 12,500/chamber	416-18	Wool	3 Rev: 3 Rot: 3 Crucible: 1 Elec: 1 Cupola:
30,000 (design and actual)	1200-5	Orlon	3 Rev:
50,000 (design and actual)	800-8	Orlon	1 Slug:

[a] Not stipulated if these are total figures for the three units.
[b] No flow diagram provided; layout of equipment not ascertainable.
[c] Not reported.

TABLE 17 (cont.)

Furnaces vented to the baghouse	Materials collected in baghouses				Frequency of bag replacement (months)
Size (tons/hr each)	Charged (lb/ton)	Produced (lb/ton)	ZnO (%)	PbO (%)	
60 2 1 @ 4, 3 @ 0.5	58	67	63	8	18
60 4	60	68	78	7.5	10-15
---[b]					
60					4
4	60[a]	88[a]	72[a]	8[a]	6
30-75 10	55	NR	58	3	12 12
80 7.5-35	NR[c]	NR	65	5-6	6
2-75 2.7 0.25 3 NR	NR	NR	55	NR	51
2 @ 30, 1 @ 12					
15-25	NR	NR	NR	NR	12

The gases leaving the melting furnaces must usually be cooled to bring the temperature down to within the temperature limitation of the filter fabrics. Direct cooling by spraying water into the combustion gases is not generally practiced because: (a) corrosion of ductwork and equipment increases; (b) vaporized water increases exhaust gas volumes; and (c) temperature of the gases in the baghouse must be kept above the dew point. Water-jacketed coolers and radiation-conversion coolers are used to cool the gases instead of water injected sprays. Duct velocities of about 3000 ft/min are used to minimize buildup of dusts in the ducts.

A Commerce Department study dealing with the economic impact of air pollution control on the secondary nonferrous metal industry indicated that the installed cost for baghouses ranges from $2.66-$8.33/acfm. The annual operating and maintenance cost ranged from $0.36-$1.07/acfm (45).

Plants engaged in secondary lead smelting require extensive systems for air pollution control and again, baghouses are considered to be the most acceptable device. Precooling is also necessary here, but the temperature is kept relatively high to prevent tar volatiles from condensing and blinding the bags. Lime may also be added as the gas enters the baghouse to help prevent this blinding action (45). When applied to reverberatory furnaces, provisions should be made to prevent sparks. Test results of secondary smelting furnaces venting to a baghouse are shown in Table 18 (44). These baghouses were equipped with Dacron fabric bags, but fiberglass bags have also been used in other lead smelting operations (46).

Air pollution control for secondary zinc retort furnaces is also achieved with baghouses. Glass bags have been found to be adequate when gas temperatures exceed the limits for cotton or orlon.

TABLE 18

Dust and Fume Emissions from a Secondary Lead-Smelting Furnace (44)

Data	Test number	
	1	2
Furnace		
Type	Reverberatory	Blast
Fuel	Natural Gas	Coke
Material charged	Battery groups	Battery groups dross, slag
Process weight (lb/hr)	2,500	2,670
Control equipment		
Type	Sectional tubular baghouse[a]	Sectional tubular baghouse[a]
Filter material	Dacron	Dacron
Filter area (ft^2)	16,000	16,000
Filter velocity (ft/min) at 327°F	0.98	0.98
Dust and fume		
Gas flow rate (scfm)		
Furnace outlet	3,060	2,170
Baghouse outlet	10,400[b]	13,000[b]
Gas temperature (°F)		
Furnace outlet	951	500
Baghouse outlet	327	175
Concentration (grains/scf)		
Furnace outlet	4.98	12.3
Baghouse outlet	0.013	0.035
Dust and fume emission (lb/hr)		
Furnace outlet	130.5	229
Baghouse outlet	1.2	3.9
Baghouse efficiency (%)	99.1	98.3
Baghouse catch (wt %)		
Particle size (μ) 0-1	13.3	13.3
1-2	45.2	45.2
2-3	19.1	19.1
3-4	14.0	14.0
4-16	8.4	8.4
Sulfur compound as SO_2 (vol %) baghouse outlet	0.104	0.03

[a]Sectional tubular baghouse. The same baghouse alternately serves the reverberatory furnace and the blast furnace.

[b]Dilution air admitted to cool gas stream.

In aluminum sweating operations, raking the metal and dross from the furnace is a critical operation from the standpoint of air pollution control and hoods should be installed to capture emissions at these locations. The fact that carbonaceous material is found in the effluent has led to the recommendation that an afterburner be used, followed by a baghouse (47). However, wet scrubbers have also been particularly suitable for the aluminum smelter. If using a baghouse system, the gases must be cooled, but spray cooling with water is not recommended because of the presence of aluminum chloride in the effluent. If the hot gases are spray cooled, the aluminum chloride hydrolyzes, producing HCl that attacks the ductwork and bags. Even condensation during shutdown may provide sufficient moisture to corrode the equipment in the presence of these chemicals (47).

J. CARBON BLACK

Most carbon black production is done by the furnace process, with a small portion done by the thermal and channel processes. Baghouses are an integral part of the furnace and thermal processes for collection of the carbon black produced. In the furnace process the baghouses are preceded by electrostatic precipitators or cyclones, which serve as primary collectors and agglomerators for the carbon black prior to final collection in a baghouse. Approximately 70% of the product black may be collected in the cyclones, with the remaining 30% collected in the baghouse.

The baghouses used to clean the gases containing carbon black are often designed so that the "dirty" gases enter the inside of the bags at the bottom and the black is deposited on the inside of the bag as the gases pass through the bag to the outside. The bags are cleaned by reverse gas flow and the collected particles fall into a collecting hopper below the bags. Fiberglass bags allow operation at temperatures up to 550°F, but various resins and silicone coatings are used to improve bag life, which is typically about 12 months.

K. COAL-FIRED POWER PLANTS

Coal-fired power plants represent a large potential user of fabric filters. However, they have only been used in pilot-scale studies. One reason that they have not been applied more widely is that the large volume of gas to be cleaned would require a very large filter installation. Also, the temperature limitations may present some drawbacks. The pressure drop of a fabric collector, although moderate, would represent a considerable increase in operating cost.

Recent emphasis on control of emissions from coal-fired power plants make the fabric filter more attractive, but present trends are toward use of wet scrubbers for removal of both particulate matter and SO_2. Some work has been done on injection of alkaline additives into the gas stream with subsequent collection on a fabric filter for removal of SO_2, but this is only in the development stages.

L. INCINERATION

Baghouses have not been used on incinerator effluent in the USA, except in pilot-scale tests, although they have been used to some extent in Europe. Part of the reason for this is the high investment and operating cost involved in their use. An EPA study showed that the investment for a fabric filter on a municipal incinerator would be $900/ton of refuse/day, and that the operating cost would be $1.10/ton (48). Unlike the wet scrubber, the fabric filter does not remove odor or gaseous pollutants. However, the potential water pollution problems associated with wet scrubbers may necessitate further consideration of fabric filters in the application.

REFERENCES

1. C. E. Billings, and J. Wilder, Handbook of Fabric Filter Technology, Contract No. CPA-22-69-38, Vol. 1, GCA Corp., December 1970, pp. 1-44.

2. American Petroleum Institute, Removal of Particulate Matter from Gaseous Wastes - Filtration, New York, 1961, p. 12.

3. Ref. 1, pp. 2-98.

4. Ref. 1, pp. 2-105.

5. Ref. 1, pp. 2-106.

6. J. W. Thomas, and R. E. Yoder, "Aerosol Penetration Through a Lead Shot Column," A.M.A. Arch. Ind. Health, 13, 550 (1956).

7. G. W. Walsh and P. W. Spaite, Characterization of Industrial Fabric Filters, Am. Soc. of Mech. Engrs., Annual Meeting, Dec. 1960.

8. Ref. 1, pp. 2-153.

9. R. H. Borgwardt et al., Filtration Characteristics of Fly Ash From a Pulverized Coal-Fired Power Plant, U. S Department of Health, Education, and Welfare, Cincinna Ohio, 1967, p. 11.

10. Ref. 1, pp. 3-5.

11. J. L. Alden, Design of Industrial Exhaust Systems, 3rd ed., The Industrial Press, New York, 1959, pp. 61, 168.

12. Ref. 1, pp. 6-82.

13. Ref. 1, Vol. 2, pp. 6.4-4.

14. Ref. 1, Vol. 4, pp. A.3-10.

15. Ref. 1, Vol. 1, pp. 3-34.

16. U. S. Department of Health, Education, and Welfare, Control Techniques for Particulate Air Pollutants, Washington, D. C., 1969, pp. 4-172.

17. R. L. Stephenson and H. E. Nixon, Centrifugal Compressor Engineering, Hoffman Industries Div., Clarkson Industries, Inc., 103 Fourth Avenue, New York, 1967.

18. R. L. Carr, Jr., "Properties of Solids," Chem. Eng., Vol. 69, 8 (Oct. 13, 1969).

19. Ref. 1, Vol. 1, pp. 3-53.

20. Ref. 1, Vol. 1, pp. 3-27

21. Ref. 1, Vol. 1, pp. 7-7.

22. Ref. 1, Vol. 1, pp. 7-9.

23. Ref. 1, Vol. 2, pp. 7.2-3.

24. Ref. 16, pp. 6-49.

25. Ref. 1, Vol. 1, pp. 7-21.

26. Ref. 1, Vol. 1, pp. 7-22.

27. Ref. 1, Vol. 1, pp. 7-29.

28. Ref. 1, Vol. 1, pp. 7-32.

29. Ref. 1, Vol. 1, pp. 7-42.

30. Ref. 1, Vol. 1, pp. 7-40.

31. L. J. Shannon et al., Engineering and Cost Study of Emissions Control in the Grain and Feed Industry, Midwest Research Institute, Kansas City, Mo., Contract No. 68-02-0213, 1971.

32. H. W. Lownie and J. Varga, A Systems Analysis Study of the Integrated Iron and Steel Industry, Battelle Memorial Institute, Contract No. PH-22-68-65, May 15, 1969.

33. W. E. Ballard,"Glass Bags From Batch to Baghouse," Rock Products, October 1962.

34. C. J. Lewis and B. B. Crocker, "The Lime Industry's Problem of Airborne Dust," Air Pollution Control Asso. 19, 31 (1969).

35. J. L. Minnick, Control of Particulate Emissions from Lime Plants, presented at 63rd Annual Meeting, Air Pollution Control Association, St. Louis, June 1970.

36. D. J. Robertson, "Filtration of Copper Smelter Gases at Hudson Bay Mining and Smelting Company, Ltd.," The Canadian Mining and Metallurgical Bulletin, May 1960.

37. "Restricting Dust and Sulfur Dioxide Emission from Lead Smelters," Verein Deutscher Ingenieure, VD 1, September 1961, p. 2285.

38. J. H. D. Hargrave, "Recovery of Fume and Dust from Metallurgical Gases at Trail, B. C.," The Canadian Mining and Metallurgical Bulletin, June 1959.

39. "Air Pollution from the Primary Aluminum Industry," A Report to Washington Air Pollution Control Board, Office of Air Quality Control, Washington State Department of Health, Seattle, Wash., October 1969.

40. R. A. Person, Control of Emissions from Ferroalloy Furnace Processing, Union Carbide Corp., Niagara Falls, N. Y., 1969.

41. American Foundrymen's Society, Foundry Air Pollution Control Manual, Des Plaines, Ill., 1967.

42. P. S. Cowen, ed., Cupola Emission Control, Gray and Ductile Iron Founders' Society, Cleveland, Ohio, 1969.

43. "Air Pollution Aspects of Brass and Bronze Smelting and Refining Industry," National Air Pollution Control Administration Publication No. AP-58.

44. U. S. Department of Health, Education and Welfare, Air Pollution Engineering Manual, Cincinnati, Ohio, Public Health Service, 1967.

45. U. S. Department of Commerce, Economic Impact of Air Pollution Controls on the Secondary Nonferrous Metals Industry, Washington, D. C., 1969.

46. P. W. Spaite, P. G. Stephan, and A. H. Rose, Jr., "High Temperature Fabric Filtration of Industrial Gases," Air Pollution Control Assoc., 11,512, (May 1961).

47. Ref. 44.

48. W. Niessen et al., Systems Study of Air Pollution from Municipal Incineration, prepared by A. D. Little, Inc. for the National Air Pollution Control Administration, Contract No. CPA 22-69-23, 1970.

Chapter 4

ELECTROSTATIC PRECIPITATORS

Sabert Oglesby, Jr.
Grady B. Nichols*

Southern Research Institute
Birmingham, Alabama

* Vice President and Director, Engineering and Applied Sciences, and Head, Particulate Control Section, respectively, Southern Research Institute, Birmingham, Alabama.

I. DESCRIPTION OF THE PRECIPITATION PROCESS

Electrostatic precipitation differs from other gas cleaning methods in that the force required to separate particulates from a gas is applied directly to the particles themselves. This force results from the electric charge on the particle in the presence of an electric field according to well-established principles of electrostatic theory. As a consequence, electrostatic precipitators accomplish the gas-solids separation with less energy than any other gas cleaning system.

Particles from industrial emission sources can acquire a charge from flame-ionization in combustion processes, triboelectric effects during transport through ducts, ionizing radiation, or from ions produced as a result of a corona discharge. Since the effectiveness of the dust separation is dependent upon the magnitude of the charge, any commercial process must ensure that the particles are charged to the maximum value consistent with other physical limitations. Charges due to other than an electrical corona are too small to be of practical significance, and for this reason, all commercial precipitators are based on particle charging resulting from corona generation.

An electric field can be established as a consequence of a concentration of charged particles, ions in the gas, or the application of a high voltage to a pair of electrodes. In a practical precipitator, all three factors contribute to the electric field.

Generation of an electrical corona requires intense local fields and breakdown of the gases which can result from the establishment of a nonuniform field, so that breakdown is confined to the zone immediately adjacent to one electrode. This can be accomplished by the use of a wire of small radius geometry as one electrode, and a plate or cylinder as the other.

The physical arrangements of precipitators vary according to whether the particle charging occurs in a separate section, or is integrated with the collection function. If particle charging and collection occur within the same section, the precipitator is termed single stage, as illustrated in Fig. 1(a). If particle charging takes place in a separate section from the collection process, the precipitator is called a two-stage precipitator, as illustrated in Fig. 1(b). The majority of industrial gas cleaning precipitators are of the single-stage type.

The precipitation process is illustrated in Fig. 2. The precipitator shown consists of a wire as the central electrode and a grounded cylinder as the receiving or collection electrode. Gas to be cleaned enters the lower

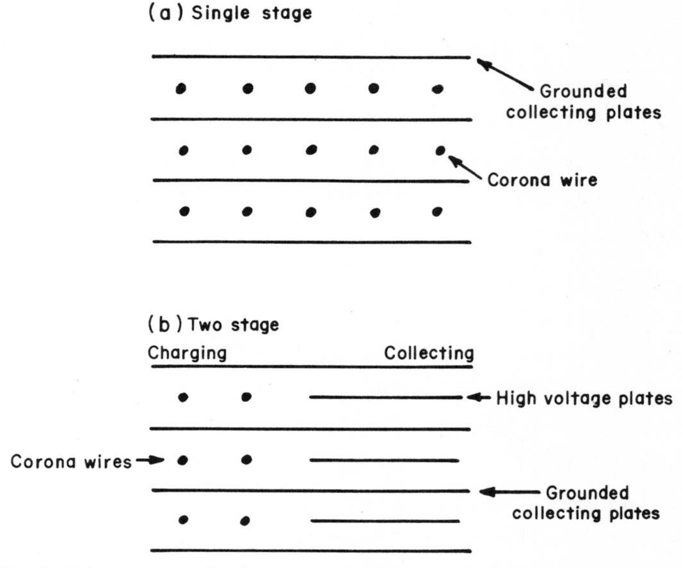

FIG. 1. (a) One- and (b) two-stage precipitator schematics.

section, where ions from the corona emanating from the central electrode region impact with and charge the particles. The charged particles move toward the collection electrode under the influence of the electric field, where they impact on the electrode or on the previously collected dust layer.

When a sufficiently thick dust layer builds up, it is removed by a sharp impact or rap, in the case of a dry-type precipitator, or it is washed from the plates, in the case of a wet precipitator.

Therefore, the principal functions of the precipitation process are: (a) generation of the corona, (b) charging, (c) collection, and (d) removal. For optimum performance, each of these functions must be performed in the most efficient manner. Physical properties of the dust, gas flow, and mechanical details of the design place con-

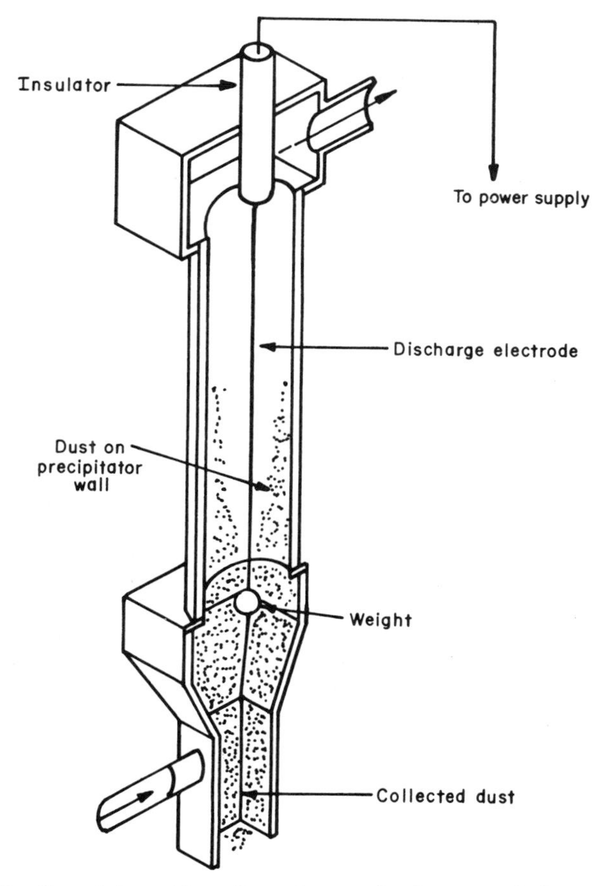

FIG. 2. Schematic of a wire and pipe precipitator.

straints on these functions and determine either the efficiency that can be achieved from a given plant or the size precipitator required to achieve a desired emission level.

The following sections discuss the theory applicable to these precipitator functions, the physical constraints that limit each, and practical problems of sizing and design of precipitators for a given dust collection problem.

II. FORMATION OF THE CORONA

Electrical conduction in gases has been investigated over a period of many years, and the basis for mechanisms of conduction is well understood as it applies to precipitator operation. The more exhaustive studies of gaseous discharges are those of Loeb (1) and von Engel (2).

Corona, as applied to electrostatic precipitators, is generated as a result of the application of a high voltage to a pair of electrodes, one of which has a small radius capable of causing a high electric field in the surrounding gas. The resulting corona is termed positive if the small radius electrode is positive, and negative if it is of the opposite polarity. Negative corona permits operation at higher voltage and higher electric field at near-standard temperature and pressure, and is used for most industrial gas cleaning operations. Positive corona is used principally in the cleaning of air in inhabited space to overcome problems with the generation of ozone inherent in negative corona precipitators.

A. ELECTRICAL CONDUCTION IN GASES

Electrical conduction in a gas can take place only if charge carriers are present. In most gases, the number of such carriers is small and the effective resistivity of the gas is extremely high. However, naturally occurring radiation present in the atmosphere causes ionization which accounts for around 20 ion-electron pairs/cm^3-sec to be formed in the gas. In the presence of an electric field, the electrons will move toward the positive electrode and the positive ions will move toward the negative electrode.

When one of the electrodes has a small radius, such as a wire, the electric field adjacent to it is high, so that electrons in the vicinity are accelerated to a velocity sufficient to cause release of orbital electrons on impact with other gas molecules. These electrons are, in turn, accelerated to a velocity sufficient to cause further ionizing collisions. The process is termed an avalanche and continues until the electric field is no longer high enough to accelerate the electrons to a velocity sufficient to ionize additional gas molecules. Fig. 3 illustrates the avalanche process.

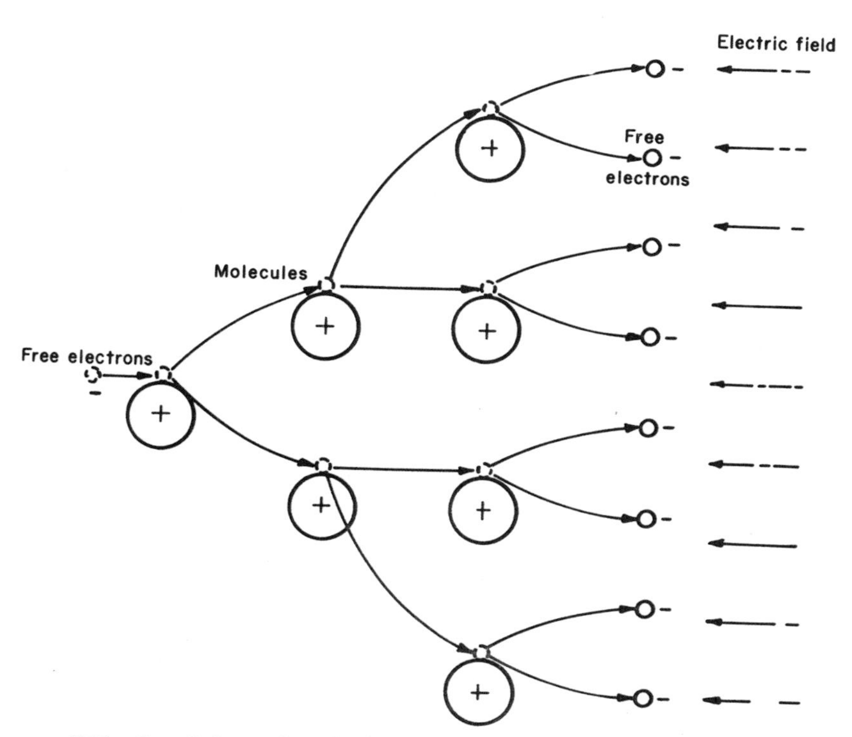

FIG. 3. Schematic of the electron avalanche phenomenon.

In a stable corona, there must be a sufficient number of initiating electrons to sustain the avalanche process. Since the number of electrons resulting from natural sources is small, additional sources must be present. In the case of the negative corona, impact of the positive ions with the electrode can result in secondary electrons being emitted from the electrode surface. Photoionization resulting from the gaseous discharge glow also furnishes additional initiating electrons. These sources provide a sufficient electron supply to sustain the avalanche process and maintain a stable corona.

In a negative corona, the electrons produced by the avalanche process move toward the collection electrode. Since the electric field diminishes rapidly inversely with the distance from the discharge electrode, the velocity of the electrons decreases with a corresponding decrease in energy, so that a collision with a gas molecule will not result in further ionization. The boundary of the corona region is therefore established by the strength of the electric field required for impact ionization.

In a positive corona, electrons produced in the ionization process move toward the positive electrode and the ions move toward the collection electrode. Otherwise, the avalanche process is similar.

At the point where impact ionization ceases, the avalanche process is quenched and no additional free electrons are released. The edge of the avalanche region is defined by the boundary of the corona glow.

The electrons passing the corona glow boundary move toward the collection plate under the influence of the electric field. If the interelectrode space contains electronegative gases, such as oxygen, sulfur dioxide, etc., which have an affinity for electron capture, electrons will be attached to the gas molecules. The negative ion thus formed will move toward the collection electrode at a velocity much less than that of an unattached electron. The presence of these slow-moving ions in the interelectrode space gives rise to a significant distributed electrical charge commonly referred to as space charge, which is necessary for the maintenance of a stable corona. If no electronegative gases were present, electrons generated in the avalanche process would be quickly swept to the anode because of their high mobility. Under these conditions, no stabilizing space charge would form and sparking would occur at approximately the same voltage required for the onset of corona. Thus, the presence of an electronegative gas is required for the maintenance of a stable negative corona. This requirement is no problem for electrostatic precipitator applications because electronegative gases are always present and are predominant in industrial gases and in the atmosphere.

The appearance of corona in a precipitator varies according to whether the corona is negative or positive. Positive corona appears as a uniform diffuse glow surrounding the corona wire. Negative corona, on the other hand, appears as tufts or discrete areas of glow along the corona wire. Fig. 4 shows the appearance of the two types of corona.

There are two steps in the development of a corona. As the voltage between electrodes is increased, a corona point will develop at a surface irregularity on the wire due to the enhanced electric field, thus initiating the avalanche process in the localized area. The highly mobile electrons are immediately swept clear of the relatively sluggish positive ions. The resulting space charge is sufficient to quench the avalanche process by the recombination of secondary electrons. The avalanche will remain quenched until the space charge moves from the

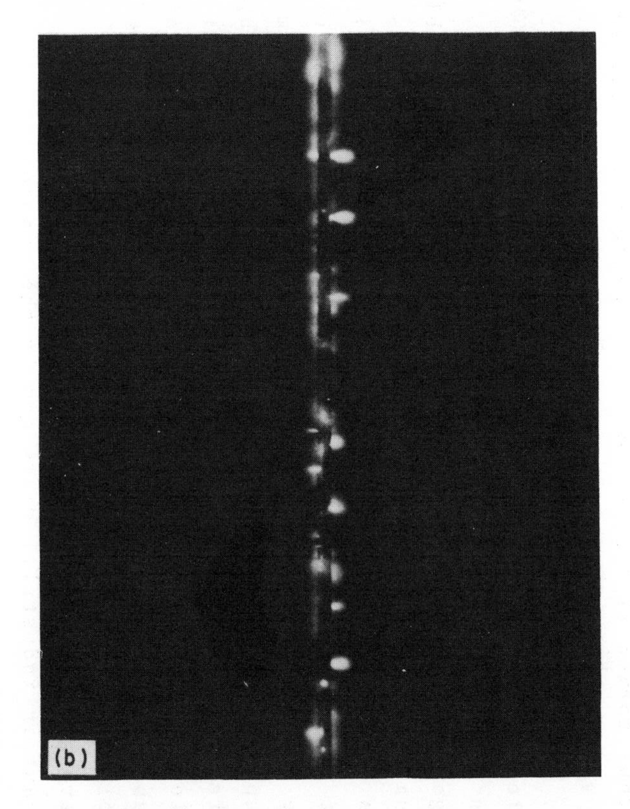

FIG. 4. (a) Photograph of positive corona and (b) negative corona.

corona wire and the process is repeated. The phenomenon is known as "Trichel Pulsing" and is the first stage of corona development. As the voltage is increased further, a stable condition is reached where the positive space charge near the wire, the negative space charge in the interelectrode region, and the avalanche electrons are in electrical equilibrium with the applied voltage.

B. CURRENT-VOLTAGE RELATIONSHIPS

The electric field strength required for the onset of corona is dependent upon the composition and density of the gas and the roughness of the wire. The field required for corona initiation in atmospheric air has been shown empirically by Peek (3) to be

$$E_c = 30\ md\left(1 + 0.30\sqrt{\frac{d}{a}}\right) \qquad (1)$$

where E_c is the field strength, kV/cm; d is the relative gas density, $\frac{T_oP}{TP_o}$; T_o is the standard absolute temperature, 293°K (20°C); P_o is the standard absolute pressure, 760 mm mercury; T and P are the temperature and pressure at operation conditions; m is the roughness factor for wire, $0.5 < m < 1$; and a is the wire radius, cm.

Once corona is established, further increases in applied voltage result in an increase in current and ultimately in a spark breakdown of the corona gap. The current-voltage relationships are governed by the precipitator geometry and by space charge effects, the latter being strongly dependent upon the nature of the gas and the gas density.

As previously discussed, space charge in a negative corona is brought about by the presence of large numbers of relatively slow-moving charges such as ions or charged particulate in the interelectrode space. This space charge reduces the potential near the corona electrodes and also repels electrons from the avalanche process as well as ions formed near the discharge electrode. The result is that for a given voltage, the greater the space charge the lower the current.

The process of electron attachment to form a space charge depends on the gas species present. In some gases, such as sulfur dioxide and oxygen, electron capture is relatively easy and the presence of even small amounts of these gases is sufficient to capture all the electrons produced in the avalanche process. In other gases, notably CO_2 and water vapor, electron attachment depends upon dissociation of the molecule by an energetic electron prior to attachment to the oxygen molecule, and attachment is less likely.

Because of these differences in the ease of electron capture, the space charge, and hence the voltage-current curves, are different for the various gases. Fig. 5 shows the data given by White (4) for carbon dioxide and various mixtures of nitrogen and sulfur dioxide for a 0.109 in. wire and 6 in. cylinder electrode.

The significant aspect of the voltage-current curves in a precipitator is the difference between the voltage required for corona initiation and that resulting in sparking. As stated previously, with no electron capture gases present, slight changes in electrode spacing or other factors would result in excessive sparking and control of the precipitator voltage would become impractical.

Most industrial gases have more than sufficient electron capture constituents for the formation of a space

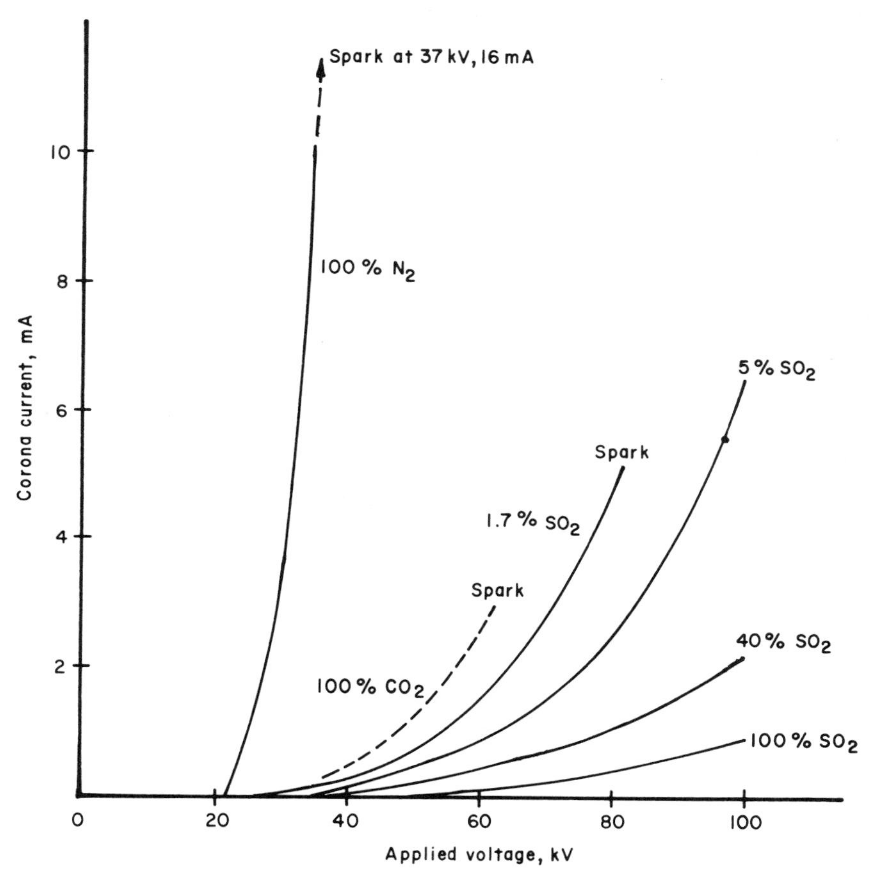

FIG. 5. Variation of negative corona current with applied voltage for various gas mixtures.

charge adequate to give a stable corona with suitable voltage-current relationship.

Voltage-current characteristics are also governed by the size of the discharge electrode, electrode spacing, and, in the case of wire and plate precipitators, by the spacing between wires. In practical precipitator design, these parameters are selected on the basis of the properties of the dust to be collected, power supply considerations, and geometrical factors.

Gas temperature and pressure also influence corona starting and sparking voltage characteristics to a considerable degree. The avalanche process is brought about by the electrons which have attained sufficient velocity for ionization of a gas molecule on impact. This velocity must be attained between inelastic collisions; thus, at higher temperatures, the gas density is lowered and the time between inelastic collisions is increased. Therefore, the field required for corona initiation would be lower at high temperatures. Conversely, lower temperatures or higher pressures would reduce the intermolecular spacing and would tend to increase the field required for the onset of corona.

For the range of temperatures used in most commercial precipitator installations, there is sufficient range between corona starting voltage and sparkover voltage to enable adequate control. However, at temperatures around 1500°F and above, sparkover voltages approach the corona starting voltage, so that operation at atmospheric pressure becomes difficult.

Studies of corona and sparking under conditions of high temperature and pressure have shown that for negative corona, effective ion mobilities increase due to the following factors (5):

1. at constant gas temperature and field strength, ion mobility increases with decreasing gas density;
2. at constant density and field strength, ion mobility increases with increasing temperature; and
3. at constant temperature and gas density, effective ion mobility increases with increasing field strength.

The effect of Factor 1 is that at lower gas densities, electrons travel further before they are attached to gas molecules to form ions. Thus, the average velocity of the current carriers is increased. The other factors (2 and 3) are reportedly due to the increase in the number of free electrons in the vicinity of the corona wire due to increased thermionic emission and to increased secondary emission from positive ion impact on the corona wire. The availability of these additional electrons and the presence of a high field again result in their traveling further toward the collection plate before being captured by inelastic collision with an electronegative gas molecule. Since the current in the interelectrode space is the product of the charge, effective mobility, and field strength, the increase in mobility results in increased current. Consequently, for negative corona, the voltage current curves would be altered toward higher currents at lower voltage.

In negative corona, the presence of negative ions near the corona wire provides a space charge which limits the field near the wire and serves to stabilize the discharge. However, with low gas densities, the highly mobile, free electrons move rapidly away from the corona wire and are not effective in providing a stabilizing space charge until they are captured by electronegative gas molecules. Since, at low gas densities, the electrons move a longer distance before capture, the space charge effect is reduced and the current rises more rapidly than at higher gas densities. Thus, sparking occurs at reduced voltage.

In positive corona, the same effects would not occur since the current carriers are positive ions. Electrons from the avalanche process are attracted to the corona wire and do not travel to the collection electrode. Consequently, positive corona should have more favorable voltage-current relationships in the high temperature region with higher sparking potentials.

III. THE ELECTRIC FIELD

The electric field influences electrostatic precipitation as it affects the particle charge and the driving force for collecting these particles. The charge on the particles is in part determined by the electric field in the vicinity of the particle during charging as is discussed in more detail in the next section. Therefore the electric field affects precipitation in a two-fold manner; first, as it contributes to the charging of the particles, and second, as it influences the collection of those charged particles.

The electric field in a precipitator results from both electrode and space charge components. The electrode component of the field is determined by the applied voltage and the geometry of the electrodes. This component of the field is augmented by the field from the space charge when ions and/or charged particulate are present in the interelectrode space.

The equations for the electric field for a wire and pipe-type precipitator are generally used because the symmetry of this configuration leads to simplified mathematical relationships. The equations for the field will be developed for the wire and cylinder electrode case.

Prior to the initiation of corona, the electric field can be described by the well known relationships of electromagnetic theory (6,7):

$$E(r) = \frac{V}{r \ln(b/a)} \tag{2}$$

where $E(r)$ is the electric field as a function of radius, r, V is the applied voltage, and a and b are the radii of the corona wire and collection electrode, respectively. With the initiation of corona, as described in the previous chapter, free electrons, ions, and charged particulate within the interelectrode space constitute a space charge that modifies the electric field configuration. Poisson's equation can be used to determine the voltage, field, and current for this dynamic situation:

$$\frac{d^2V}{dr^2} + \frac{1}{r}\frac{dV}{dr} + \frac{\rho}{\varepsilon_0} = 0 \text{ (cylindrical coordinates)} \quad (3)$$

where ε_0 is the permittivity of free space, (f/m). The space charge, ρ (c/m^3), given in terms of charge per unit volume, is related to the current density, j (A/m^2), carried by the motion of the various charge carriers, which is related to the carrier mobility, μ (m /v-sec), expressed as an equation

$$j = \rho\mu E \quad (4)$$

When multiple carriers are present, then the current from each carrier species must be considered. Thus, Eq. (4) becomes

$$j = (\rho_e\mu_e + \rho_i\mu_i + \rho_p\mu_p)\ E \quad (5)$$

Since for typical installations utilizing electrostatic precipitators the gases contain an electronegative gas, essentially all of the free electrons are quickly attached to neutral molecules to form negative ions. The term for free electron space charge is neglected so that Eq. (5) can be simplified to

$$j = (\rho_i\mu_i + \rho_p\mu_p)\ E \quad (6)$$

If the term in brackets is replaced with an expression which contains the space charge and an equivalent mobility, we have

$$j = \rho \mu_{eq} E \tag{7}$$

which yields an expression for the space charge of

$$\rho = \frac{j}{\mu_{eq} E} \tag{8}$$

The right-hand portion of Eqs. (6) and (7) can be manipulated in order to find an expression for this mobility, which is approximately

$$\mu_{eq} \approx \frac{\rho i}{\rho} \mu_i \tag{9}$$

where ρ_i is the ionic space charge density and ρ is the total ionic and particulate space charge density. Using these relations in Poisson's equation, an expression for the electric field as a function of the radius can be shown to be

$$E(r) = -\left[\left(\frac{r_0 E_c}{r}\right)^2 - \left(\frac{r_0{}^2}{r^2} - 1\right) \frac{j}{2\pi\varepsilon_0 \mu_{eq}}\right]^{1/2} \tag{10}$$

Figure 6 illustrates the variation in electric field with and without corona current. Breakdown field strengths can be computed on the basis of Eq. (1). For atmospheric air at standard conditions, the breakdown strength is of the order of 30 kV/cm for parallel plate electrodes (but is higher for nonuniform fields).

The voltage current characteristics can be determined by integrating Eq. (10) from near the surface of the

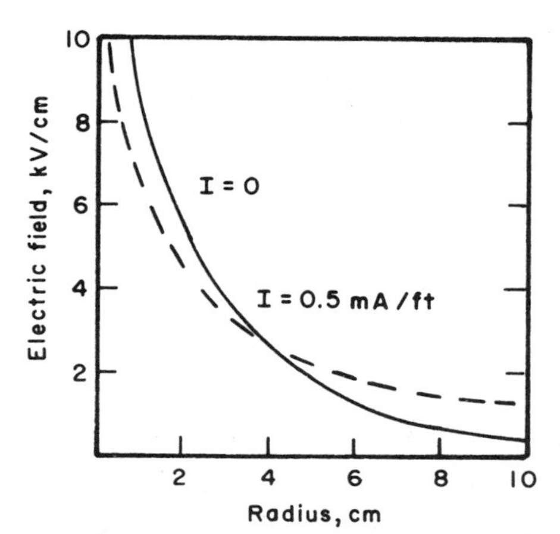

FIG. 6. Electric field vs radius for no current flow and for a current per unit length of wire of 0.5 mA/ft.

wire to the collection electrode

$$V = \int_a^b -E(r)\,dr = -\int_a^b \left[\left(\frac{r_0 E_c}{r}\right)^2 - \left(\frac{r_0^{\,2}}{r^2} - 1\right)\frac{i}{2\pi\varepsilon_0\mu_{eq}}\right]^{1/2} \tag{11}$$

The radius of the corona glow region is very nearly equal to the radius of the wire, a. Substituting a for r in Eq. (11) and integrating and applying the boundary conditions yields Eq. (12)

$$V = V_0 + aE_c\left[\sqrt{1 + \frac{i\rho}{2\varepsilon_0\mu_i\rho_i}\cdot\frac{b^2}{a^2E_c^2}} - 1 - \ln\sqrt{\frac{1 + \dfrac{i\rho}{1 + 2\mu_i\varepsilon_0\rho i}\cdot\dfrac{b^2}{a^2E_c^2}}{2}}\right] \tag{12}$$

Further simplifications can be made for moderate and high currents such that a reasonably close approximate equation for full-scale installations is

$$V = V_0 + aE_c\left[\sqrt{1 + \frac{i}{2\pi\varepsilon_0\mu_{eq}} \cdot \frac{b^2}{a^2E_c^2}} - 1\right] \qquad (13)$$

Equation (13) contains a term, μ_{eq}, which is the equivalent mobility of the composite space charge, charged particulate, and the ionic current. An approximate value can be determined for this term by neglecting the current carried by the particulate and solving Eq. (4) for the space charge due to ions. The space charge due to the particulate can be determined by computing the saturation charge expected for the particle concentration with the average electric field. A determination for this factor is required for solving Eq. (19).

For reasonably accurate estimates, the space charge from the particulate may be neglected. For this case, the mobility of the ion can be used as the equivalent mobility. This, of course, will cause an error in the space charge of the order of a factor of 2 or 3, but will demonstrate the order of magnitude of the voltage, current, and electric field.

An equation for the wire and duct electrodes has been derived (8); however, due to the added complexity of the mathematics, this equation is not commonly used.

IV. PARTICLE CHARGING

The corona process leads to the formation of large numbers of ions in the interelectrode region. These ions move toward the collection electrode at a velocity determined by the mobility of the ions and the magnitude of the electric field. Particles introduced into the interelectrode region must receive their charge by attachment of these ions. This attachment of ions is brought about by one of two processes; field charging or diffusion charging. In field charging, the ionic flow is caused by the directed flow of ions along field lines which intercept the particles. In the case of diffusion charging, the random motion of the ion due to its thermal energy causes it to impact with a particle and impart its charge to it.

In practical precipitators, both field charging and diffusion charging are significant in terms of particle charging. For large particles, field charging is the predominant mode, whereas diffusion charging predominates in the case of very fine particles or fume.

A. FIELD CHARGING

The sequence of events associated with field charging is suggested by the illustrations in Fig. 7. In 7 (a), an uncharged spherical particle with a dielectric constant significantly greater than 1 is suspended in a uniform electric field. The presence of this particle causes a distortion in the electric field in the vicinity of the particle. If ions are now introduced in the region, they will tend to follow the electric field "lines" or maximum field gradient. This causes the ions to impact with and be retained by the particulate.

After a period of time, the ionic flow to the particulate will cause a net charge to accumulate on the particle. This charge produces an electric field that decreases as the reciprocal of the square of the radius as depicted in Fig. 7 (b). This self field superimposed on the uni-

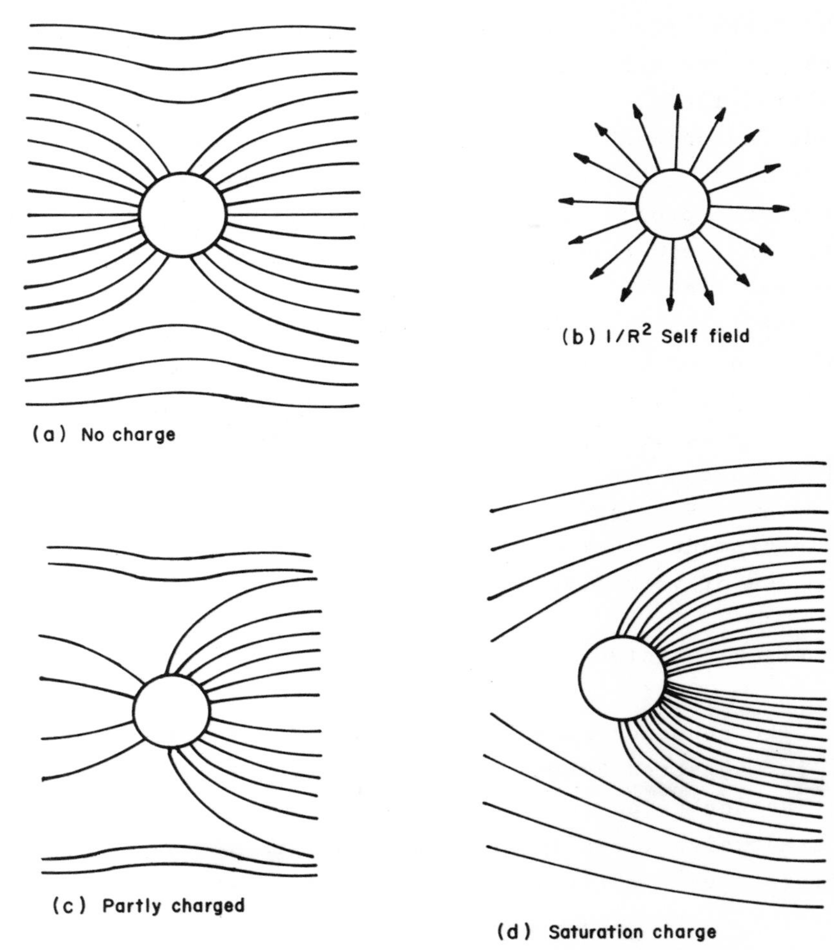

FIG. 7. Electric field in vicinity of particle.

form applied field yields a resultant field as shown in 7 (c).

Finally, the particle will acquire a saturation value of charge such that the electric field from the charged particulate just balances the applied field and charging ceases. This condition is depicted in 7 (d). Derivations for the spatial distribution of charge and the charging rate are covered by several authors (4,9,10,11,12). In these derivations the current flow to the particle is related to the free ion density, mobility, and applied field. The expression for current flow is then integrated with respect to time to yield a value for the charge. This expression is

$$q(t) = q(s) \frac{1}{1 + \frac{\tau}{t}} \tag{14}$$

and q_s is the saturation value of charge from field charging

$$q_s = \frac{12\pi\varepsilon\varepsilon_0 a^2 E_0}{\varepsilon + 2} \quad \text{(in MKS units)} \tag{15}$$

where τ is the charging time constant defined as that time required for a particle to acquire 50% of the final value of charge. This time constant is related to the free ion density and the mobility of the carrier species by

$$\tau = \frac{4\varepsilon_0}{N_0 e\mu} \tag{16}$$

where N_0 is the ion density (number per cubic meter).

B. DIFFUSION CHARGING

Diffusion charging differs from field-dependent charging in that the driving force for the motion is from the thermal energy that gives rise to random velocities for the ions. The average velocity for the ionic cloud is proportional to the square root of the temperature. Within the ionic cloud, individual velocities range in value up to several orders of magnitude greater, and less, than the RMS velocity. The velocity distribution is thought to be either approximately Gaussian or perhaps log-normal distribution. Thus, there is a nonzero probability of finding an individual ion with some value greater than any particular value of velocity. Thus, there is a nonzero probability of finding an ion with sufficient kinetic energy to overcome the energy barrier caused by the field from a charged particle. This leads to a continued flow to charged particulate, the rate of which decreases with increasing charge, but never reaches a zero flow rate.

This mechanism is known as diffusion charging. The thermal velocity is in addition to any velocity caused by the applied field. An applied electric field causes a bias on the component of velocity in the direction of the field which is often neglected in theoretical descriptions of diffusion charging.

In contrast with the field charging mechanism, theory does not predict a saturation value of charge for the diffusion charging process. There is always a nonzero probability of finding an ion with sufficient kinetic cnergy to climb the potential hill from the charged particulate. However, a saturation value of charge does exist even though it may be only of academic interest. This saturation value of charge is that value of charge that causes an electric field at the surface of the particulate to cause the expulsion of an electron by field emission.

Various authors (12,13) have derived expressions for diffusion charging rates. In general, the charging rate is proportional to the absolute temperature and the free ion density raised to some fractional power. If one neglects the influence of the electric field, the expression for the charging rate for diffusion charging is

$$q(t) = \frac{akT}{e} \ln \left[1 + \frac{\pi a v N_0 e^2 t}{kT} \right] \qquad (17)$$

where k is the Boltzmann constant (J/°K), T is the temperature, °K, and ln is the natural logarithm.

C. COMBINATION FIELD AND DIFFUSION CHARGING

Both field and diffusion charging are active for the charging process. However, for large particles, the component of charge due to field charging is so great in comparison to the diffusion component that the diffusion charging is neglected. Conversely, for the very small particles, the saturation value of field charging is so small that only the diffusion charging process need be considered. For the intermediate range of particles (0.1 - 0.8 μm), both diffusion and field charging mechanisms must be considered. Before the saturation value of field charging is attained, both mechanisms are active. In this case, the charging rates for both systems must be added. After the saturation value for field charging is attained, only the diffusion charging component needs to be considered. Thus, the charging behavior for this intermediate size range can be described as augmented field charging initially, followed by diffusion charging. Several authors have considered these factors extensively (13,14).

D. PRACTICAL ASPECTS OF CHARGING

Particle charging theories have been developed for highly idealized conditions. The assumptions in these theories include such factors as a uniform and constant electric field, spherical isolated particles, and constant free ion densities. The actual situations encountered in practice include heavy dust loads of uncharged particles introduced into the inlet of a precipitator that is powered by either a half-wave or full-wave rectified voltage waveform. The only filtering applied to this power supply is from the distributed capacitance of the precipitator electrode system and power distribution system. Thus, the applied voltage, as well as the free ion density, changes with time. Hence, the charging rate changes with time, and in fact, the charging may be interrupted during portions of the voltage cycle. The time varying secondary voltage and an estimate of the charge vs time for a range of conditions are shown in Fig. 8.

A further complication is brought about by the introduction of large quantities of dust into the precipitator inlet. The available charge is quickly bound to the particulate, which causes an immediate reduction in the current. The highly mobile gas ions are quickly attached to the relatively sluggish particulate. Since the charging rate is proportional to the free ion density, this current quenching in the inlet section causes a decrease in the charging rate. For low current densities, such as are encountered in precipitators collecting high-resistivity dust, charging times can be significant and can reduce collection efficiency. If dust resistivity is low, charging times may be low enough to be insignificant.

V. PARTICLE COLLECTION

Fundamental electrostatic theory establishes that the magnitude of the force acting on a charged particle under the influence of an electric field is dependent upon the charge on the particle and the strength of the electric

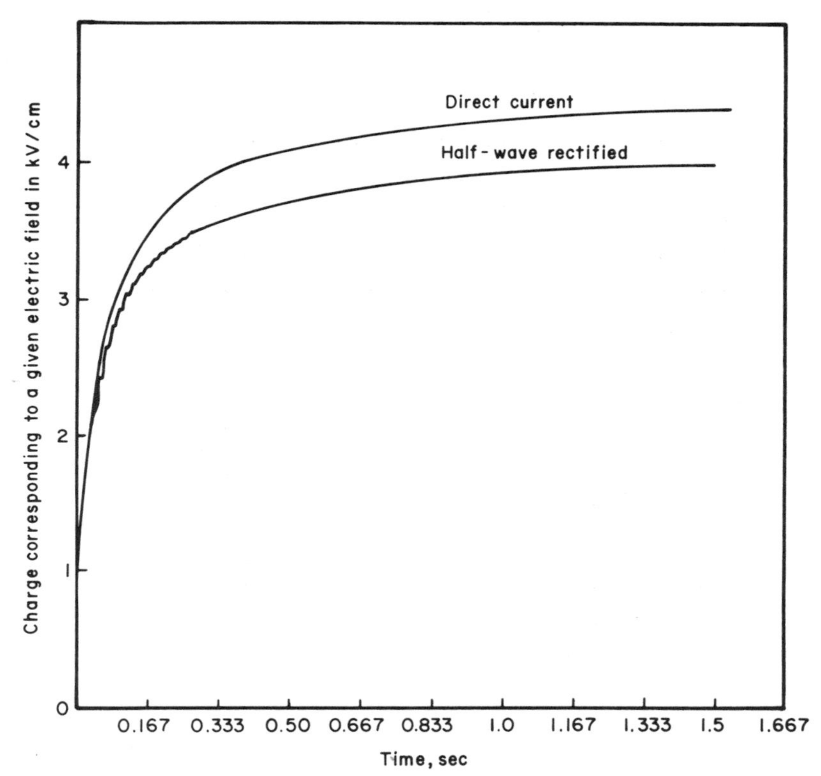

FIG. 8. Charging characteristics for half-wave rectified and direct current supplies with 4.35 kV/cm peak values.

field. The direction of the force depends upon the polarity of the particle charge and the direction of the electric field.

Simplified concepts of particle collection often consider that the electrostatic forces predominate, and that motion of the particle toward the collection plate is governed primarily by the electrostatic forces. However, in a full-size precipitator, the aerodynamic forces associated with the highly turbulent gas flow predominate.

This causes the motion of particles smaller than about 10 μm in diameter to be almost completely determined by the gas stream motion. Near the collection plate, the influence of gas turbulence is reduced, and electrostatic forces predominate. The collection of an individual particle therefore depends on the probability that it will enter the region where the electrostatic forces result in its deposition on the collection surface.

In addition to the domination of the particle motion by the gas flow in the central regions of the precipitator other factors influence particle collection. Once a dust layer is formed on the collection electrode, impingement of a particle being precipitated can reentrain those previously collected. Scouring of the dust layer, reentrainment of dust during rapping, and unusual electrical conditions, including sparking and back corona, can also alter the basic collection process. In developing a fundamental theory of particle collection these factors are neglected in order to simplify the derivation and facilitate understanding of the collection process. The effects of these factors on collection are discussed following derivation of the basic collection theory.

A. FORCES ACTING ON THE PARTICLES

The forces acting upon charged particles, other than the electrical force and the viscous drag force caused by the motion of the particle through the gas, can be neglect The electrical force accelerates the particle toward the collection electrode, while the viscous drag force of the gas opposes this flow. The final value of velocity attain will be the velocity that causes the viscous drag force to just balance the electrical driving force.

In the region adjacent to the collection electrode, the turbulent flow is reduced by the frictional force from the collection electrode to a value of the same order of magnitude as that of the electrical velocity of the intermediate and smaller particles. The trajectory of charged particulate in this boundary region is governed by the electrical velocity toward the collection plate and by the frictionally reduced gas velocity through the precipitator. Possible gas velocity distributions across the duct in a precipitator are shown in Fig. 9, and the vector sum of the electric and gas velocities at the edge of the boundary zone is shown in Fig. 10.

The velocity with which a charged particle is driven to the collection electrode is related to the charge on the particle, the dimensions of the particle, the value of the electric field driving the particle, and the viscosity of the host gas. The velocity of the particle relative to the gas can be expressed mathematically by equating the electrical and viscous forces acting on the particle. The electrical and viscous forces are given respectively on the left and right sides of Eq. (18)

$$qE = 6\pi a \nu w \tag{18}$$

which can be solved for the electrical migration velocity, w:

$$w = \frac{qE}{6\pi a \nu} \tag{19}$$

where q is the charge, C; E is the electric field, V/m; a is the particle radius, m; and ν is the viscosity, kg/m-sec.
(Note: 1 kg/m-sec = 10 poise.)
Gravitational and inertial forces have been neglected in the discussion. These factors are not significant for

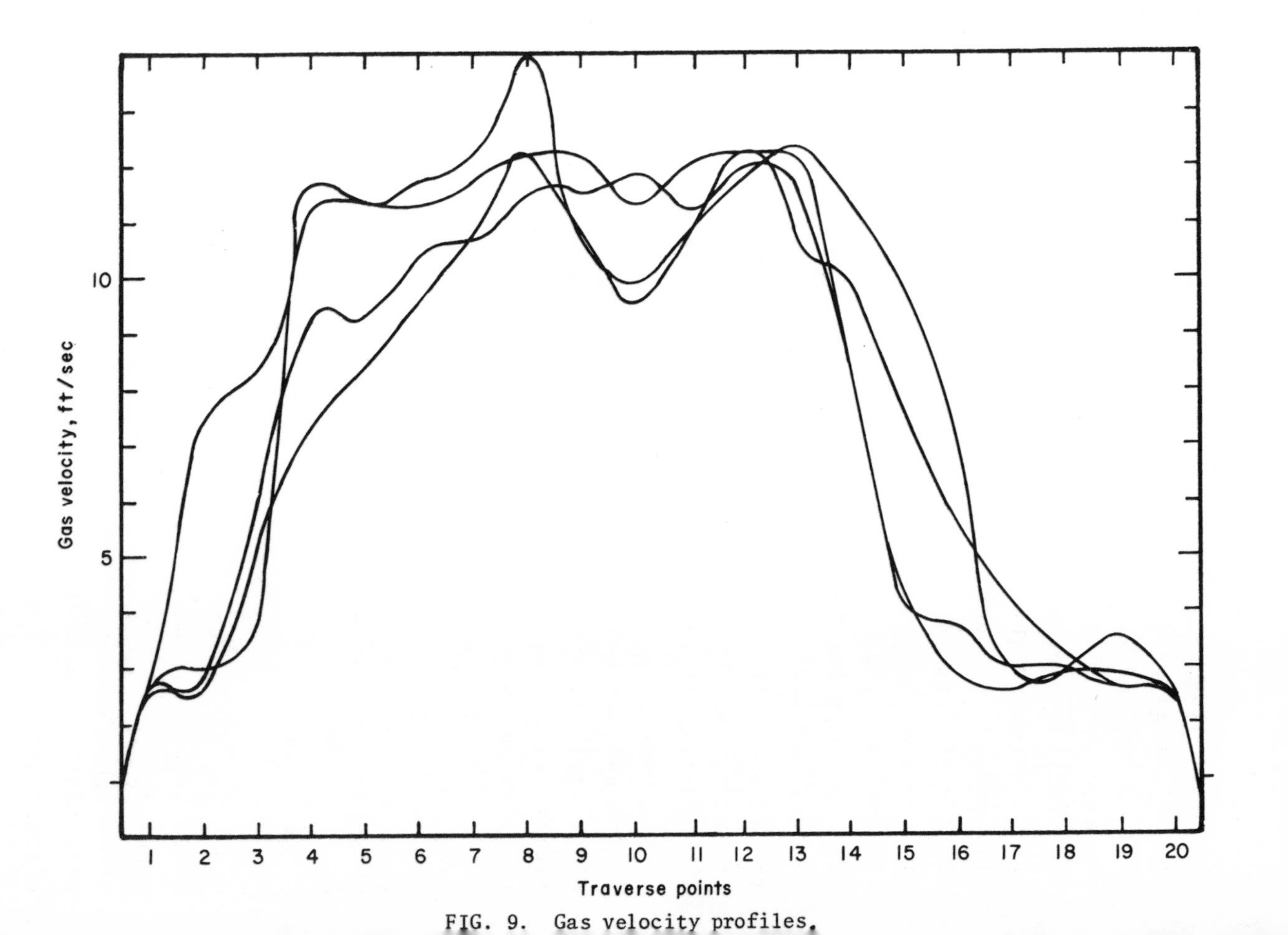

FIG. 9. Gas velocity profiles.

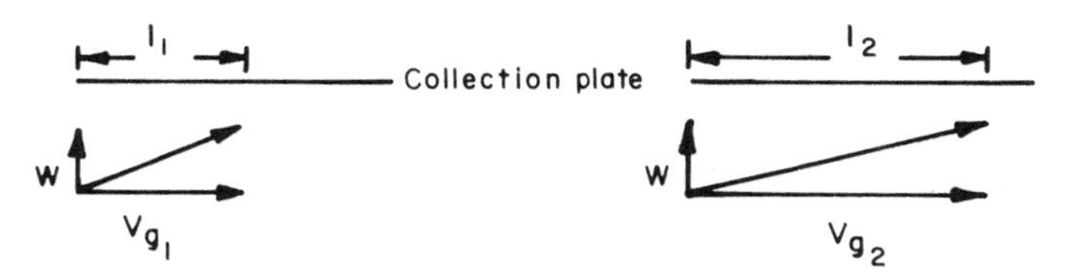

FIG. 10. Sketch showing the variation in length required for collection for two gas velocities with a constant migration velocity (laminar flow case).

particles less than about 100 m in diameter which is the range of interest in electrostatic precipitation.

The equation for the migration velocity for a particle is dependent upon which of the two charging mechanism dominates. For the larger particle where field charging dominates, the migration can be expressed by Eq. (20a)

$$w = \frac{2\varepsilon\varepsilon_0 E_0 E_p a}{(\varepsilon + 2)\nu}\left(\frac{1}{1 + \frac{\tau}{t}}\right) \tag{20a}$$

which for particles for steady-state conditions with a rather high dielectric constant becomes

$$w = \frac{2\varepsilon_0 E_0 E_p a}{\nu} \tag{20b}$$

When diffusion charging is the dominant charging mode, the migration velocity becomes

$$w = \frac{E_p(1 + A\frac{\lambda}{a})\ kT}{6\pi\nu\qquad e}\ \ln\left(\frac{1 + \pi a\nu N_0 e^2 b}{kT}\right) \tag{21}$$

The term $\left(1 + A\,\frac{\lambda}{a}\right)$ is the well-known Cunningham correction factor that is applied to Stokes' law for particles of a size comparable to the mean-free path length in the gas stream. This factor can generally be neglected for particulate of the size range associated with field charging.

B. PARTICLE COLLECTION WITH LAMINAR FLOW

For simplicity the discussion of electrical collection begins with the highly idealized case of laminar flow and then extends to the more complex situation of turbulent flow, which exists in essentially all commercial installations. For the laminar flow example, consider that the gas stream and entrained particulate moves through the precipitator with a uniform gas velocity, v. As described above, the particulate is driven toward the collection plate with an electrical velocity, w. The particulate trajectory will then be the resultant from the vector sum of these two velocities, as shown in Fig. 11. Also shown in the figure is the development of a dust-free zone at

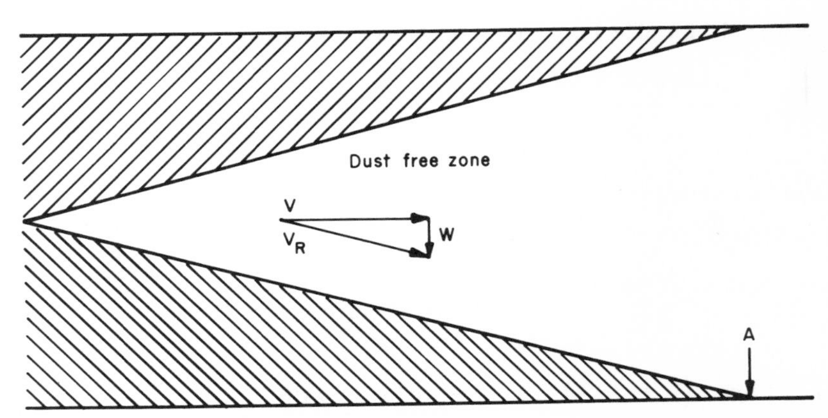

FIG. 11. Collection schematic for laminar flow precipitator.

the center of the precipitator that progressively grows as the collector is traversed until at point A, all the particulate has been removed for 100% collection. This development assumes that the particles were previously charged to saturation, no reentrainment of the collected dust occurs, and all particles are of the same size with identical migration velocities.

Since the migration velocity for field charging is proportional to the particle radius, a composite dust with a range of particle sizes would behave somewhat differently from the single particle size. The distance required for the total collection of the small particles, 1 μm diameter, for example, would be ten times as great as for a particle with a diameter ten times greater. Thus, for a particle size range that includes particles from the larger to the smaller, the dust concentration would decrease as the precipitator was traversed, with the larger particles removed first and progressively smaller ones collected subsequently. Thus, a concentration gradient, as shown in Fig. 12, would develop in the longitudinal as well as the transverse direction.

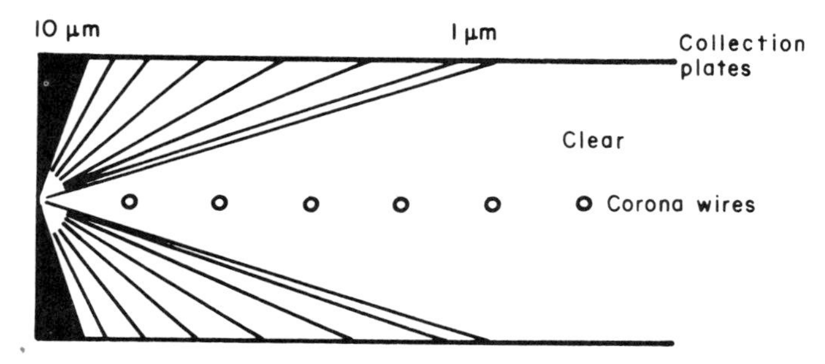

FIG. 12. Concentration gradient that should develop for laminar flow collection with a range of particle sizes.

C. PARTICLE COLLECTION WITH TURBULENT FLOW

In all practical precipitators the gas flow is turbulent, and this turbulent flow fundamentally determines the trajectory of the dust particles in the interelectrode space. Near the wall, however, the turbulent eddies are damped by the friction. Within this boundary region, the magnitude of the electrical velocity is of the same order of magnitude as the gas velocity. Within this region, where precipitation actually takes place, the trajectory of the charged particulate is determined by the vector sum of the electric and gas flow velocity vectors.

Within the interelectrode space outside the boundary region, the magnitude of the electrical migration velocity is small compared with the magnitude of the gas turbulent velocity. In this region, the effect of the electrical velocity is to apply a small bias toward the collection electrode to the overall turbulent flow. The electrical velocity can essentially be neglected in all but that region of space adjacent to the collection electrode.

The consequence of this turbulent flow situation is that the collection efficiency that could be expressed as a linear function of the precipitator length for laminar flow is converted to an exponential relation in the case of turbulent flow. In the distance where 100% collection occurred for laminar flow, the concentration is reduced to a value of $1 - \frac{1}{\varepsilon}$ for the turbulent flow, where ε is the base of the natural logarithm. The collection efficiency is given by:

$$\eta = 1 - \exp\left(- \frac{A}{V} w\right) \tag{22}$$

where η is the efficiency, A is the collection surface area, V is the gas volume, and w is the precipitation rate paramet

This exponential efficiency equation is known as the Deutsch-Anderson equation. The exponential relationship, which was discovered experimentally by Anderson in 1919, was derived from theoretical calculations by Deutsch in 1922. Several simplifying assumptions were made by Deutsch in his original work. These include:

1. The particles are considered to be fully charged immediately on introduction into the collection system.
2. Turbulent and diffusive forces cause the particles to be distributed uniformly in any cross section. (This assumption is more restrictive than necessary in actual practice.)
3. The velocity of the gas stream does not affect the migration velocity of the particle.
4. Particle motion is governed by viscous drag where Stokes' law applied.
5. The particle always moves at its electrical terminal velocity.
6. Dust particles are separated enough for their mutual repulsion to be neglected.
7. The effect of collision between ions and neutral gas molecules can be neglected.
8. There are no disturbing effects present such as erosion, reentrainment, uneven gas flow distribution, or back corona.

Derivations of this equation are included in several texts (4,10).

Equation (22) pertains specifically to the efficiency of the collection of particles with a given migration velocity and hence of a given size. Since precipitators typically collect particulate with a wide range of sizes, no one equation is sufficient to describe the composite collection efficiency for the system. A range of particle sizes will have a range of associated migration velocities. The collection efficiency for the composite could be expressed as a summation of the products of the dust load within an increment of size distribution with the collection efficiency associated with that size interval, or by a definite integral for continuous particle size distributions. Reasonably close approximations can be made by dividing the specific particle size distribution into a small number of increments and calculating the overall collection efficiency. An example of this calculation is given in Table 1.

D. FACTORS MODIFYING PARTICLE COLLECTION

The collection efficiency as described by Eq. (22) is based on the highly idealized case described by the original assumptions. These idealized factors seldom exist in practice. The particle charging time is seldom negligible, which results in the particle traversing a significant percentage of the precipitator before acquiring a saturation value of charge. The net result is that the migration velocity is less than would be predicted based on the assumption of instantaneous charging.

The assumption of a uniform particle concentration may be significantly in error. Preliminary tests indicate that the electrical forces tend to cause an increase in the particle concentration in the vicinity of the collection plate. This factor should cause an increase in the collection efficiency.

Reentrainment can occur as an erosion during the collection process, as reentrainment during plate rapping, or as a pickup by the gas stream from the material in the dust hoppers. This reentrainment constitutes a reduction in the collection efficiency.

Uneven gas flow causes a two-fold reduction in the collection efficiency. First, the Deutsch-Anderson equation shows that the collection efficiency is inversely related to the gas velocity or volume flow rate. Second, the region experiencing the high gas velocity also carries a greater percentage of the total dust emission. These two factors result in reductions in collection efficiency.

The particular particle size distribution that is introduced into the precipitator also has an effect on the collection efficiency. If the precipitator is preceded by a mechanical collector, the larger particles will be removed ahead of the precipitator. In the absence of a mechanical collector, the large particles, which are easier to collect, will still be present in the precipitator. Two examples will demonstrate the effect of particle distribution.

For illustrative purposes, assume that the inlet particle size for an installation with no mechanical collector can be approximated by the distribution shown in Table 1. The electrical conditions are such that the migration velocity for each size increment is as stated, with an area-to-volume-flow ration of 12 ft^2/cfs, or 100 ft^2/1000 cfm. The overall collection efficiency and effective migration velocity (EMV) for this installation is as shown.

Next, consider the situation where a mechanical collector is installed that modifies the inlet particle size distribution to that shown in Table 2. The efficiency within each size range remains the same, but the shift in the particle size distribution causes a decrease in the overall collection efficiency and EMV as shown. This change in the particle size causes a change in the EMV from 15.5 cm/sec to 11.4 cm/sec.

Table 1

Assumed Inlet Particle Size Distribution

Precipitator Performance Without a Mechanical Collector

Particle size m	% by weight	Migration velocity, cm/sec	Efficiency this interval	Uncollected material
0.5	1	4	79.2	0.21
1.5	2	12	99.1	0.018
3.0	4	24	99.99	0.0004
6.0	6	48	100	0
12.0	12	96	100	0
24.0	25	192	100	0
48.0	50	384	100	0
TOTAL	100			0.2284

Efficiency = 100 - 0.23 = 99.77%

EMV = 15.5 cm/sec

Table 2

Assumed Inlet Particle Size Distribution

Precipitator Performance with a Mechanical Collector

Particle size μm	% by weight	Migration velocity cm/sec	Efficiency this interval	Uncollected material % by weight
0.5	5	4	79.2	1.05
1.5	10	12	99.1	0.09
3.0	20	24	99.99	0.002
6.0	30	48	~100	0
12.0	20	96	~100	0
24.0	10	192	~100	0
48.0	5	384	~100	0
TOTAL	100			1.142

Efficiency = 100 - 1.142 = 98.86%

EMV = 11.4 cm/sec

Now, consider the effect of changing the gas velocity through the system while maintaining a fixed inlet particle size distribution. This change in velocity results in a change in the $\frac{A}{V}$ that varies inversely with velocity. If we utilize the particle size distribution shown in Table 2 with the associated migration velocity as an example, we can compute the collection efficiency and effective migration velocity for a range of gas velocities. The results of this exercise are shown in Fig. 13 and Table 3, where the effective migration velocity for gas velocities ranging from 4 to 12 ft/sec is given. An assumed area-to-volume flow ratio of 12 ft /cfs, at a

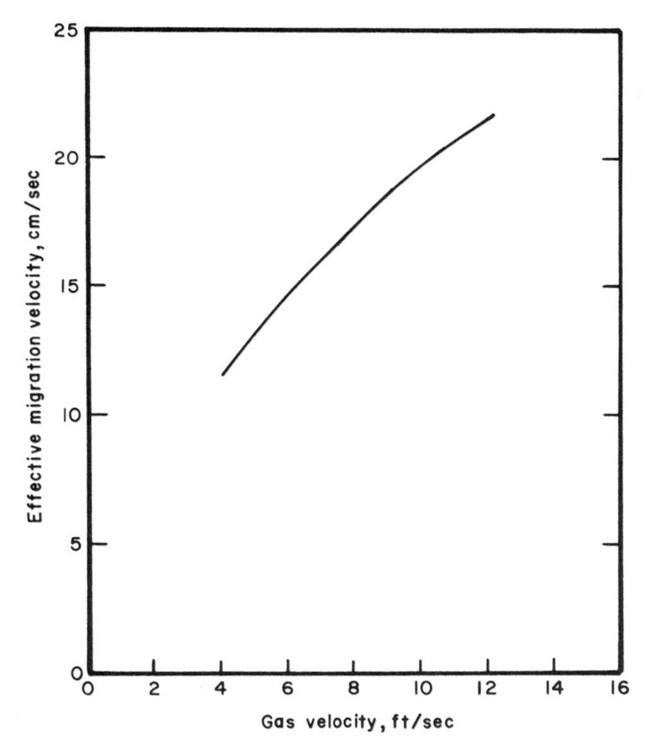

FIG. 13. Variation in precipitation rate parameter with gas velocity showing effect of particle size distribution.

Table 3
Incremental and Total Collection Efficiencies
With Effective Migration Velocities as a Function of Gas Velocity

Particle size μm	Migration velocity cm/sec	Percent by weight this increment	Gas velocity, v = 4 ft/sec			6 ft/sec		
			$\frac{A}{V}$ w	η	loss	$\frac{A}{V}$ w	η	loss
0.5	4	5	1.57	79.2	1.05	1.05	65	1.75
1.5	12	10	4.72	99.1	0.09	3.14	95.7	0.43
3.0	24	20	9.45	99.99	0.002	6.30	99.8	0.04
6.0	48	30	18.9	100	0	12.6	100	0
12.0	96	20	37.8	100	0	25.2	100	0
24.0	192	10	25.6	100	0	50.5	100	0
48.0	384	5	151.2	100	0	101	100	
			η = 98.86			η = 97.78		
			w = 11.4			w = 14.5		

gas velocity of 4 ft/sec, was used for the example. From this example, we see that conventional electrostatic precipitation theory predicts an increase in the effective migration velocity with gas velocity for realistic particle size distributions.

VI. PARTICLE REMOVAL

Dust deposited on the collection electrodes must be removed by a wet or dry process. In the case of liquid aerosol collectors, such as tar separators, acid mist collectors, etc., the material coalesces and drains from the collection plates. Solid material, on the other hand, requires an external method of removing dust. Wet

v =								
8 ft/sec			10 ft/sec			12 ft/sec		
$\frac{A}{V}$ w	η	loss	$\frac{A}{V}$ w	η	loss	$\frac{A}{V}$ w	η	loss
0.785	54.3	2.28	0.627	46.6	2.67	0.52	40.6	2.97
2.36	90.6	0.91	1.88	84.6	1.54	1.57	92.2	2.08
4.72	99.1	0.18	3.78	97.7	0.46	3.15	95.7	0.86
9.45	99.994	0.002	7.55	99.95	0.015	6.3	99.7	0.09
18.9	100	0	15.1	100	0	12.6	100	0
37.8	100	0	30.3	100	0	25.2	100	0
75.6	100	0	60.4	100	0	50.4	100	0
η = 96.7			η = 95.315			η = 94		
w = 17.35			w = 19.4			w = 21.5		

collectors utilize water or some other liquid to wash the plates carrying the dust with it. Several types of wet removal systems are in current use. One type utilizes a flooded header with cylindrical collection electrodes. The top of the electrodes form a weir and water flows inside the cylinder carrying the collected dust with it. Alternative types of wet collectors utilize water sprays to atomize water into the precipitator section where the water droplets precipitate onto the collection surface and subsequently drain, carrying the collected dust with them. Figs. 14 and 15 illustrate the various configurations of precipitators with wet removal systems.

Principal advantages claimed for wet removal systems are that: (a) reentrainment losses are kept to a mini-

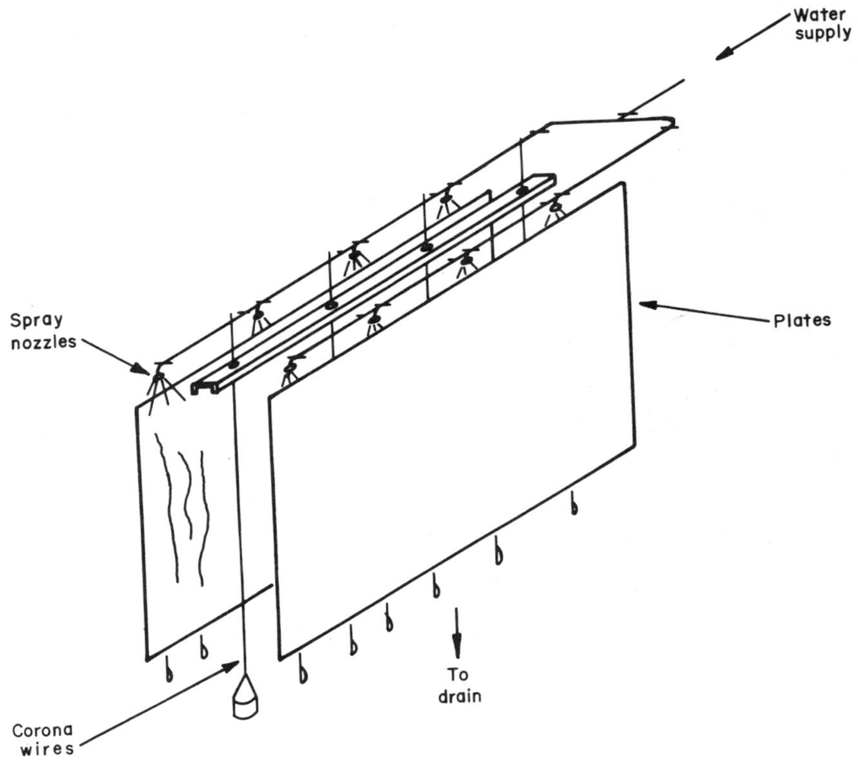

FIG. 14. Spray-type wet-wall precipitator.

mum, and (b) resistivity problems are eliminated. These factors are in part related to the "wettability" of the dust. Other claims are made for wet systems, including condensation on small particles for better collection, space charge enhancement, etc. The principal difficulties with wet collection systems are corrosion and scaling problems, loss of plume bouyancy, and problems of handling the slurry.

Dry dust removal is accomplished by periodic or continuous rapping of the collection electrodes. If the dust deposit on the electrodes is allowed to accumulate,

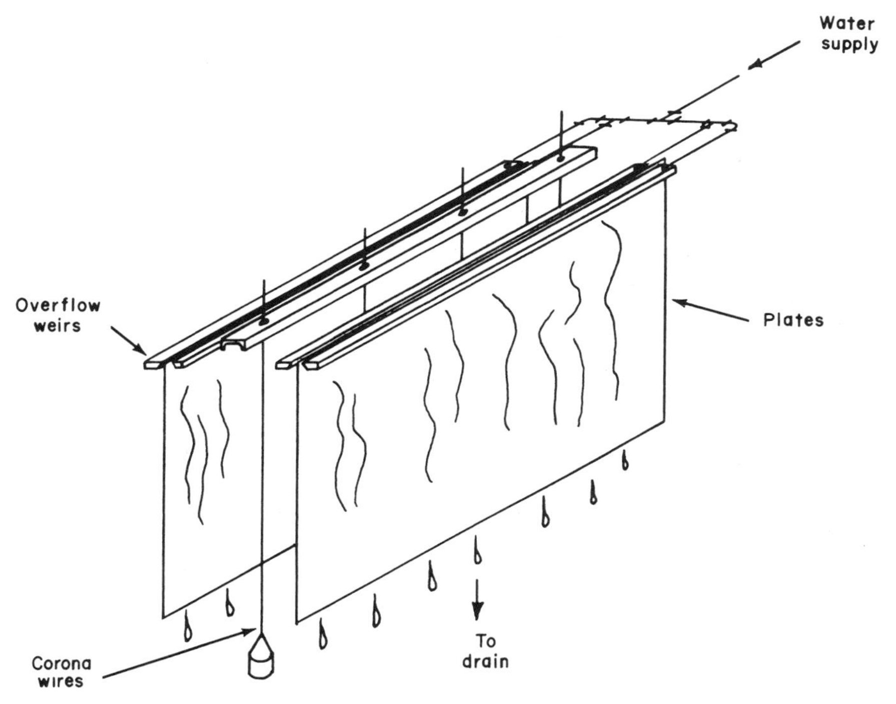

FIG. 15. Weir-type wet-wall precipitator.

it will fall in sheets into the hopper. The majority of precipitators collecting solid particulates are of the dry dust removal type. The principal problem with dry dust removal is reentrainment of the collected dust, which must be maintained at a minimum for good collection efficiency.

Dust reentrainment can occur as a result of several factors. If gas velocity is high, dust can be scoured from the surface. However, recent studies indicate that scouring is not a serious problem at gas velocities below around 12 ft/sec (15). Large particles tend to be scoured more easily than small ones, so that some scouring of large particles may occur if gas velocity distribution is very nonuniform.

Reentrainment losses, however, are mainly associated with dust removal, which requires careful attention to rapping frequency and intensity. Rapping frequency is of importance in preventing too thick a deposit from accumulating on the plate. Under these conditions, the dust layer will fall of its own weight. When dust voluntarily falls from the plate, it can achieve a relatively high velocity in free fall, and reentrainment losses can be quite severe. If, on the other hand, rapping is too frequent, the dust layer will be too thin and rapping will tend to powder the dust, where it will be picked up and carried away by the gas stream.

Rapping intensity is a second variable that needs to be controlled for optimum performance. Too soft a rap can fail to dislodge the dust at the proper time, so that on the succeeding rap, the dust can fall freely because of the weight of the collected dust layers. Too severe a rap also causes poor performance, since the dust layer will be thrown out into the gas stream where some dust will be picked up and reentrained.

Figure 16 shows the relationship between rapping intensity and efficiency. The curve shows that the optimum rapping intensity can be altered by rapping frequency and by dust properties. Because of the exponential collection efficiency of a precipitator, the rate of dust layer buildup on the first sections will be higher than on subsequent ones. Consequently, rapping cycles are often varied between sections, the first sections being rapped two to three times for each rap of the last section.

Dust properties determine optimum rapping conditions. A low-resistivity dust is held to the plates

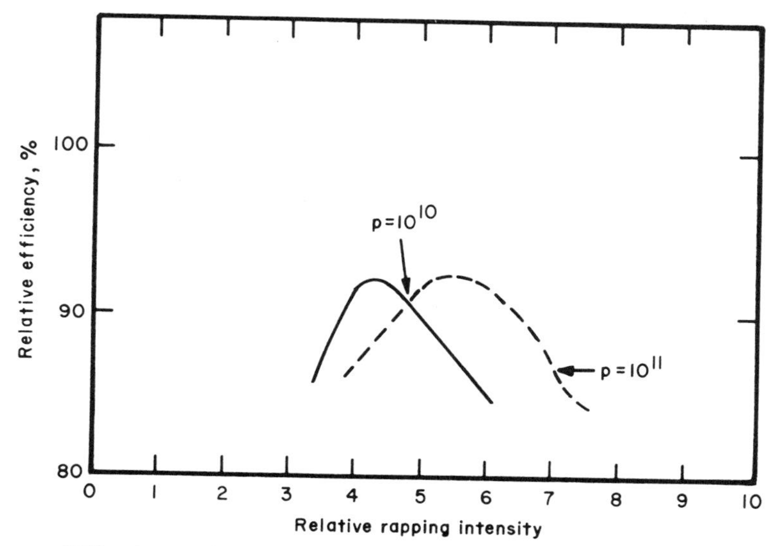

FIG. 16. Effect of rapping intensity on efficiency.

principally by mechanical and molecular adhesive and cohesive forces and is relatively easy to dislodge. A high-resistivity dust, on the other hand, is held primarily by electrical forces, in addition to the above, and the rapping intensity required to dislodge the dust can be quite high. Very high-resistivity dust often cannot be removed by conventional rapping, and power-off rapping is used. If, say, five or six series sections are used, power-off rapping could be used with very little loss if the gas velocity is sufficiently low. Otherwise, the reentrained dust would not be collected and a rather heavy discharge of dust would follow such a power-off rap.

Rapping losses for a very low-resistivity material cannot be completely controlled and can be higher than those for a dust with a resistivity in a more favorable range.

Adjustment of rapping frequency and intensity is generally accomplished in the field. The technique utilized is to minimize rapping puffs, which can be observed visually by use of a light and observation port near the plates of the last field or by obscuration or other types of instruments in the duct near the precipitator.

Reentrainment losses can adversely affect precipitator performance if they are unacceptably high. White (4) points out several methods of detecting abnormally high reentrainment losses by: (a) particle size analysis, (b) analysis of the charge on the particles, and (c) an abrupt change in the precipitation rate parameter with gas velocity. Since large particles are more easily reentrained, their presence at the precipitator exit indicates reentrainment losses. Reentrained particles of intermediate-to-low resistivity tend to take on the opposite charge polarity due to the pith-ball effect, so that the appearance of large numbers of oppositely charged particles tends to indicate reentrainment. Plots of the precipitation rate parameter as a function of gas velocity can indicate reentrainment if there is a sudden decrease in migration velocity.

Reentrainment losses can be kept to within very acceptable limits if the rappers are properly designed and adjusted. However, poor gas flow quality, poor adjustment, or unfavorable dust properties can cause severe degradation of precipitator performance.

VII. MECHANICAL AND ELECTRICAL COMPONENTS

The major components that make up a precipitator are the power supply or high-voltage source, electrodes, precipitator housing, and gas flow control equipment. The specific design and layout of precipitators vary in accordance with the type of service and with suppliers.

A. POWER SUPPLIES

The power supply system for an electrostatic precipitator is composed of a high-voltage transformer, a rectifier, a control element, and a sensor for the control system. Transformers for modern precipitators are capable of delivering secondary voltage in the range of 30 to 100 kV, with the specific voltage requirements governed by the electrode spacing, geometry, dust resistivity, and gas temperature, pressure, and composition.

The high-voltage rectifiers used in present day precipitators are predominately of the silicon type because of their high conversion efficiency and high reliability. However, vacuum tube rectifiers are used on many precipitators now in service.

Control equipment for power supplies is of varying design depending upon a particular manufacturer's philosophy of control. Current-limiting controls are required on all power supplies to protect the rectifiers and transformers under arcing conditions and, under some conditions, to serve as the primary control element. During an arc, the current would become excessively high without a limitation device. This limit can be provided by means of resistive, reactive, and active control elements in the power supply.

The voltages and currents to the precipitator are established by control circuits, the function of which is to maintain optimum precipitation conditions.

One of the most common types of control circuits utilizes spark rate to sense proper operating conditions. The average value of the voltage in a precipitator continues to increase after sparking begins and continues to increase until after a certain spark rate occurs. Further sparking causes the average voltage to decrease because of the increase in the number of secondary short circuits and excessive current drain. The optimum spark rate can be set for a given condition and can vary from 0 to around 200 sparks/min.

This automatic spark rate control circuit is acceptable for low-current power supplies that deliver around 100 mA. Larger power supplies, with their inherently lower internal impedances, sometimes develop power arcs rather than sparks. Spark rate controllers may not operate acceptably on these devices.

An alternate method used by some precipitator suppliers is to control to maximum secondary voltage. In this case, the input voltage is caused to fluctuate through a narrow range and the control seeks the point of maximum secondary voltage.

Judgment is required in setting the controls for a power supply for dusts of widely varying characteristics. For example, a back corona situation may develop prior to sparking, and this may require adjusting to a current limit as opposed to a spark rate limit.

One of the principal variables in power supply design is the degree of sectionalization to be provided. For many reasons, it is preferable to supply power by a number of small power supplies as opposed to a few large ones. First, if the precipitator is to operate in a sparking mode, spark quenching can be more effective if the power supply impedance is high. Large power supplies have inherently low impedances and can prevent quenching a spark during low-voltage conditions. Also, smaller electrical sections can permit operation at higher average voltage. Since the voltage limit is determined by the worst condition in a precipitator, a condition of poor electrode spacing control would limit the voltage for a smaller portion of the precipitator. Finally, the more sections that are used, the less the effect on emission of the failure of one section. The latter is of particular significance in the case of wire-type electrodes where wire breakage can disable an entire section.

B. VOLTAGE WAVEFORM

A precipitator presents a resistive-capacitive type load to the power supply, and a typical voltage waveform displays the rise in voltage characteristics of a rectified 60-cycle supply. As the voltage from the power supply decreases, the voltage on the precipitator would tend to decrease; however, due to the capacitive effect resulting from the large electrode surface area, the voltage decays as a typical resistive-capacitive discharge. The resistance of the system is determined by the dynamic voltage-current characteristics of the corona discharge. The capacitance of the precipitator usually maintains the voltage above the level of corona onset so that corona current is continuous in a normal precipitator.

Use of unfiltered, rectified voltage in a precipitator is generally better than filtered dc voltage, since higher peak voltage can be maintained before sparkover, and more effective spark quenching is possible.

Abnormal precipitation conditions can often be detected by examination of the voltage waveform. Extinguishing of the corona between cycles can be observed by alteration of the voltage waveform. This takes the form of a discontinuity in the discharge portion of the curve with the voltage holding constant for a portion of the cycle. Voltage waveforms for this condition and for normal conditions are shown in Fig. 17.

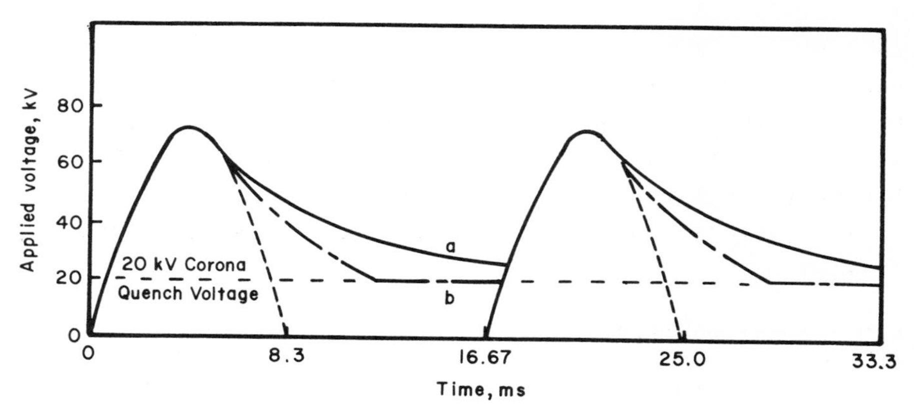

FIG. 17. Precipitator 60 Hz half-wave secondary voltate vs time waveform: (a) without corona current quenching; (b) with corona current quenching.

Variations in voltage waveform have been made by the use of special voltage shaping circuits and through the use of pulsed power supplies. These approaches attempt to maintain higher peak voltages through the use of rapid voltage rise rates. At this time, very limited quantitative data are available on the possible improvement in precipitator performance due to voltage shaping techniques.

C. ELECTRODE CONFIGURATION

The type of electrode structures varies according to manufacturer and type of service. The discharge electrodes can be of the weighted type, which is used by the majority of American manufacturers, or the supported type, common to many European precipitators. A third type structure is the spiral wire electrode used by one European manufacturer. Figure 18 demonstrates three electrode designs typical of the three approaches. Specialized applications are also used involving horizontal rods and other structures.

The principal requirement of the discharge electrode is to provide a source of very high electric field for

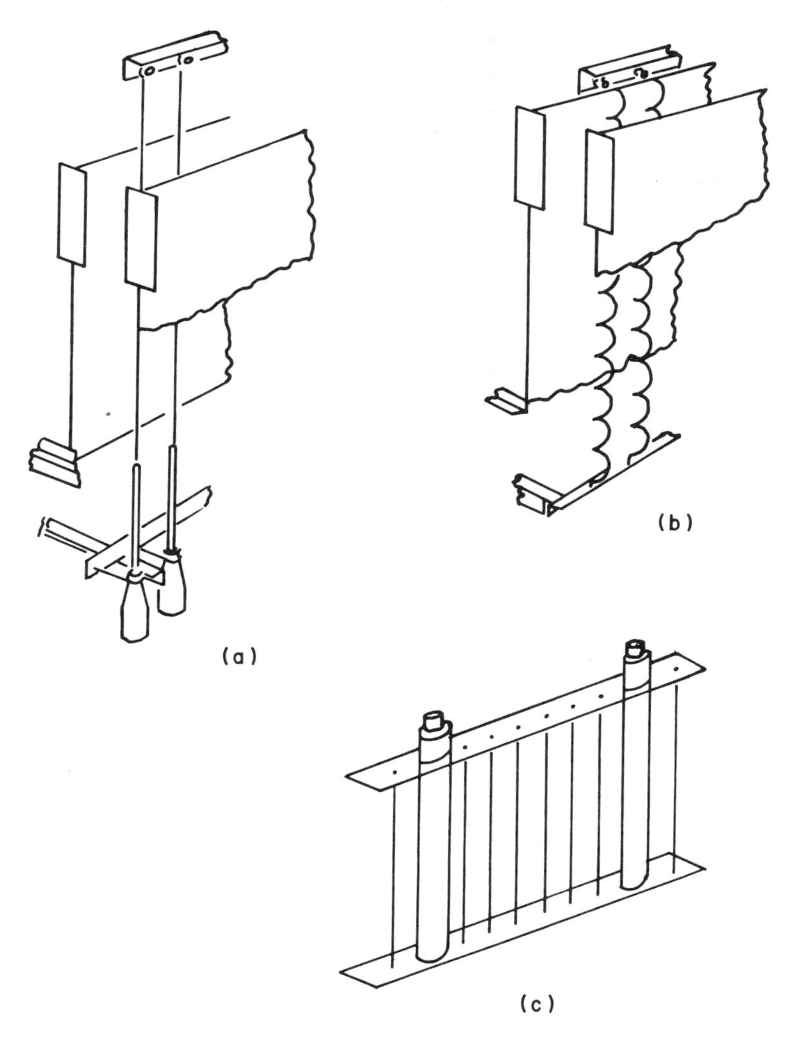

FIG. 18. Typical corona electrode designs utilized in industrial electrostatic precipitators.

generation of a corona. This, in turn, requires a small radius of curvature which can be provided by a wire or by a shaped electrode. Shaped electrodes in common use are square wires with or without additional forming, punched sheet metal, barbed wires, etc.

The size of the wire or the particular geometrical configuration of formed electrodes can be varied to change the voltage-current curves. The smaller the wire or the sharper the radius, the greater the current for a given voltage. In cases where high dust loadings of fine fume are encountered, the space charge tends to quench corona current and high-current electrodes are used.

In addition to providing the corona current, the discharge electrode structure must maintain accurate interelectrode spacing. Otherwise, sparking will occur at lower-than-normal voltage. Since sparkover will generally occur at the point of minimum spacing, variations in spacing must be kept to a minimum. In wire-type electrodes, spacing is maintained by weights attached to the lower ends of the wires and guides to keep the lower ends from moving under the influence of aerodynamic and electrical forces that are present. The supported electrode structure maintains alignment by means of the support frame itself.

One advantage of the supported electrode structure is that mechanical failure of the discharge electrode may not cause shorting of the precipitator section and hence electrode failure, so often a problem with the weighted wire electrode, does not present as great a problem with the supported electrode system.

D. COLLECTION ELECTRODES

Collection electrodes vary primarily in the type of baffles used with flat plate electrodes to prevent reentrainment and in the height of the electrodes. The trend in precipitator design has been toward higher electrodes ranging from 24 to 40 ft.

Collection plates can be of other than flat sheet design with baffles. Heavy gauge screens of the perforated or expanded metal type, as well as others, have been used in commercial practice. Cylindrical electrodes are also used for many applications.

Baffles are used in collection electrodes for two purposes. First, they provide stiffness needed for support of the electrode and second, they provide a region of low turbulence to minimize reentrainment of the dust, especially during rapping. Baffles take the form of channels, shaped protrusions, etc., as illustrated in Fig. 19.

Studies have been made by White(4) and others of the effect of collection plate geometry on reentrainment losses. Although there is wide variation in the type of plate geometry used in commercial practice, the functional differences in the various designs appear to be small.

The types of precipitators in use vary depending upon whether the collection electrodes are plates or cylinders, whether the gas flow is vertical or horizontal, the type of dust removal system, etc. Figure 20 shows a typical horizontal-flow, wire-and-plate type precipitator. This type of precipitator is widely used for control of fly ash, cement kiln dust, metallurgical fume, recovery boiler emissions, and similar applications. Figure 21 shows a precipitator used for removal of tar from a manufactured gas processing plant. These figures illustrate the range of design that can be found for control of a wide variety of emissions.

Rappers used in removal of dust from dry process collection plates can be of the pneumatic, electromag-

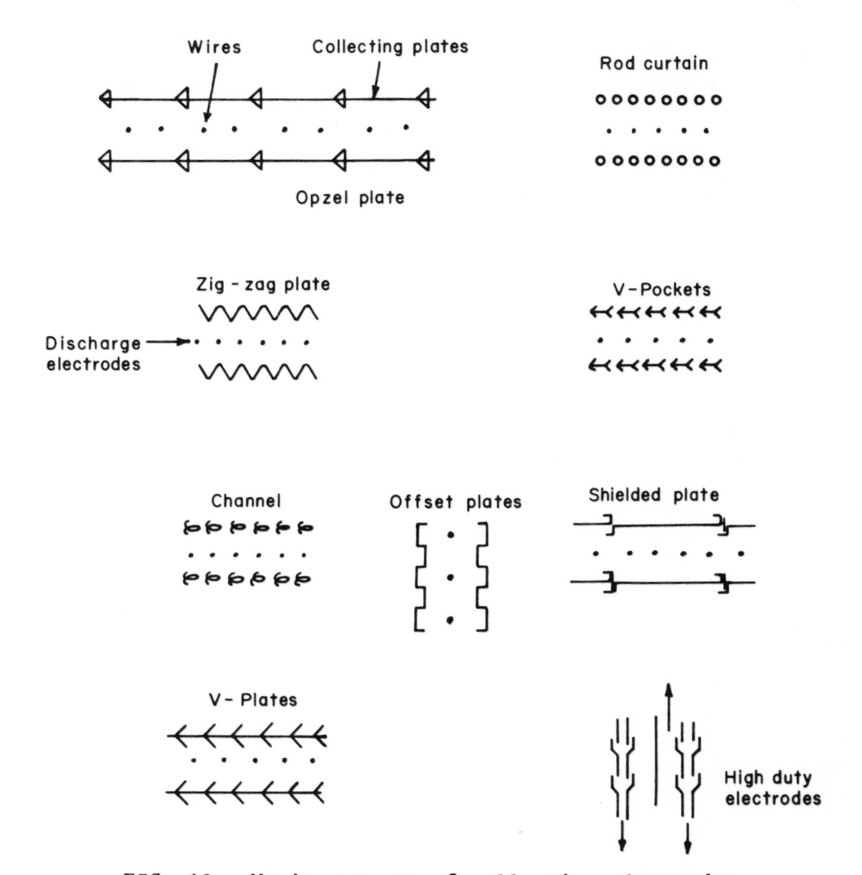

FIG. 19. Various types of collection electrodes.

netic, or mechanical types. For some applications, rapping is accomplished by vibration, which can be generated by unbalanced motors or electromagnetic vibrators. Vibrator-type rappers are often used on discharge wires.

Single impact rappers are the most common type used on present-day precipitators. The impact is provided by lifting a weight and allowing it to fall on an anvil coupled to the electrode support structure. Electro-

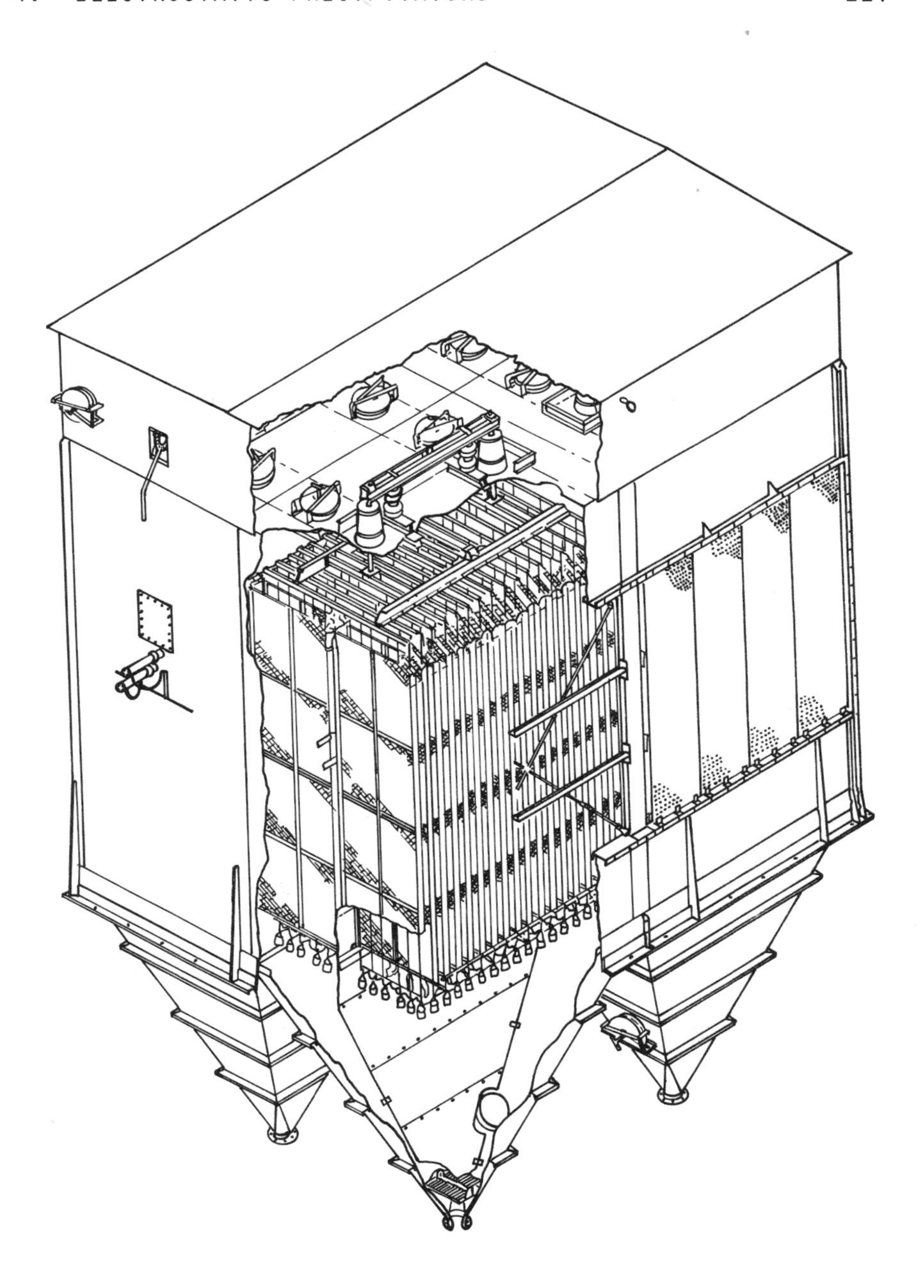

FIG. 20. Parallel plate precipitator.

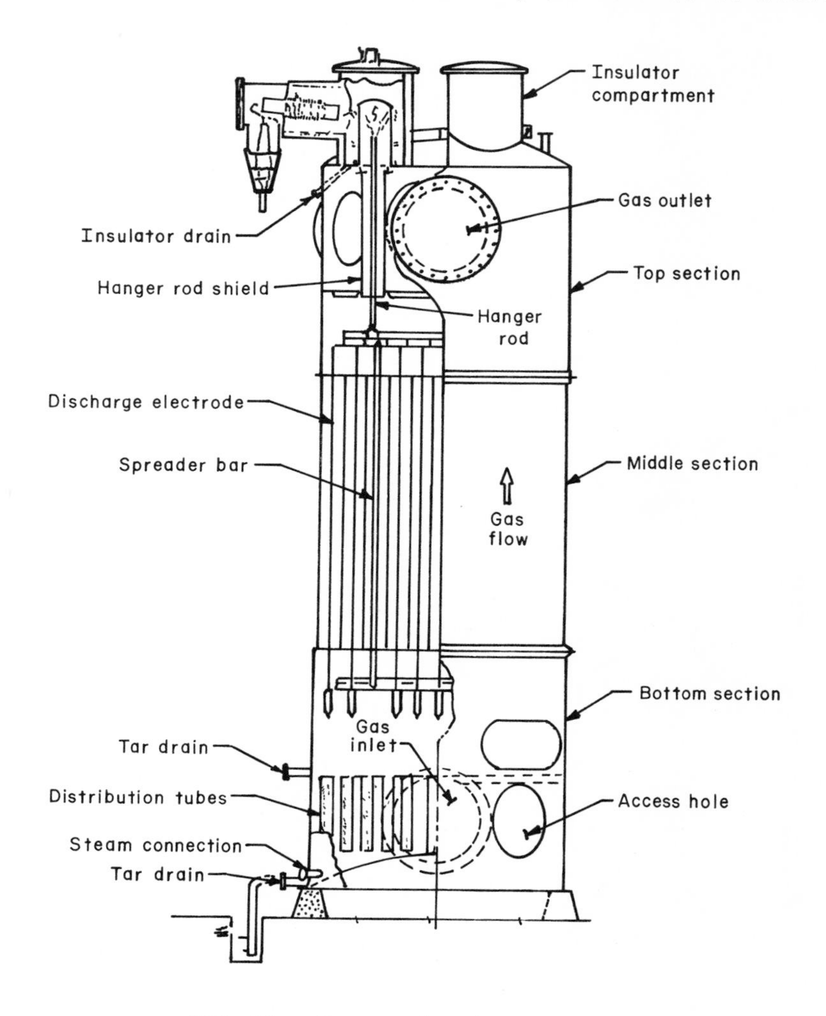

FIG. 21. Concentric ring detarrer.

magnetic or pneumatic rappers are often located on the top of the precipitator and the weights fall by gravity onto the anvil. They can also be spring-activated, in which case the rappers can be located on the side of the precipitator, and rapping can be accomplished perpendicular to the plate. Rapping intensity can be controlled by adjusting the height that the weights are lifted. Control panels are often provided with means for controlling the voltage to the electromagnetic lifters or the air pressure to the pneumatic lifters. Additional variations can also be achieved by changing the weights of the hammers.

Figure 22 shows a mechanical-type rapper, in which the rapping hammers are lifted by means of a shaft common to a number of rappers. Rappers can be arranged to provide impact in the plane of the sheet. Rapping adjustments include the changing of the speed of the

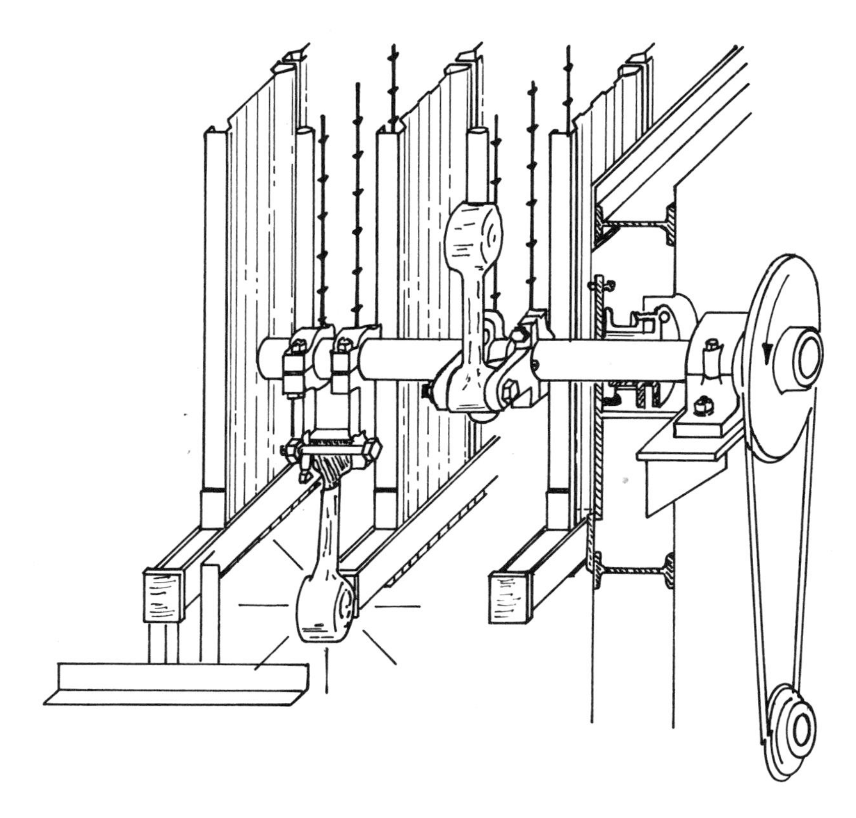

FIG. 22. Mechanical type rapper.

shaft for modifying the rapping frequency and adjusting the rapping intensity by replacing the hammers with others of a different mass.

The plate area rapped at any one time varies between applications. In general, the smaller the plate area rapped, the lower the visibility of the rapping puffs.

E. GAS FLOW

In many instances, precipitators are purchased for a given application and the control of duct design is beyond the responsibility of the precipitator manufacturer. Nevertheless, the quality of the gas flow can make a substantial difference in precipitator performance.

Poor gas flow leading to the precipitator can lead to dust fallout ahead of the precipitator, resulting in large deposits which can seriously alter the gas flow pattern. Large variations in gas velocity can lead to excessive reentrainment as well as poor utilization of various sections of the precipitator.

Temperature uniformity of the gas is also prerequisite to good performance. In some instances, temperature from heat exchangers, or from the process itself, can vary by as much as 100°F across the duct. Such variations can cause resistivity values to be widely different and thereby seriously impair performance. Good system design, therefore, requires that the gas be properly mixed to reduce the temperature gradient and that the velocity distribution be corrected to give a reasonable degree of uniformity.

On large installations handling high gas volumes, physical modeling of the gas flow system is generally required to achieve the desired gas flow quality.

The use of diffusion entrance plates as a part of the precipitator shell to improve gas flow quality is a common practice. The diffuser serves to reduce the scale of the turbulence to the order of magnitude of the holes. Multiple diffuser plates in series are also used in an effort to control gas flow quality. However, if the entering gas flow is extremely nonuniform, diffuser plates alone do little to correct the problem.

VIII. PRACTICAL LIMITATIONS OF PRECIPITATOR PERFORMANCE

From a theoretical standpoint, the efficiency of an electrostatic precipitator when collecting a given dust is determined by the magnitude of the charge on the dust particles, the viscosity of the gas, and the losses due to reentrainment of the dust. These factors are governed principally by properties of the dust, quality of gas flow, design of the electrical energization equipment, mechanical condition of the precipitator, etc.

A. EFFECT OF DUST RESISTIVITY

Resistivity of the dust is one of the most common factors limiting precipitator performance. If the resistivity of the collected dust layer is high, the current density will be limited by breakdown of the interstitial gases in the dust layer. The electric field in the dust layer is determined by the product of the dust resistivity and current density

$$E = j\rho \qquad (23)$$

Electrical breakdown of the dust layer occurs at fields of around 10 to 20 kV/cm in most industrial applications.

The effect of the dust resistivity can be shown by the curves in Fig. 23. Curve 1 is for a clean plate with no dust deposit, while curves 2 through 5 are for layers of thickness of 2.3 mm with various resistivities as shown. An increase in current follows a voltage increase until, ultimately, electrical breakdown

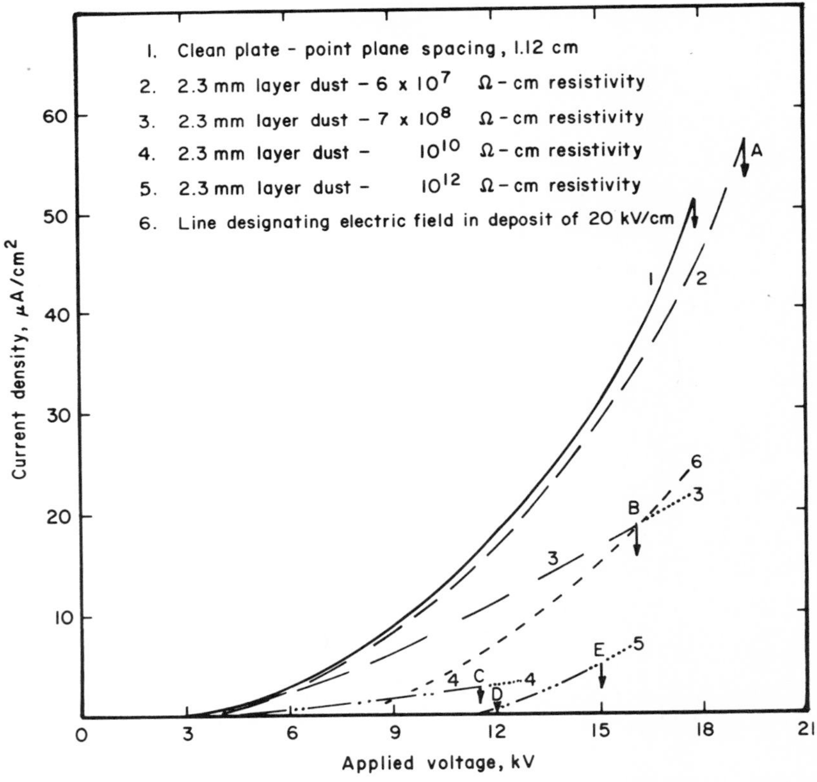

FIG. 23. Experimental volt-current curves for point-plane device with a variety of dust resistivities.

of the gas in the interelectrode space occurs. This breakdown mechanism is from a sparkover initiating at the anode and propagating across the interelectrode space. Conditions at the collection electrode, therefore, limit the voltage at which a precipitator can operate.

Curve 2 shows the voltage-current curve for a dust resistivity of 6×10^7 Ω-cm applied to the plate of the point-plane precipitator. The voltage drop across the dust layer causes a shift of the V-I curves as shown. For this value of resistivity, a field strength necessary to cause breakdown in the dust layer is not reached within the normal current density range. The voltage can therefore be increased until point A is reached, at which time the voltage across the interelectrode space is sufficient to propagate a spark. Curve 3 shows the voltage-current relationship for a 7×10^8 Ω-cm resistivity. With this resistivity, a field of 10 kV/cm is reached at a current density of 20×10^{-6} A/cm^2, and electrical breakdown of the gases occurs as indicated by point B. If the combined voltage drop across the dust layer and the interelectrode space is sufficient to propagate a spark, sparking will occur. The voltage required to propagate a spark is lower following breakdown of the dust layer than it is for a clean plate.

Curve 4 shows the voltage-current curve for a resistivity of 1×10^{10} Ω-cm, with sparking occurring at point C. Curve 5 is for a resistivity of 1×10^{12} Ω-cm. With this resistivity, breakdown in the dust layer would occur at point D. The voltage at this point is insufficient to cause a spark to propagate across the interelectrode space, and sparking would not occur. The continuous breakdown of the dust layer produces positive ions which move toward the negative electrode.

The effect of this back corona is to discharge the previously charged dust, seriously hampering precipitator performance.

When a condition of back corona is reached, the shape of the voltage-current curve is altered from the resistive characteristic. The curve can parallel the clean plate curve, indicating a constant voltage drop across the dust layer, or it can reverse itself, giving higher current for lower voltage. Increasing voltage above the back corona point may lead to sparkover, as shown by point E.

Both sparking and back corona, therefore, constitute limits on precipitator voltage and current. When constrained to operate at low current densities, the effect is to increase the time required for particle charging and to reduce the electric field. All of these factors combine to reduce the effectiveness of precipitation.

B. FACTORS AFFECTING RESISTIVITY

Electrical conduction through dusts or granular materials can take place through the bulk of the material itself or through adsorbed layers of material on the surface. If conduction is through the bulk of the material, it is termed volume conductivity, and if a conduction occurs through the surface layer, it is termed surface conductivity. Both surface and volume conductivity are important in terms of electrical precipitation.

The mechanism of conduction by each of these methods has not been fully resolved; however, it has been established that surface conduction takes place as a result of adsorbed layers of moisture on the dust surface. When other materials, such as SO_3, are present in the flue gases, they can serve as effective conditioning agents to enhance surface conduction. Conduction through the adsorbed layer is probably ionic.

In the case of volume conduction, the charge transfer takes place by the migration of alkali metal ions, principally sodium. An extensive investigation of the volume conduction process and quantitative relationships between resistivity and fly ash composition has been conducted by Bickelhaupt (16).

Both surface and volume conduction are dependent upon flue gas temperature, as shown in Fig. 24. Volume conduction, represented by the right-hand position of the curve (above about 400°F), is dependent upon material composition. The data shown are for fly ash from electric power boilers. Except for rather extreme differences in composition, the resistivity falls within the bands indicated. However, if widely different compositions are encountered, volume conductivity can be altered considerably.

The surface conductivity portion of the curve (below around 350°F) is highly dependent upon flue gas composition as well as temperature. In the case of fly ash, the variation in surface conductivity is primarily dependent on the amount of sulfur trioxide (SO_3) present in the flue gas. As a general rule, the quantity of SO_3 present increases with the sulfur content of the coal. However, other factors, such as the amount of

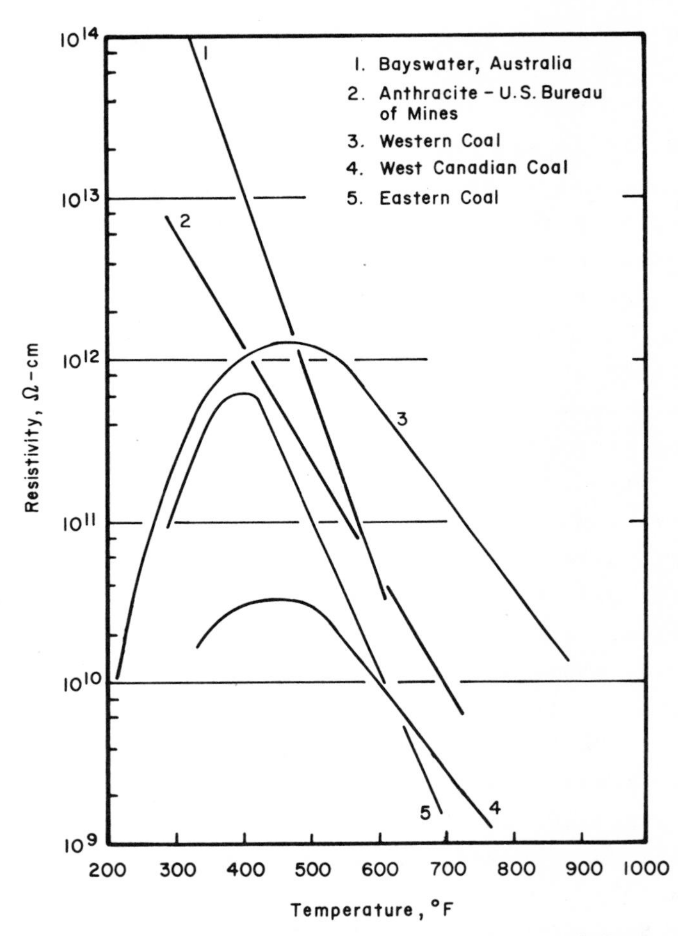

FIG. 24. Variations of electrical resistivity of fly ash from different sources. The straight-line portions are from laboratory data while the left-hand curving parts are in situ data.

oxygen present, the time available for oxidation of the SO_2, possible catalytic effects, etc., determine the extent of the increase. The composition of the ash can also have a considerable influence on the effectiveness of the SO_3. An ash high in lime (CaO) content can give a basic ash surface that can react with the normal conditioning agent, requiring larger quantities of conditioning agent to effect resistivity change.

Since dust resistivity can be influenced by so many variables, trying to predict it from fuel or ash properties is impractical. The most direct and, at present, the only reliable method of determining resistivity is by in-situ measurement under typical operating conditions.

C. MEASUREMENT OF DUST RESISTIVITY

There is considerable confusion about the data reported for the electrical resistivity of a given dust and operating condition. Comparative tests have shown variations as great as two orders of magnitude attributable to differences in the apparatus and techniques used for measurement. Consequently, interpretation should be based on a single method of measurement or allowance should be made for differences in the measurement methods.

The first rule of resistivity measurement is that it should be made in situ in the gas environment if the temperature is in the range where surface conduction is significant. Laboratory measurements are faced with the considerable uncertainty of reproducing flue gas conditions, the data are generally higher than in-situ values, and direct correlation with in-situ data is not practical.

The various types of in-situ resistivity probes differ in the manner in which the sample is collected and in the geometry of the measurement cell. Collection of the sample can be by means of a mechanical cyclone or an electrostatic collector. The latter can be of the point-plane or wire and cylinder type. Measurement cells can be of the parallel disk or concentric cylinder type with various design differences.

Figures 25 and 26 show two typical types of resistivity probes. The first type consists of a cyclone collector for sample collection and a concentric cylinder for resistivity determination (17). The cyclone collector can be inserted directly into the flue gas, or a probe can be inserted into the duct to withdraw a sample of the gas to the resistivity probe located in a heated chamber external to the duct. In the latter case, isothermal conditions must be maintained throughout the sample train during the test. The second type of probe illustrated utilizes a point-plane precipitator for

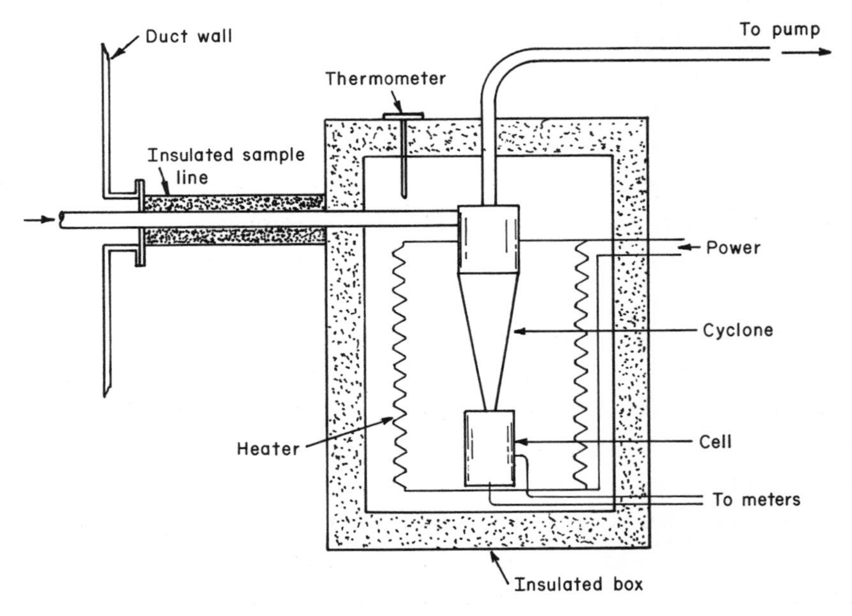

FIG. 25. Cyclone collector - cylindrical electrode cell for collection external to the duct.

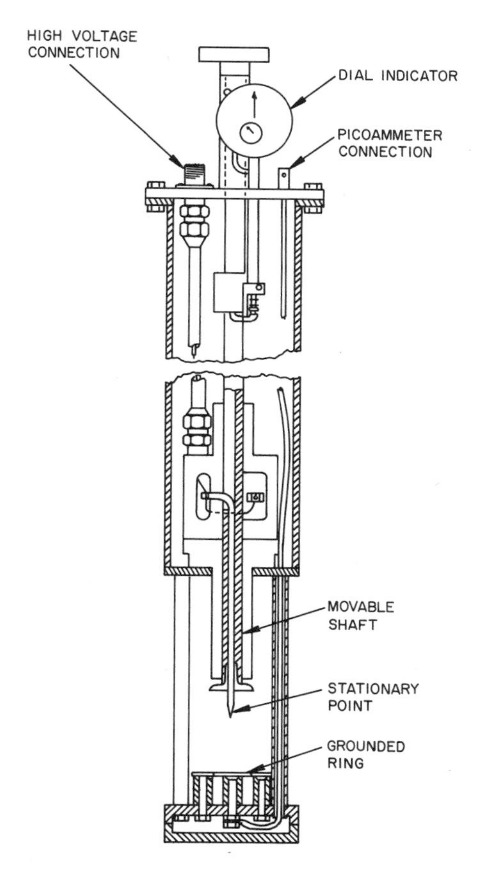

FIG. 26. Point-to-plane resistivity probe equipped for thickness measurement.

depositing a dust sample on a disk (18). Following collection, a second disk is lowered onto the dust surface and resistivity is measured by measuring the voltage across the disk and the current through the dust layer.

There are many variations of the two basic types of probes, depending upon details of mechanical design. Extensive tests with the two types of probes show two significant factors to be taken into consideration in measuring resistivity. First, there is a considerable variation in resistivity with time, especially with the cylindrical cell electrode when measuring the resistivity of some types of dust. This time variation is thought to be due to current absorption or polarization effects at the electrodes. Second, resistivity changes with the magnitude of the electric field in the dust deposit, i.e., the current-voltage relation is nonlinear. One significant difference in the normal practice with the two probes is to measure the resistivity with the parallel disk probe at a point just prior to electrical breakdown of the dust, whereas the cylindrical cell probes are normally operated at a constant field of around 2 to 3 kV/cm. In terms of precipitator operation, the electric field in the dust deposit will be at near breakdown for high-resistivity dusts, but lower for low-resistivity dusts.

Figure 27 shows a comparison of resistivity data measured on the same dust, using the point-plane probe with a parallel disk cell and the cyclone probe with a cylindrical cell. Data with the parallel disk cell were taken over a wide range of voltages, which permitted comparison on the basis of the same average electric field. Comparison was based on the peak value of current (minimum resistance) obtained with the cylindrical electrode cell. The data indicate that resistivities measured with the cyclone collector-cylindrical cell probe were somewhat higher than those measured with the point-plane, parallel-disk cell probes. The primary differences probably are due to differences in density of the dust, particle size variations in the collected sample, method of deposition, etc.

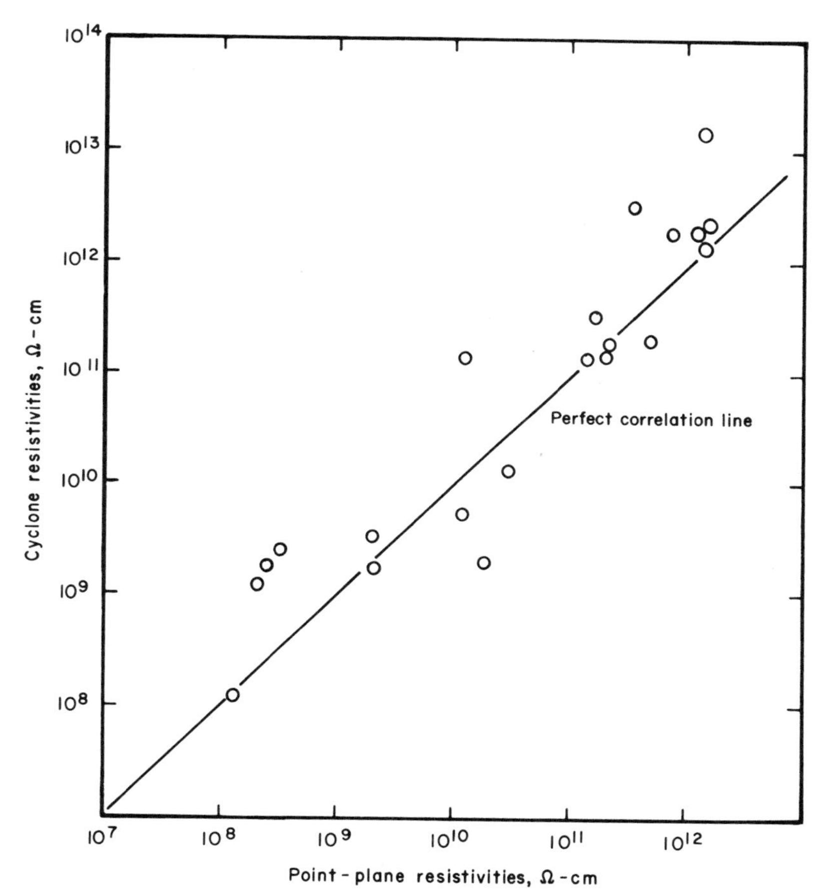

FIG. 27. Comparison between resistivity measured with the cyclone and the point-plane probes. Peak current values used for cyclone probe.

D. METHODS OF OVERCOMING HIGH RESISTIVITY

When high resistivity dust is encountered, it can be overcome by several techniques. First, since resistivity is dependent upon temperature, the option exists of operating at a temperature which will give favorable resistivity. In many applications, such as fly ash collection, normal operating temperatures are in the vicinity of 300°F, which can be the peak resistivity for many conditions. The choice of operating temperatures is based generally upon process economics and operating considerations. In the case of fly ash, the minimum temperature is governed primarily by corrosion and fouling conditions of the air heater. The average cold-end temperature of the air heater must be maintained above a temperature governed by the composition of the flue gases. The maximum temperature is determined by heat recovery and boiler economy considerations.

If low-sulfur coal is being burned, corrosion and fouling are not so much of a problem, and lower temperatures usually can be tolerated. Considerable operating experience has been accumulated at flue gas temperatures as low as 220-230°F when burning low-sulfur coal. Within the temperature range of 220-230°F, resistivity of fly ash can generally be brought to within an acceptable value and this constitutes one possible option to overcoming the resistivity problem.

A second option is to operate at temperatures in the range of 500-900°F where volume conduction is sufficiently high to bring dust resistivity to within acceptable limits. In many applications, such as metallurgical fume, cement kiln dust, etc., precipitators are operated within this temperature region. In the case of fly ash, the precipitators can be located between the economizer and air heater, where a temperature of 600-800°F can be maintained. These so-called "hot precipitators" can be effective in overcoming the high dust resistivity problem.

The primary considerations in selecting the alternative approaches to altering temperature for resistivity control are those associated with the process and economics of the options. In the case of low-temperature operation, variations in fuel composition, ease of temperature control, plume buoyancy, etc., are the primary considerations. High-temperature considerations are the increased gas volume, necessity for more ductwork, increased insulation, possibility of corrosion following ash removal, etc. Each of these considerations must be analyzed for each application.

An alternate approach to temperature control is the use of chemical conditioning agents added to the flue gas. The major attempts at conditioning have been in connection with fly ash precipitators, although successful conditioning of synthetic catalyst dust from petroleum catalytic crackers has been achieved.

Since the principal cause of high-resistivity in fly ash is a low level of SO_3 in the flue gases, one technique for conditioning has been to artificially increase the SO_3 concentration by additions to the flue gas in quantities ranging from around 5 to 20 ppm.

Methods of SO_3 conditioning vary according to the starting material. One method used successfully has been the vaporization of stabilized SO_3. Figure 28 is a diagram of a system described by Whitehead (19) and used successfully at the Kincardine Plant of the South of Scotland Electricity Generating Board. The plant consists of a means for storing the SO_3 in a heated tank, a supply of dry, heated air for converting the SO_3 to vapor, and heated lines to convey the air-SO_3 mixture to the injection point.

A second type of conditioning plant utilizes evaporation of concentrated aqueous sulfuric acid in the temperature range of 400-500°F. In this type plant, the maximum concentration of acid vapors that can be pro-

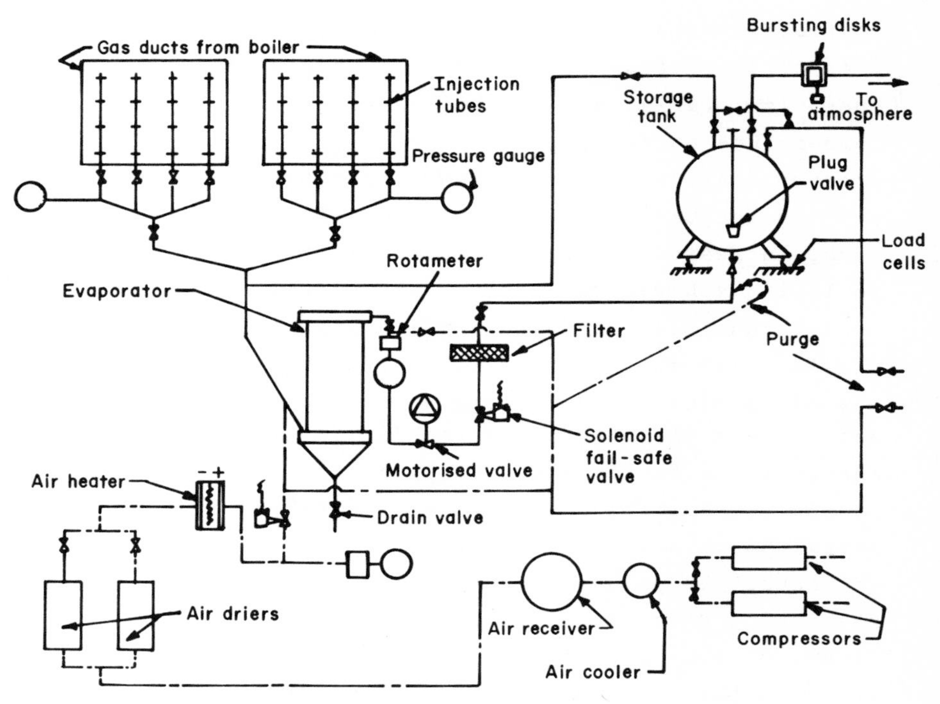

FIG. 28. Conditioning plant using stabilized SO_3.

duced is limited by the amount of heat available for evaporation. If the only heat source available for evaporation is heated air at about 500°F, concentration of the acid is limited to around 5000 ppm in the injection lines.

In a high-temperature injection system, heat for evaporation of the acid is supplied from combustion of natural gas. In this system, adequate heat is available to maintain much higher concentrations in the injection lines. Two problems that must be avoided in the high-temperature H_2SO_4 system are:

1. injection lines must be heated to prevent condensation of the SO_3 and H_2O as H_2SO_4-H_2O liquid aerosol prior to injection; and

2. consideration should be given to the prevention of conversion of SO_3 to SO_2, which can occur in a gas at high temperatures and low O_2 concentrations.

The principal constituents of an H_2SO_4 conditioning system are the acid storage tank, metering pump, air blower, vaporizer, and heat-traced ducts and injection manifolds. A schematic of an acid conditioning plant is shown in Fig. 29.

The feasibility of SO_3 conditioning in reducing fly ash resistivity has been demonstrated on a variety of types of ash, ranging from highly basic to neutral, and at temperatures ranging from 270 to 325°F.

Figure 30 shows the resistivity of various types of fly ash as a function of the concentration of conditioning agent. Changes in the effectiveness, as measured by the change in resistivity, are due to several factors, among which are:

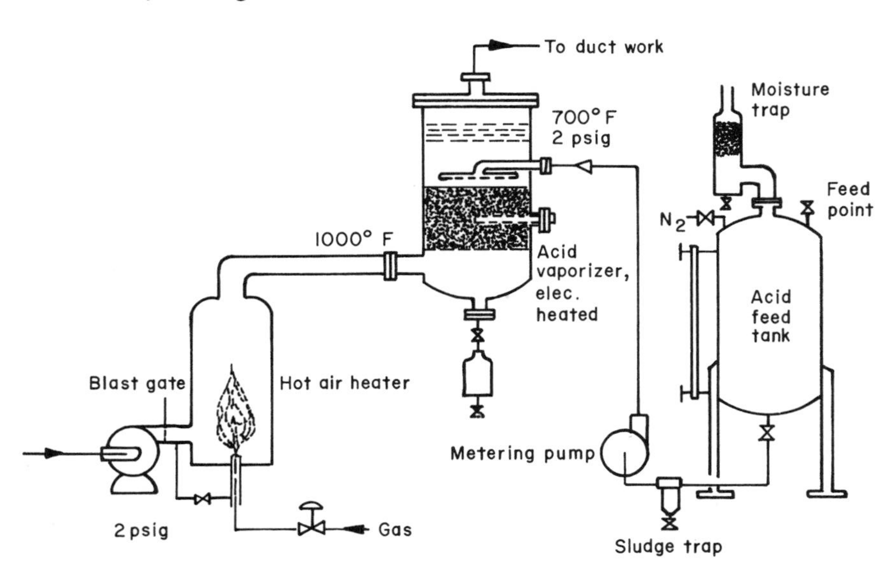

FIG. 29. SO_3 acid conditioning plant.

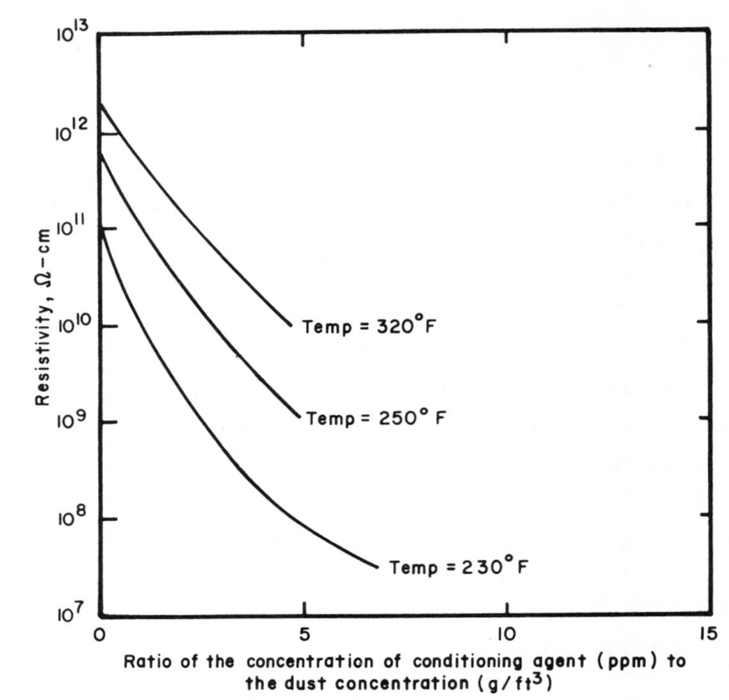

FIG. 30. Resistivity of fly ash as a function of the relative concentrations of conditioning agent and fly ash.

1. whether the injection is ahead of the mechanical collector, and hence required to treat a larger concentration,
2. whether the ash is highly basic or neutral, and
3. temperature of flue gases.

Conditioning with full-scale and pilot units has demonstrated that it is effective in reducing resistivity. However, conditioning will not correct the malfunctioning of a precipitator where other problems are encountered such as poor gas flow, inadequate power or sectionalization, excessive reentrainment, or inadequate design.

Acceptance of flue gas conditioning has not been very good because of maintenance, control, and safety problems associated with the conditioning plant. These can be overcome to some extent as experience is gained with operating these plants. However, for new precipitators, where other options are available, they are often preferred over flue gas conditioning.

IX. DESIGN AND SIZING OF PRECIPITATORS

Current methods of design and sizing of electrostatic precipitators are based primarily upon analogy with similar plants or empirical relationships between fuel or dust characteristics, desired efficiency, and gas volume.

Although there are several variations as to the details of design methods, they are based generally upon the exponential relationship between efficiency and the specific plate area as related by the familiar Deutsch equation. The factor, w, is the important consideration in arriving at a suitable design. This factor is empirical and most precipitator manufacturers maintain files of precipitator applications with the corresponding values of w, and these form the basis for design for new installations.

The precipitation rate parameter, as used here, differs from the theoretical migration velocity discussed in a previous section. Although the same general parameters influence both, the precipitation rate parameter takes into account reentrainment, uniformity of gas flow, electrical sectionalization, and other factors; specifically design, construction, and operation defects not included in the theoretical migration velocity.

One of the most significant variables in dust properties influencing precipitator performance is the particle size of the dust. Figure 31 shows the variation in the specific collecting area required for a group of applications with widely varying particle sizes. As the graph shows, the value of w decreases rather severely with decreasing particle size. The spread in the values of w indicates the range of precipitation rates experienced in each of the application areas.

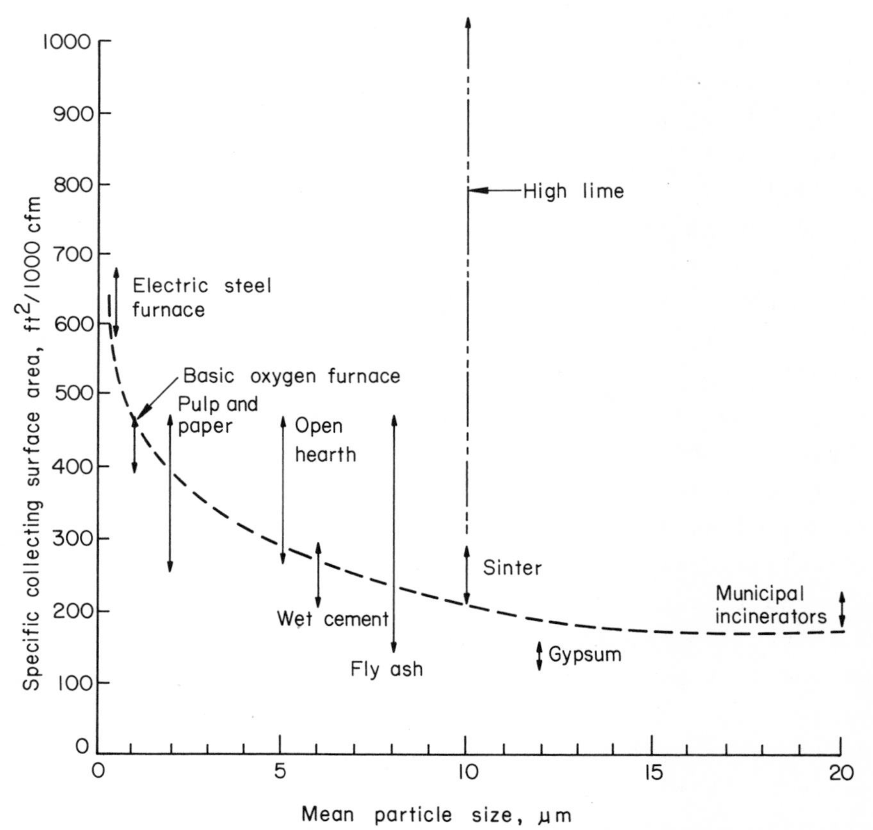

FIG. 31. Variations in specific collecting area with dust particle size.

From the design standpoint, the spread in many applications is too great to permit rational design based upon these numbers alone. Consequently, some method of reducing the uncertainty in precipitation rate parameter is required.

The principal factors, other than particle size, influencing precipitation rate parameter are the charge on the particles, the collection field, and reentrainment losses. Both particle charge and electric field are dependent upon the voltage and current density at which the precipitator operates. These, in turn, are limited by dust resistivity in some instances and gas breakdown in others.

Variations in precipitation rate parameter are greater for fly ash than for any other single application due principally to variation in resistivity resulting from differences in fuel composition, particle size distribution, and good or poor gas flow conditions. Generally speaking, high-sulfur (3-4%) coals produce ash with relatively low resistivity that precipitates well at normal operating temperatures, whereas low-sulfur coals produce ash of high resistivity that is relatively difficult to precipitate.

White (4) has developed an empirical relationship between dust resistivity and precipitation rate parameter for fly ash as shown in Fig. 32. This relation also has a theoretical basis. Such relationships as this permit narrowing of the design precipitation rate parameters.

Where flue gas temperatures are not varied, sulfur content of the fuel can be used to determine an approximate precipitation rate parameter. Ramsdell (20) has developed empirical relationships based upon experience of one company. Curves derived from Ramsdell's data

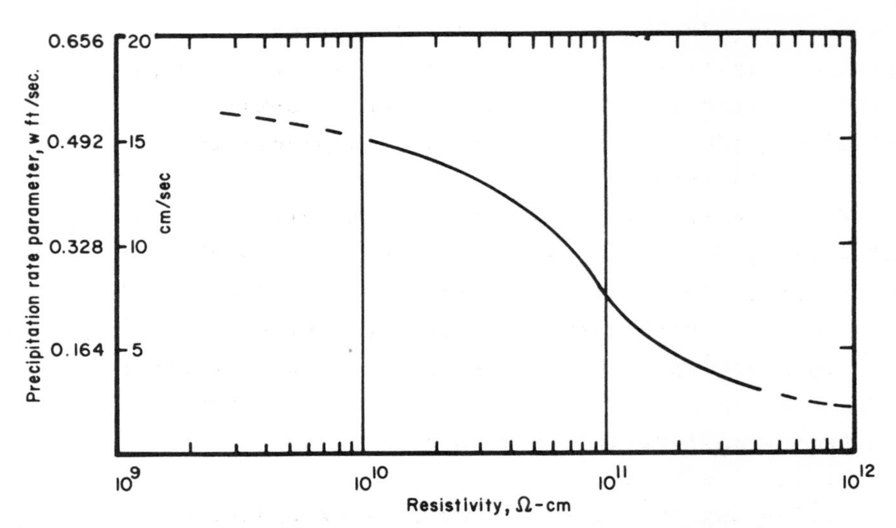

FIG. 32. Empirical relationship between precipitation rate parameter and resistivity for fly ash [from White (4)].

are shown in Fig. 33. These relationships have been widely used as a basis for design and design analysis of fly ash precipitators.

Although the value of w is taken as constant in the Deutsch-Anderson equation, it does vary with a number of parameters because of the range of particle sizes in a typical dust. Larger particles are collected easily in initial sections of the precipitator, whereas smaller particles tend to be caught in the latter sections or to not be captured at all. Consequently, if high efficiencies are to be allowed, the additional precipitator surface added must collect the smaller fraction of the dust. The design w must therefore be reduced somewhat for very high collection efficiencies.

In other applications than fly ash, relationships between w and process parameters have not been so well

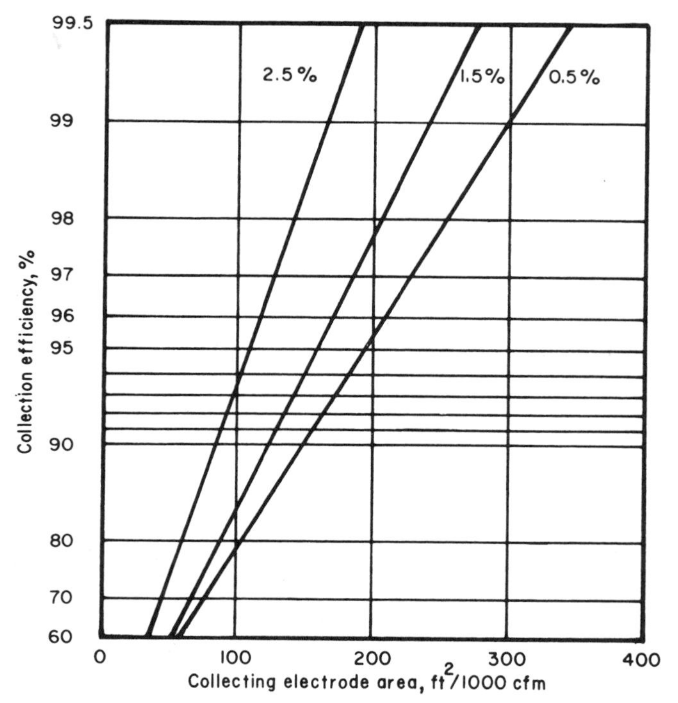

FIG. 33. Collecting electrode area vs collection efficiency - sulfur content of the coal is the parameter.

developed, and more dependence upon analogy with existing plants is required. As with fly ash, dust resistivity plays an important role, and attention to those factors affecting resistivity can result in more reliable design procedures.

Other factors in the design of precipitators include the power requirements and the degree of sectionalization required. Power input is determined empirically and varies according to the application. Figure 34 shows the variation in precipitation rate with power input for fly ash precipitators. The level of power input is determined principally by dust resistivity. Power den-

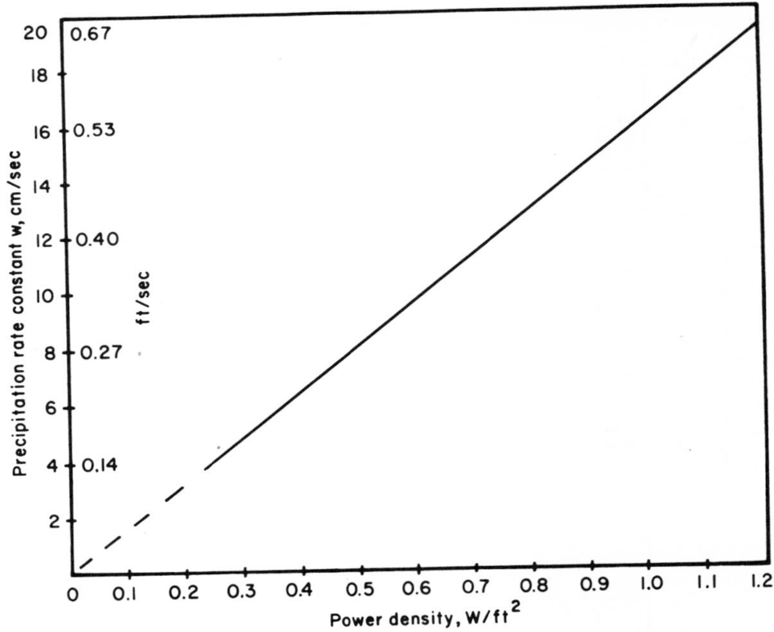

FIG. 34. Linear relationship between precipitation rate parameter and power density for fly ash collectors.

sities of 1 W/ft^2 are not uncommon for low resistivity dusts, whereas power densities in the range of 0.3 to 0.4 W/ft^2 are the limit for high dust resistivities. From the standpoint of power supply design, therefore, the power requirements can be computed based on the design w and total plate area.

Another design option is the number of independently powered bus sections. As a rule, a greater degree of electrical sectionalization results in higher voltage, better control of sparking, less reentrainment, and generally better operation.

Ramsdell has determined empirically that the number of independently powered bus sections for fly ash precipitators is related to efficiency as indicated in Fig. 35.

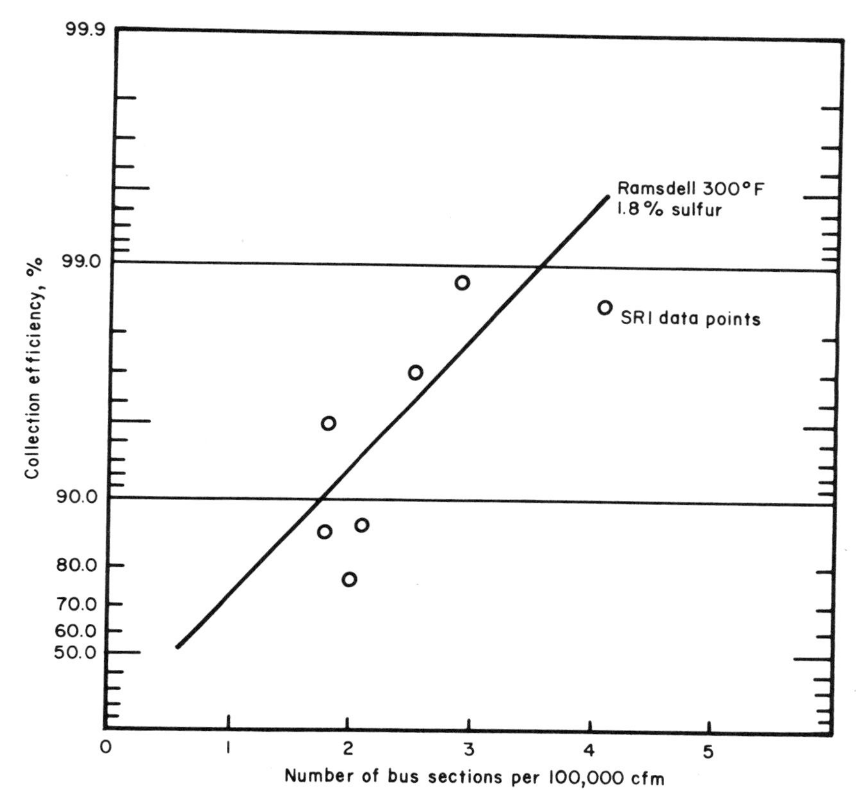

FIG. 35. Variation in efficiency with degree of sectionalization.

Again, power densities, sectionalization, etc., are not nearly so well developed or publicized for other applications as they are for fly ash precipitators. The principal reasons for the attention to fly ash are:

1. the large number of fly ash precipitators and the large size of each precipitator, and
2. the relatively large variability of the ash due to variations in coal composition.

A. COST

The principal variation in cost of an electrostatic

precipitator is its size in terms of collecting surface area. Gas volume, efficiency, and precipitation rate parameter all influence the collecting surface area. This area, in turn, determines the overall size and weight of the precipitator, foundation requirements, etc. A general average installed cost for a precipitator in 1970 was around $6 per square foot of collection surface area.

Costs for specific applications have been developed based on efficiency ranges and gas volume handled. These are given in detail in the Manual of Electrostatic Precipitator Technology (9).

Costs of precipitators like those of other control techniques can vary widely depending upon temperature, whether the installation is new or a back-fit, local labor costs, contractual guarantees, etc.

Electrostatic precipitators have been applied to a wide variety of dust control problems ranging from collection of fume from precious metal forming operations to control of heavy dust concentrations in large gas volume applications. The principal advantage of electrostatic precipitators is the ability to separate small particles from a gas stream with relatively low energy requirements. In comparison with other dust control methods, precipitators are somewhat higher in first cost, but lower in operating costs than competitive dust collection methods. Properly engineered and maintained, precipitators offer a reliable and effective method of dust control. Their principal uses have been in those applications involving high gas volume where savings in energy costs are appreciable.

Table 4 lists the applications of precipitators with the number of precipitators in each of the application areas.

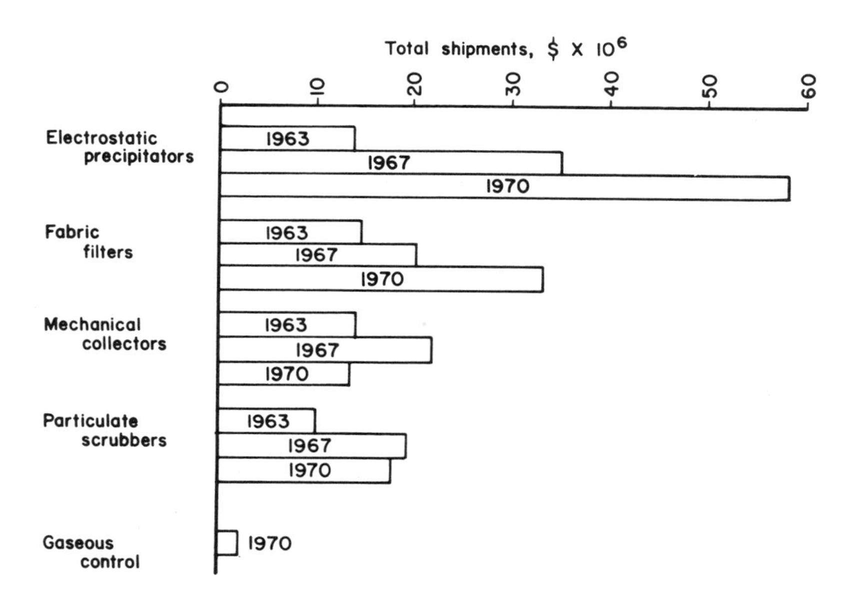

FIG. 36. Total shipments for various types of particulate control devices for 1963, 1967, and 1970.

Table 4
A Summary of Electrostatic Precipitator Applications

Application Area	1920-1924	1925-1929	1930-1934
Fly Ash	1	11	12
Metallurgical			
Ore roasters, zinc or lead zinc		9[a]	2
Ore roasters, molybdenum sulfide		2[a]	1
Ore roasters, pyrites		2	1
Aluminum (carbon plant)			
Soderberg pots			
Aluminum remelt furnaces			
Aluminum prebake potline			
Cryolite recovery			
Cadmium recovery zinc sinter machine			
Copper			
Lead blast furnace		3[a]	3
Gold and silver			1
Tin smelting		1	
Scarfing machine			
Sintering machine			
Blast furnace			4
Basic oxygen furnace			
Open hearth			
Sinter machine gas			

1935- 1939	1940- 1944	1945- 1949	1950- 1954	1955- 1959	1960- 1964	1965- 1969	Total
56	74	90	63	90	87	171	655
8	8	2	2	1		1	33
1	1						5
3	1		1				8
	2		3				5
	2			1			3
			3	2			5
			3	2			5
			1				1
5	5	4	4				18
1						1	2
	1						7
2			1			2	6
	2						3
				4	3	1	8
			1	7	7[b]		15
17	29	24	26	21	6		127
					2	5	7
			2	9	8	4	23
			2	2		3	7

Table 4 (Continued)

Application Area	1920-1924	1925-1929	1930-1934
Chemical			
Sulfuric acid			
Acid mist			
Phosphorus electric furnace			
Phosphate			
Phosphoric acid mist		2	
Disodium metaphosphate			
Rock products			
Alumina calciner			
Cement kiln (wet)			
Cement kiln (dry)			
Cement plant dryers and mills			
Gypsum			
Petroleum			
Catalytic crackers			
Pulp and paper			
Pulp mill			
Paper mill			
Tar recovery		63[a]	20

[a]Prior to 1930. [b]1960-1969. [c]1930-1939.

1935-1939	1940-1944	1945-1949	1950-1954	1955-1959	1960-1964	1965-1969	Total
			15	10	1	4	30
			3	12	5	8	28
	2	2	2	1	2	1	10
			2				2
1	3	1	2	1			10
			1				1
	3		3			1	7
7[c]		1	7	24	17	17	73
14[c]			3	17	6	3	43
1[c]				2	3		6
4	3	3	18	19	10	1	58
	24	13	3	2			42
	12	28	20	37	2	15	114
		2	2	11	25	31	71
33	62	82	35	14	6		282

The use of electrostatic precipitators has increased rather steadily since the first commercial applications. Figure 36 shows the sales as compared with other dust control methods for the years 1963-1970.

REFERENCES

1. Loeb, L. B., "Fundamental Processes of Electrical Discharge in Gases," in Gaseous Discharge, Wiley, New York, 1939.
2. von Engel, A., Ionized Gases, Clarendon, 1955.
3. Peek, F. W., Jr., Dielectric Phenomena in High Voltage Engineering, 3rd ed., McGraw-Hill, New York, 1929.
4. White, H. J., Industrial Electrostatic Precipitation, Addison-Wesley, Reading, Mass., 1963.
5. Shale, C. C., et al, "Feasibility of Electrical Precipitation at High Temperatures and Pressures," Bureau of Mines Report of Investigations 6325, U. S. Department of the Interior, 1963.
6. Frank, N. H., Introduction to Electricity and Optics, McGraw-Hill, New York, 1950.
7. Stratton, J. A., Electromagnetic Theory, McGraw-Hill, New York, 1941.
8. Cooperman, P., "The Dependence of Electrical Characteristics of Dust Precipitators on Their Geometry," Research Report 46, Research Corp., Bound Brook, N.J., unpublished.
9. Oglesby, S. and Nichols, G. B., A Manual of Electrostatic Technology, Part I. Fundamentals (1970), PB 196389; Part II. Application Areas (1970), PB 196381; Selected Bibliography of Electrostatic Precipitator Literature (1970), PB 196379; generated under Contract CPA 22-60-73 for the National Air Pollution Control Administration.
10. Rose, H. E. and Wood, A. J., An Introduction to Electrostatic Precipitation in Theory and Practice, Constable & Company, London, 1956.
11. Pauthenier, M. and Moreau-Hanot, M., "The charging of spherical particles in an ionized field," J.Phys. Radium, 3, 590 (1932).
12. Arendt, P. and Kallman, H., "The mechanism of charging cloud particles," Z. Phys., 35, 421 (1926).

13. Liu, B. Y. H. and Yeh, H. C., "On the theory of charging of aerosol particles in an electric field," J. Appl. Phys., 39, (3), 1396 (1968).

14. Murphy, A. T., Adler, F. T. and Penney, G. W., "Theoretical analysis of the effects of an electric field on the charging of fine particles," Paper 59-102, AIEE Winter General Meeting New York (February 1959). Published in Communications and Electronics, September, 1959.

15. Southern Research Institute Final Report to the Environmental Protection Agency on Contract CPA 70-166, 1972.

16. Bickelhaupt, R. E., "Electrical volume conduction in fly ash," APCA Journal, 24, (3), 251 (1974).

17. Cohen, L. and Dickenson, R. W., "The measurement of the resistivity of power station flue dust," J. Sci. Instrum. (London) 40, 72 (1963)

18. Nichols, G. P., "Technique for Measuring Fly Ash Resistivity," Report to the Environmental Protection Agency (Submitted August 5, 1974).

19. Whitehead, C. L., "Gas conditioning," Proceedings, Electrostatic Precipitator Symposium, Birmingham, Alabama, 1971.

20. Ramsdell, R. G., "Design criteria for precipitators," paper presented at the American Power Conference, Chicago, Illinois (April 1968).

Chapter 5

WET GAS SCRUBBERS

John J. Kelly

Chemical Engineering Department
University College
Dublin, Ireland

I. INTRODUCTION

This chapter presents a "state of the art and science" survey on wet gas scrubbers. The range of equipment in use for cleaning dirty gases is very extensive. There are about 15 categories of gas scrubbers and anywhere from 20 to 30 design variations within each category. Each equipment manufacturer has his own "unique" design and thus the selection of one scrubber for a specific job is a formidable task. The equipment catalogs that are supplied by manufacturers vary greatly in the quantity and detail of information given and, with few exceptions, they do not give the performance curves for the equipment, which is otherwise well described and lauded. The technical and trade literature contains a vast amount of information on the gas cleaning equipment that is currently available. Numerous review articles have been written by Stairmand (1-5) and others (6-10) which provide useful introductory material on this subject.

For particulate collection duty, equipment is judged by its performance in removing particulate solids over a given size range. With the universal tightening up of permissible emission limits in recent years, interest has now focused on the collection efficiencies of particles in the range below 5 microns [1 micron (μm) = 10^{-6} m]. The average collection efficiencies of 18 varieties of gas cleaning equipment are given in Table 1. This equipment may be subdivided into five principal groups, which are listed in Table 2 together with expenditure data in the United States for the year 1967.

The value of shipments of industrial gas cleaning equipment for the United States in 1967 was twice that for 1963. This reflects the growing pressures from local and national legislation and from public opinion.

TABLE 1

Average Collection Efficiencies of Gas Cleaning Equipment (3)

Equipment type	Percentage efficiency at		
	50μm	5μm	1μm
Inertial collector	95	16	3
Medium-efficiency cyclone	94	27	8
Low-resistance cellular cyclones	98	42	13
High-efficiency cyclone	96	73	27
Impingement scrubber (Doyle type)	98	83	38
Self-induced spray deduster	100	93	40
Void spray tower	99	94	55
Fluidized bed scrubber	> 99	99	60
Irrigated target scrubber (Peabody type)	100	97	80
Electrostatic precipitator	> 99	99	86
Irrigated electrostatic precipitator	> 99	98	92
Flooded disk scrubber, low energy	100	99	96
Flooded disk scrubber, medium energy	100	99	97
Venturi scrubber, medium energy	100	> 99	97
High-efficiency electrostatic precipitator	100	> 99	98
Venturi scrubber, high energy	100	> 99	98
Shaker-type fabric filter	> 99	> 99	99
Reverse-jet fabric filter	100	> 99	99

TABLE 2

Manufacturers' Shipments of Industrial Gas Cleaning Equipment in the United States for 1967 (10)

Equipment type	Expenditure ($1000)	Percentage of total expenditure
Mechanical collectors	22,381	20.2
Wet scrubbers	25,999	23.5
Electrostatic precipitators	36,509	33.0
Fabric filters	21,730	19.7
Gas incinerators and absorbers	3,976	3.6
Totals	110,595	100.0

Within the subheadings of Table 2, wet scrubbers differ from the others in that they represent a much wider variety of equipment. This arises from the number of different collection mechanisms that may operate when a liquid is used to scrub a dirty gas; some scrubbers use only one mechanism, but most use a combination of those mechanisms, as is described in the sections below.

II. SURVEY OF TYPES AND PERFORMANCES

A. CLASSIFICATION OF SCRUBBERS

Any equipment classification for the range of wet scrubbers in use must of necessity by an arbitrary one.

They come in many shapes and forms and no one unique collection mechanism clearly distinguishes one group from another. The objective in each case is essentially the same; that is, to bring about intimate contact between the dirty gas stream and the scrubbing liquid, which in most cases is water. The removal from the gas stream of particulate pollutants is readily achieved once the particles have been wetted. The removal of gaseous pollutants is a gas-absorption process but depends also on intimate liquid/gas contact. In such processes, the scrubbing liquid may not be water, but some solvent liquid that has a high solubility for the gaseous constituent to be removed. However, the most usual duty for wet scrubbers is the removal of particulate solids from a gas stream. In some instances, the same item of equipment simultaneously removes both particulate and gaseous pollutants.

The major consideration for scrubbers is to obtain effective utilization of the water in terms of energy input and collection efficiency while at the same time avoiding problems of corrosion and erosion. Scrubbers may present liquid waste disposal problems and often emit a condensed water plume from the stack in cool weather. Wet scrubbers may be broadly classified as low- or high-energy scrubbers. Low-energy scrubbers having 1-6 in. water gauge of pressure drop include simple spray towers and wet cyclones. Water requirements typically run from 3-6 gal/1000 ft^3 of gas and collection efficiencies can exceed 90-95% when the particles are above 10 μm in diameter. The low-energy scrubber has found frequent application in incinerators, fertilizer manufacturing, lime kilns, and iron foundries, but is not capable of meeting current emission requirements in these industries. High-energy scrubbers, such as the Venturi type, have collection efficiencies up to 99.5% and greater, with pressure drops within the range 10-35 in. of water. High-energy scrubbers are used with steel furnaces, wood-paper and pulp mills, and foundry cupolas where the particulate size in the effluent fumes may fall below 0.25 μm.

The simplest type of scrubber is a spray chamber, which is an empty tower in which the gas flows through a shower of water droplets produced by suitably positioned spray nozzles. These are effective for the collection of large particles (>50μm) and also for cooling, humidifying, and conditioning the gas. The flow configurations of spray chambers are cocurrent, countercurrent, and cross flow. In cocurrent flow, the gas and water drops flow in the same direction, so that relative velocities and thus, impingement rates are low. Countercurrent flow is achieved by spraying the water in at the top of the chamber with upward gas flow. The gas flow rate is generally restricted to rates below that which will entrain most of the water drops. In cross flow, the liquid spray is introduced at right angles to the gas flow. Mixed flow occurs in nearly all cases because of turbulence, droplet inertia, air resistance, and gravitational force. A fuller description of these effects is given in Sec. III of this chapter.

The efficiency of collection in a spray tower may be increased through the use of centrifugal force. A spiral motion is given to the gas stream by having a tangential inlet port at the tower bottom and/or by fitting internal vanes and impellers. The increased performance is achieved at the cost of additional pressure drop and thus power requirements (see Sec. IV).

A different method of gas-liquid contacting and spray formation is used in the self-induced spray scrubber. The inlet gas is introduced at high velocity onto or just under the liquid level surface at the tower bottom, causing considerable turbulence, spray formation, and effective gas/liquid contact. The main advantage of self-induced spray scrubbers lies in their ability to handle high dust concentrations in the gas and water slurry. Sludge removal may be achieved by drag chains or by sluicing. Mist carryover is minimized by baffles or swirl chambers (see Sec. V). Impingement plate scrubbers have turbulent gas-liquid contact zones, since the gas flows at high velocity through the orifices in each plate into the liquid cross flow. One or more plates may be fitted as required in this type of unit (see Sec. VI).

Cylindrical and rectangular towers packed with materials such as raschig rings, saddles, tile, and marbles may be used for dust and mist collection and for gas absorption. Packed beds are operated in either cocurrent, countercurrent, or cross flow modes. The bed may be fixed or free to move in the manner of a fluidized bed chemical reactor. The scrubbing liquid serves to wet, collect, and wash the particulate matter from the bed (see Sec. VII).

High-energy scrubbers of the Venturi type impart high velocity to the gas stream by means of a converging duct section and contact the stream with low-pressure injected water. Venturi scrubbers may have single or multiple throats in which the gas flows in parallel or in series through a number of identical units contained in a single housing. The scrubbing liquid is introduced at right angles to the high-velocity gas flow in the Venturi throat. The velocity of the gas causes the disintegration of the liquid into fine droplets. An alternate method is to direct a high-velocity liquid jet along the axis of the Venturi throat. This ejector scrubber uses the water jet both to pump and scrub the gas. The flooded disk scrubber has a variable orifice throat and finds application in cases where slurry plugging and variable gas flows are a problem. Its performance is similar to that of a Venturi scrubber (see Sec. VIII).

In mechanically induced spray scrubbers, sometimes referred to as disintegrators, mechanical energy is used to improve efficiency and to reduce water and space requirements. Some units use a partially submerged rotor, others disintegrate water injected between a rotor and stator. Centrifugal wet fan scrubbers are of this type. Erosion, abrasion, and corrosion considerations limit the range of applications of mechanically induced spray scrubbers (see Sec. IX).

B. PERFORMANCE AND EFFICIENCY

Two efficiencies are commonly used for describing the performance of pollution control equipment. "Grade efficiency" (η_{ig}) is the efficiency of collection at the mean size of a fractional size range. It may be calculated from the equations

$$\eta_{ig} = \frac{M_{i1} - M_{i2}}{M_{i1}} (100) \tag{1}$$

$$= \left(1 - \frac{M_{i2}}{M_{i1}}\right) (100) \tag{2}$$

where M_{i1} and M_{i2} are the mass flows of particulates in the fractional size range "i" at the inlet and outlet respectively. Expressed in terms of concentrations (gr/scf) in the size range "i," this equation becomes

$$\eta_{ig} = \left(1 - \frac{C_{i2}}{C_{i1}}\right) (100) \tag{3}$$

In these equations, C_i and M_i are generally interchangeable, since with most scrubbers the total inlet and outlet gas flow rates expressed in standard units are the same. The grade efficiency for a scrubber (for example, see Fig. 3) is a plot of η_{ig} against particulate size and is, perhaps, the most relevant curve in assessing a scrubber's performance.

"Collection efficiency" (η) is the over-all efficiency of collection measured in units of weight per cent removed. It may be calculated from the equation

$$\eta = \left(1 - \frac{C_2}{C_1}\right) 100 \tag{4}$$

where C_1 and C_2 are the total solids' concentrations into and out of the scrubber respectively. It is related to the grade efficiency by the summation

$$\eta = \sum_{i=1}^{N} (W_i \ \eta_{ig}) \tag{5}$$

where W_i is the weight fraction of particulates in the fractional size range "i" at the scrubber inlet, and N is the number of ranges selected. η is thus the weighted average value of the grade efficiencies of the scrubber over the size range in question and it may be readily calculated from Eq. (5) when the grade efficiency curve is available. A simple calculation example of these efficiencies for a hypothetical scrubber is given in Table 3.

While a 95% collection efficiency clearly indicates that only 5% by weight of all the particulates flowing into the gas cleaner escape to the stack, it could also mean, since this 5% will contain a high proportion of the smallest sized particles, that some 40% of the total surface area flowing into the cleaner escapes to the stack. This could mean a visible and unsatisfactory stack plume. Where the grade efficiency curve of a collector is known, the efficiency based on the surface area may be approximated by assuming a spherical particle shape. Since it is the surface area that reflects the light, and thereby makes a stack gas

TABLE 3

Calculations of Emission Rates and Collection Efficiency

Size range (μm)	Given: Inlet M_{i1} (gr/s)	Given: Inlet w_i (wt. fraction)	η_{ig}[a] (%)	Calculate: Outlet M_{i2} (gr/s)	Calculate: Scrubber collection efficiency
0-10	100	0.125	80	$100-\frac{80}{100}(100) = 20$	$\eta=0.125(80) + 0.375(90) + 0.5\ (95)$
10-25	300	0.375	90	$300-\frac{90}{100}(300) = 30$	=91.25% (or)
25-50	400	0.5	95	$400-\frac{95}{100}(400) = 20$	$=\frac{800-70}{800}$
Totals	800	1.0		70	=91.25%

[a]Alternatively, where the inlet and outlet concentrations are known, the grade efficiency curve for that operation may be drawn.

visible, the "surface area efficiency" may be the more relevant efficiency where a clean stack is important. It is not, however, a common term in usage for describing gas cleaning equipment.

For a given dust, higher removal efficiency requires greater energy input to the scrubber. The collection efficiency η is a function of the gas-water contacting power and can be expressed in terms of the dimensionless transfer unit N_t, discussed in Sec. VII of this chapter.

$$N_t = -\int_{C_1}^{C_2} \frac{dC}{C} = -\ln\left(\frac{C_2}{C_1}\right) \tag{6}$$

Thus

$$\eta = \left(1 - \frac{C_2}{C_1}\right) 100 = (1 - \mathrm{Exp}(-N_t))\ 100 \tag{7}$$

In other words, the collection efficiency increases exponentially with the number of transfer units.

The total power consumption, P_t, is the sum of the power input to the gas P_g, and to the liquid P_ℓ. These may be determined from the following relationships: (11,12)

$$P_g = 0.157\Delta P \tag{8a}$$

$$P_\ell = 0.583\Delta P_\ell (Q_\ell / Q_g) \tag{8b}$$

$$P_t = P_g + P_\ell \tag{8c}$$

where P_t, P_g, and P_ℓ are in horsepower per 1000 cfm; ΔP is the gas pressure drop in inches of water gauge; ΔP_ℓ is the pressure drop for the water input in psi; Q_ℓ is the water rate in gpm; and Q_g is the gas rate in cfm.

In many cases, data for the number of transfer units and total power consumption fall on a straight line when plotted on log-log coordinates. Thus they have been correlated empirically by the equation: (12)

$$N_t = \alpha P_t^{\beta} \tag{9}$$

where α and β are characteristic parameters that depend upon the nature of the dust being collected and the type of scrubber used. Some representative values are given in Table 4. The seven general types of scrubber are shown in Table 5 with a summary of their performances and ranges of operation.

Calvert (13) presents expressions for the collection efficiency of packed beds and spray scrubbers with different types of flow. The final expression in each case is Eq. (7) with N_t interpreted in terms of the type of scrubbing. For packed beds

$$N_t = C\left(\frac{z}{D_c}\right)\frac{V_r \rho_p d_p^2}{18\mu R_c} \tag{10}$$

where z is the height or length of the scrubber; D_c is its diameter or width; V_r the drop velocity relative to the gas; R_c is the radius of the particle collector packing materials. The constant C is a representative value for typical packing; ρ_p is the particle density; d_p its diameter, and μ is the viscosity of the liquid.

For the cross flow spray scrubber, and dust particles in the size range "i",

$$N_t = \frac{3Q_w h}{4Q_g r} \cdot \frac{\eta_{ig}}{100} \tag{11}$$

where Q_w and Q_g are the liquid and gas volumetric flow rates; h is the scrubber height; r is the drop radius; and η_{ig} is the particle grade efficiency.

For vertical countercurrent flow scrubbing

$$N_t = \frac{3Q_w V_t z}{4Q_g r v_d} \cdot \frac{\eta_{ig}}{100} \tag{12}$$

where V_t is the particle terminal settling velocity; and v_d is the drop velocity relative to the duct.

For cocurrent flow scrubbers

$$N_t = [13{,}500(L') + 1\cdot 2L'^{2\cdot 5}V_g]\ \eta_{ig}\ (2 \times 10^{-5}) \tag{13}$$

TABLE 4

Scrubber Efficiency Parameters (12)

Scrubber	Aerosol	Parameters for use is Eq. (9) (α)	(β)
Venturi	Talc dust	2.97	0.362
	Phosphoric acid mist	1.33	0.647
	Foundry cupola dust	1.35	0.621
	Open hearth steel furnace fume	1.26	0.569
	Odorous mist	0.363	1.41
Venturi evaporator	Hot black liquor gas	0.522	0.861
Venturi and cyclonic spray	Lime kiln dust (raw)	1.47	1.05
	Black liquor furnace fume	1.75	0.620
	Ferrosilicon furnace fume	0.870	0.459
Venturi, pipe line, and cyclonic spray	Lime kiln dust (pre-washed)	0.915	1.05
	Black liquor fume	0.740	0.861
Venturi condensation scrubber with:			
1. Mechanical spray generation	Copper sulfate	0.390	1.14
2. Hydraulic nozzles	Copper sulfate	0.562	1.06
Orifice and pipe line	Talc dust	2.70	0.362
Cyclone	Talc dust	1.16	0.655

TABLE 5

Summary of Scrubber Types, Performance, and Ranges of Operation

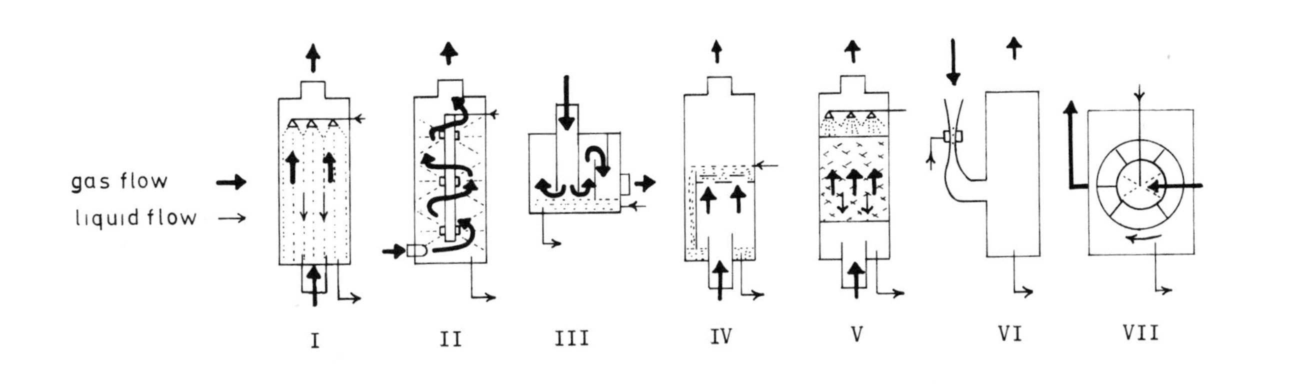

	Design	Approximate grade efficiency at 5 μm	Pressure drop range (in. water gauge)	Liquid consumption range (gal/1000 ft^3)
I	Gravity spray	80[a]	0.5 - 2.0	5 - 20
II	Centrifugal or wet cyclone	87	1 - 15.0	2 - 15
III	Self-induced spray	93	2 - 15	0.5 - 1

IV	Impingement plate	97	1 - 8[b]	3 - 5
V	Packed bed	99	0.2 - 1.0	8 - 20
VI	Venturi	99+	5 - 35[c]	2 - 10[d]
VII	Mechanically induced spray	99+	1.5 - 4.0	4 - 5

[a] Very approximate. The literature contains widely differing values; for example, see Stairmand (2) and Strauss (11).

[b] Venturi slot plates give rise to much higher pressure drops - see text.

[c] Pressure drops of 70 in. water gauge have been cited (9, 10).

[d] For the venturi jet scrubber, this range goes up to 50 gal/1000 ft^3 (10).

where L' is the ratio of liquid to gas flow in gal/1000 ft^3 and V_g is the gas velocity relative to the duct. Equations (10-13) serve to illustrate the role of design variables on performance as related through Eqs. (7) and (8). This provides the basis for analysis and further development of scrubbers. Other factors in considering the use of scrubbers include corrosion, loss of plume buoyancy, and water disposal.

Water usage and waste disposal can be critical factors in the selection of a scrubber. Settling tanks and ponds, continuous filtration, liquid cyclones, centrifugation, and chemical treatment are ways of handling the slurry of collected material. The method used depends upon the recovery value, the need for chemical treatment, the space available, and such other local factors as may be relevant. The possibility of replacing an air pollution problem with a water pollution one is, of course, a factor that must be considered at the initial stages of equipment selection.

C. COLLECTION MECHANISMS

Inertial impaction and interception are the predominant mechanisms for particulate capture in wet gas scrubbers.

Collection by inertial impaction occurs when a particle in the gas stream approaching a droplet separates from the path of its gas streamline around the droplet and continues on to collide with it [see Fig. 1(a)]. The eventuality of this collection mechanism depends more on the mass or inertia of the particle than on its size. It is therefore the predominating mechanism for the collection of dense particles. On the other hand, the mechanism of collection by interception depends more on the size of the particle than on its mass or inertia. In this case, the particle follows the path of the streamlines around the droplet and interception occurs when a streamline carrying a particle passes within the distance dp/2 from the droplet's surface, where dp is the diameter of the particle [see Fig. 1(b)].

Other mechanisms of lesser effect are due to diffusion, thermal effects, and electrostatic attraction. Diffusion is a contributory mechanism with very small particles (<0.5μm) due to the effect known as Brownian movement, whereby the particles are small enough to be moved from their paths by molecular bombardment, thereby causing collisions

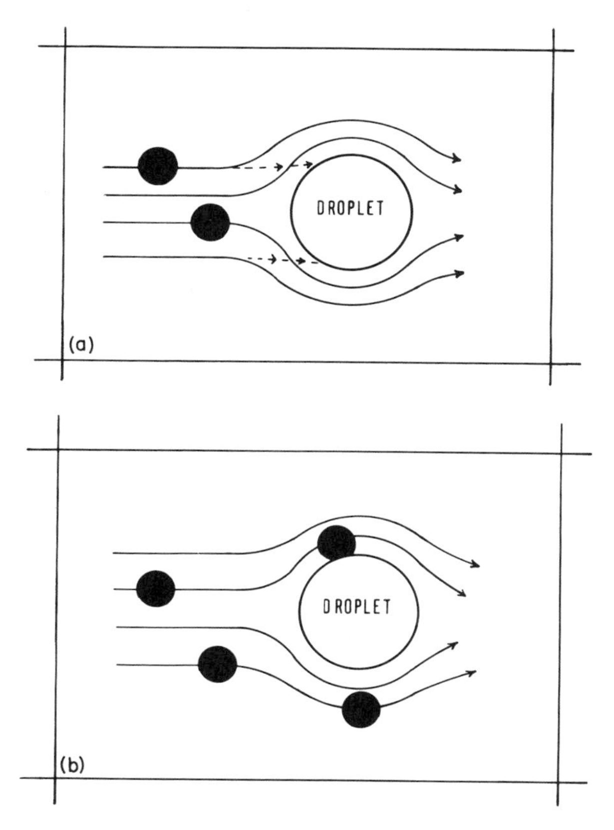

Fig. 1. Collection mechanisms: (a) inertial impaction (b) interception.

with the scrubbing droplets. Similarly, thermal effects influence only very small particles that will tend to travel down a thermal gradient because of the greater bombardment by the gas molecules on the warmer side of the particles. The electrostatic charge produced on droplets during atomization is generally too small to influence collection, except when the droplets are charged by an external source (11). Liquid condensation on particles, thus increasing their size, and ease of collection are further mechanisms used very effectively in the Venturi scrubber (13). In this process, the gas is saturated before entering the reduced pressure zone in the Venturi throat, where more liquid is added. Condensation on dust particles takes place in the diffuser as pressure rises and velocity decreases. The particles thus increase in size, agglomerate, and are more easily collected.

For the removal of gaseous pollutants, the gas is contacted with a liquid having a high solubility for the particular pollutant. The process is known as gas absorption and for many pollutants of high solubility, such as hydrogen fluoride or hydrogen chloride, water is used. In other cases with less soluble gases, acid or salt solutions are used because they react chemically with these gases. For example, for sulfur trioxide removal, dilute sulfuric acid is used, producing a more concentrated acid.

1. Combined Mechanisms

The over-all collection efficiency in any wet scrubber is generally the result of a number of these mechanisms working simultaneously. The contribution of any one mechanism depends on the particle and droplet sizes and on the velocity of the gas stream relative to the liquid droplets. In a typical air pollution problem, where a range of particle and spray droplet sizes are encountered, all these mechanisms, and perhaps others unaccounted for, have a role to play. An excellent review of the separate and combined mechanism theories has been given by Strauss (11).

D. SCRUBBER SELECTION

The following criteria are suggested as a basis for the selection of a scrubber for a particular air pollution problem:

1. Grade Efficiency Curves

The grade efficiency curve is the most important item of information that should be provided. With a knowledge of the dust concentration and particle sizes, it is a simple calculation, as illustrated in Table 3, to estimate the dust emission to the atmosphere.

It is important that the grade efficiency curves are applicable to the gas flow rates at the conditions of temperature and pressure, and to the particular pollutant under consideration. Particle collection efficiency may depend on the density of the particle as well as on its size, so it is necessary to check that the grade efficiency curve was obtained for the dust in question or for a similar dust. The saturation level of the gas is dependent on temperature and pressure, which conditions therefore have a direct bearing on the particle collection efficiency.

2. Operational Flexibility

With any new item of plant, it is always recommended to consider its potential as a process bottleneck in the event of increasing the main plant throughput. Therefore, it may be of importance to know the dependence of the collection efficiency on the total gas throughput above, and indeed below, the design value. Similarly with the dust loading, it may be important to know how the plant will behave with unsteady dust concentrations in the gas flow or at sustained higher than design dust loadings.

Some scrubber designs provide for this by having control over the unit pressure drop, thus allowing for variation in the severity of the collection process.

3. Solids Disposal Facilities

It is, of course, important that the air pollution problem is not replaced by a water pollution problem. Units that have the advantage of low water consumption often have a problem of sludge removal. Where no convenient river exists, or where the nature of the solids is such that it may not be simply dumped in a river, the provision of settling tanks and associated equipment has to be considered.

4. Ease of Operation and Maintenance

Apart from their contribution to the running costs, which are discussed below, the considerations of easy operation and trouble-free maintenance are of importance in scrubber selection. In general, moving or rotating parts inside the scrubber should be avoided, since bearings have a short life in the gritty atmosphere prevailing. Also, small flow areas can give rise to plugging problems. Equipment with these difficulties may have compensating factors such as high efficiency operation and/or easily replaceable key parts.

5. Cost Data

Where a number of units satisfy the four operational points discussed above, the next consideration is, of course, the cost of the unit. In order to arrive at the total capital cost figure, it is necessary to add to the basic equipment cost the additional cost of plant auxiliaries such as fans, pumps, and motors and an allowance for site clearance and plant erection. For wet gas scrubbers, an approximate rule of thumb for this additional cost is to take it as 100% of the basic equipment cost. Some units, because of their particular design, have a figure somewhat less than this, whereas others could be more costly. For electrostatic precipitators or mechanically induced spray scrubbers, because of their compactness, a figure of 50% would be more accurate. Running costs are made up of four principal items: (a) power requirements related to gas pressure drop; (b) power requirements related to water pressure drop; (c) the cost of water; and (d) the maintenance costs. From such data, the cost per annum or the cost per ft^3 of gas may be readily calculated for comparison purposes to assist in the final selection of a scrubber (5).

Approximate cost figures applying to Great Britain (1971) for 14 different types of particulate collection equipment are given in Table 6. These figures were calculated for a gas containing 30% by weight below 10 μm. An extensive analysis of the economic considerations in air pollution equipment relating to the United States is given in Ref. 10.

III. GRAVITY SPRAY SCRUBBERS

The simplest type of wet gas scrubber is the gravity spray scrubber. It is an empty tower where water is sprayed down countercurrent to the rising dirty gas. A perforated

TABLE 6

Approximate Costs of Various Dust-Arresting Systems Treating 100,000 m^3/h[f] of Dusty Gases at 20°C (26)

Equipment	Capital cost[a] Total installed (£)[e]	(£/m^3/h capacity)	Pressure drop (mbar)[g]	Power[b] cost (£/annu
Medium-efficiency cyclones	9,400	0.094	9.2	1,69
High-efficiency cyclones	18,300	0.18	12.2	2,26
Impingement scrubber (Doyle type)	35,800	0.36	20.0	3,76
Self-induced spray deduster	25,100	0.25	10.0	1,85
Void spray tower	52,500	0.53	3.5	2,38
Fluidized bed scrubber	19,900	0.20	12.5	2,83
Irrigated target scrubber (Peabody type)	31,000	0.31	15.2	2,90
Electrostatic precipitator, medium efficiency	88,300	0.88	2.2	1,00
Irrigated electrostatic precipitator	113,400	1.13	1.5	1,11
Venturi scrubber, medium energy	40,600	0.41	49.8	9,41
Electrostatic precipitator, high efficiency	148,000	1.48	5.0	2,27
Venturi scrubber, high energy	44,100	0.44	78.5	14,87
Shaker-type fabric filter	62,100	0.62	6.2	1,87
Reverse-jet fabric filter	87,100	0.87	7.5	3,96

[a]Includes fans, pumps, motors, and erection.

[b]Cost of electrical energy as taken 0.5 p/kWh, fan and motor efficiency 60%.

[c]Cost of water assumed to be 0.14 p/m^3.

[d]Capital charges taken as 30% annum of total installed cost. 8,000 h/annum operation assumed.

[e]£1.00 = 100p = $2.32 (mid 1974 equivalents)

[f]100,000 $m^3/h \doteq 58,850\ ft^3/min$.

[g]1 mbar $\doteq$ 0.4 in. water gauge.

Water[c] cost (£/annum)	Maintenance and replacement (£/annum)	Total running cost (£/annum)	Capital[d] charges (£/annum)	Total cost including capital charges	
				(£/annum)	(p/1000m^3)
--	60	1,750	2,820	4,570	0.57
--	60	2,320	5,490	7,810	0.98
140	320	4,220	10,740	14,960	1.87
110	210	2,170	7,530	9,700	1.21
3,300	320	6,000	15,750	21,750	2.72
2,550	750	6,130	5,970	12,100	1.51
550	320	3,770	9,300	13,070	1.63
--	400	1,400	26,490	27,890	3.49
440	420	1,970	34,020	35,990	4.50
1,210	320	10,940	12,180	23,120	2.89
--	920	3,190	44,400	47,590	5.95
1,210	320	16,400	13,230	29,630	3.70
--	3,200	5,070	18,630	23,700	2.96
--	6,000	9,960	26,130	36,090	4.51

plate with uniform hole distribution or a shallow bed of tower packings is sometimes used at the bottom of the tower to ensure even gas distribution. A mist eliminator at the tower top is necessary if high gas velocities are used.

Pressure drops are normally less than 1 in. water gauge. Collection efficiency is low for particles less than 10 μm; it is for the removal of particles greater than 50 μm that spray towers are used industrially. They are rarely used for the removal of gaseous pollutants. Spray towers are most commonly used in conjunction with more efficient scrubbers where they function as precleaners, removing the bulk of large-sized liquid or solid particles.

A study on the effects of the particle and spray droplet sizes on the collection efficiency for the simple gravity spray tower has been made by Stairmand (2). Figure 2 is a plot of the results of this work. It shows that the

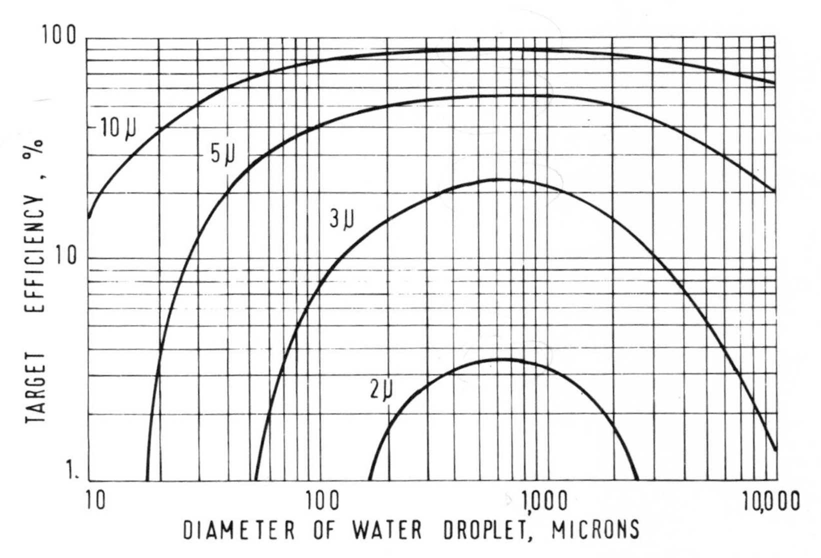

Fig. 2. Impaction collection in a gravitational spray tower (2).

maximum collection efficiencies for all dust particle sizes fall within the range of water droplet diameter 500-1000 μm. Thus a nozzle producing a coarse spray with droplets just below 1 mm is satisfactory. Impingement-type nozzles are widely used. An extensive survey of spray nozzle types and performances has been made by Marshall (14).

In practice, the gas flow rate up the tower is between 2-5 ft/sec. The droplet size must be such that its terminal settling velocity is greater than this velocity, otherwise excessive water entrainment over the tower top will take place. For water droplets in the range 1-50 μm in diameter, Stoke's Law may be used for the calculation of terminal settling velocities with negligible error (15):

$$V = \frac{d_p^2 (\rho_p - \rho_g) g}{18\mu} \tag{14}$$

where d_p is the droplet diameter; ρ_p and ρ_g are the droplet and gas densities, respectively; g is the acceleration due to gravity; and μ is the gas viscosity. For droplet sizes in excess of 50 μm, calculations must be made from empirically derived drag relationships, such as that of Schiller and Naumann (16, 17). For droplet diameters less than 1 μm, allowance must be made for "intermolecular slip," which phenomenon has the particles falling faster than the prediction from Eq. (14). This allowance takes the form of a multiplying factor, such as the Stokes-Cunningham correction factor, to the prediction from Eq. (14) (11).

Consistent with the theory of interception and inertial impaction, the collection efficiency of a spray tower is dependent on the droplet size and on the relative motion between the gas and the water droplets. The occurrence of an optimum droplet size may be reasoned as follows: for a given water flow rate to the spray nozzles, the finer the spray the greater is the fraction of tower cross-section traced out by the falling droplets and therefore the greater is the probability of particle capture by the mechanism of interception. This is clearly so as the ratio of the surface area to volume of a sphere giving the surface area generated per unit volume of liquid flow is $6/d_p$.

On the other hand, because of the lower settling velocity of the finer spray, the relative motion between the gas and the descending droplets will be lesser than with larger droplets. Thus the probability of particle collection by the mechanism of inertial impaction, which is dependent on the value of the relative motion, will fall off with reduction in droplet size.

The addition of the effects of these two rivaling phenomena gives rise to an optimum droplet.

A further, if somewhat lesser, consideration is the time of fall of the water droplets in the tower. This time, fixed by the actual velocity of the droplet relative to the tower wall, is dependent on the droplet size and the velocity of the ascending gas. For a given spray operation, that is, fixed water flow rate and droplet size, this time of fall increases with an increase in gas velocity, as also does the tower holdup of water droplets. This phenomenon, taken in isolation, will increase the probability of particle capture simply because of the increased loading of the tower volume with water droplets. However, this effect is limited by droplet entrainment in the gas stream, which occurs when the gas velocity approaches that of the settling droplets. In practice, the gas velocity up the tower is set at about 50% of the terminal settling velocity of the droplets.

Thus, for a droplet size of 500 μm, a terminal settling velocity is calculated to be 6 ft/sec, and a gas velocity of 3 ft/sec would be suitable. This calculation provides the first estimation of the tower diameter that is required. As with most other types of wet scrubbers, close control over the spray formation to ensure uniform droplet size is required for efficient operation.

With cocurrent spray towers, there is no such restriction on the gas velocity, since flooding or excessive entrainment is less likely to occur. It is therefore possible to operate with much higher velocities than in countercurrent operations, with the result that smaller diameter towers may be used, thus saving in capital costs and in plant space. However, since the gas and droplets travel in the same general direction, the probability of particle capture by either inertial impaction or interception is low, and therefore the collection efficiency is also low.

IV. CENTRIFUGAL OR CYCLONE SCRUBBERS

The collection efficiency of a dust particle by a liquid droplet depends on two principal factors: (a) the particle/droplet relative velocity (inertial impaction); and (b) the spray density (interception).

From such considerations, it is seen that the modification to the simple gravity spray tower, in which the gas is introduced tangentially at the bottom of the tower, greatly improves its scrubbing operation. A typical grade efficiency curve for irrigated and dry cyclone scrubbers is shown in Fig. 3.

The centrifugal force given to the gas stream causes the particulate solids to be thrown outwards to the chamber wall where they are trapped in the wall liquid flow. Inlet gas velocities are normally in the range 50-150 ft/sec. For example, the outward accelerating force on a particle moving at 50 ft/sec at a radius of 1 ft is 95 times that of gravity, or 95 g; in practice, values of several hun-

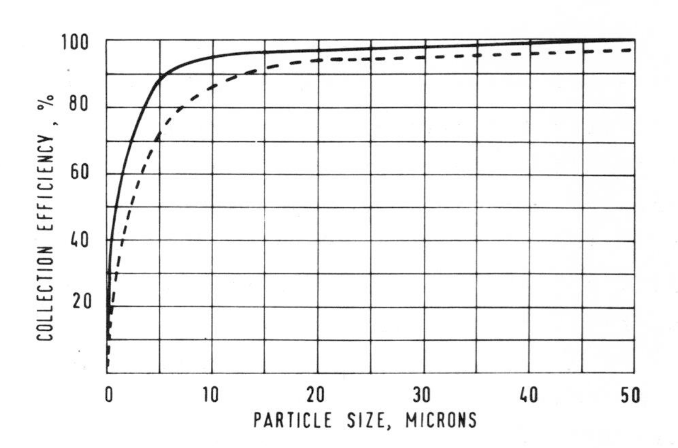

Fig. 3. Grade efficiency curves for large diameter irrigated cyclones. Efficiency at 5 μm = 87% (——irrigated; – – – dry).

dred g are realized. With the higher gas velocity, the particle/droplet relative velocity is increased and collection by inertial impaction is enhanced. Furthermore, the spiral path followed by the gas is a longer one through the tower than in the simple spray tower; impingement opportunities are therefore greater. However, since the gas will not maintain its spiral motion, preferring to take the path of least resistance (and hence minimum pressure drop) up the tower, internal devices such as guide vanes, baffle plates, etc., are often incorporated to sustain this motion.

A finer spray may be used in the centrifugal scrubber than in the simple gravity scrubber, since the chances of entrainment or liquid carryover are less likely. Because of the centrifugal action imparted to the gas, the spray droplets are thrown outwards to the tower wall where they constitute the wall flow to the bottom outlet of the scrubber. The effective life of a spray droplet is therefore quite short. The efficiency with which a liquid droplet will collect dust particles has been shown to be related to the dimensionless group (Dg/Vf), referred to here as the Stairmand Number (see Fig. 4) where D is droplet diameter in microns; g is acceleration due to gravity, ft/sec^2; V is relative velocity of approach of droplet and dust particle, ft/sec; and f is the terminal settling velocity of the dust particle, ft/sec.

In order to maximize their effectiveness, sprays are generally directed against or across the motion of the gas spiral; thus the relative velocity, V, is increased with a consequent reduction in the Stairmand Number and an increase in the collection efficiency by impaction. Furthermore, with a finer spray, there is a probability of particle/droplet collision by interception, though less by impaction. The finer the spray, the shorter is the time during which a droplet maintains an independent velocity from the gas stream and an effective collection capability.

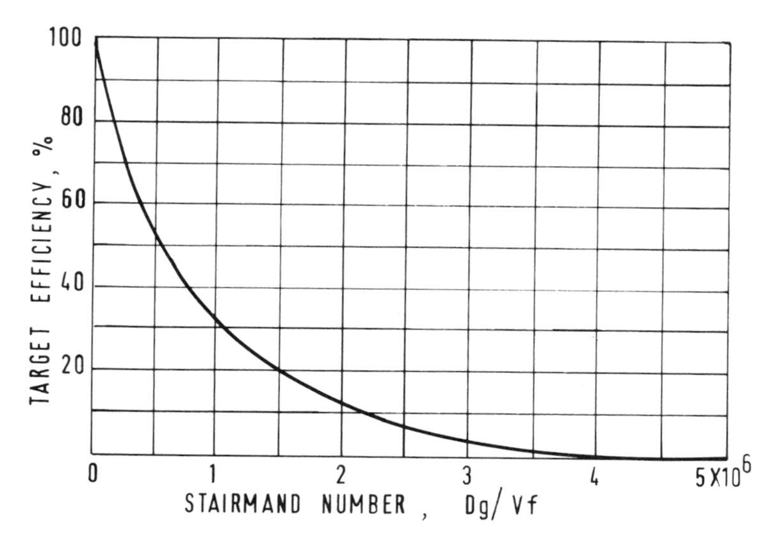

Fig. 4. Individual particle collection efficiency related to Stairmand Number (Dg/Vf).

For a given spray droplet, the Stairmand Number is at a maximum immediately after its formation from the spray nozzle. The droplet/particle's relative velocity then falls off inversely as the droplet's size. The optimum droplet size has been theoretically estimated at 100 μm in diameter (42) (see Fig. 5). In practice, atomizing sprays of 100-200 μm range are used (20). Spiral spray, spinning disk, jet impingement, and sonic nozzles are used to obtain these fine droplets (11, 19).

Centrifugal scrubbers are used effectively for particle removal down to the 5-μm range. For the removal of particles in the submicron region, they are sometimes used in an integral type of contact system after, for example, a high-energy jet or venturi scrubber; their function in such

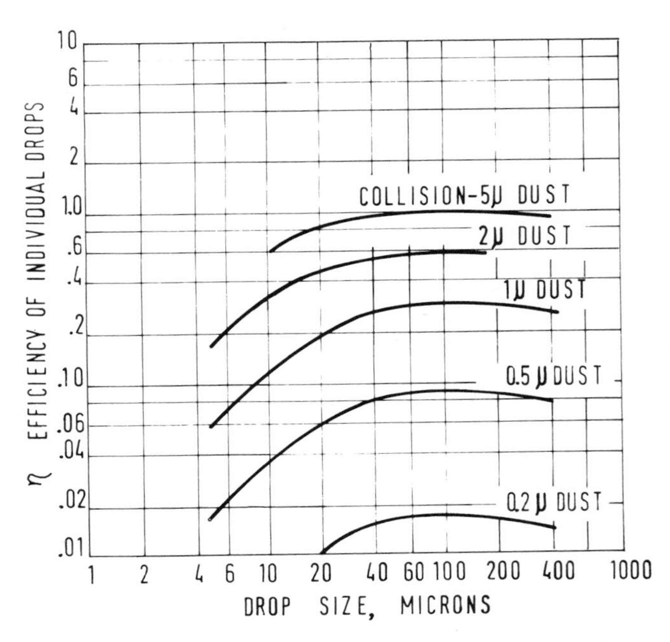

Fig. 5. Impaction collection for centrifugal force of 100 times gravity.

instances is to remove the particles that have been previously wetted in the more effective contactor (9,21,22,23). They are also used for limited gas absorption duties with scrubbing solutions other than water (9, 24).

Pressure drops fall within the range 1-4 in. water gauge. Cyclone scrubbers are particularly suitable for handling large gas volumes with high dust contents; the likelihood of downtime due to plugging is small. The high pressures required by the spray nozzles mean greater pumping costs than the gravity spray scrubber. In some designs a low-pressure liquid supply is used, the spray being formed by the action of the swirling gases.

A. INDUSTRIAL DESIGNS

A centrally positioned spray manifold is used in the Pease-Anthony design (see Fig. 6) (25). Operational control is provided for by the damper on the gas inlet, which allows for regulation of the inlet gas velocity and the scrubber pressure drop. Further control may be effected on the inlet water pressure to the spray manifold. Pressure drops of 1-4 in. of water gauge and liquid rates of 4-5 gal/1000 cfm are normal. Removal efficiencies in the range 95-98% have been obtained with a variety of dust types with particle sizes as low as 0.5 μm. It has also found application for the gas absorption of SO_2 from boiler flue gases, using a weak alkali solution scrubbing liquid with absorption efficiencies in excess of 94% (26).

A wall-mounted spray jet assembly is shown in Fig. 7, for the series 500 Mikro/Airetron Cyclonic Scrubber. This design has been developed to handle particles of 1.0 μm in size or larger, with recovery efficiencies in excess of 98%. It is also used as a gas absorber for readily soluble gases such as HCl, NH_3, SO_2, and H_2S; for the latter two, alka-

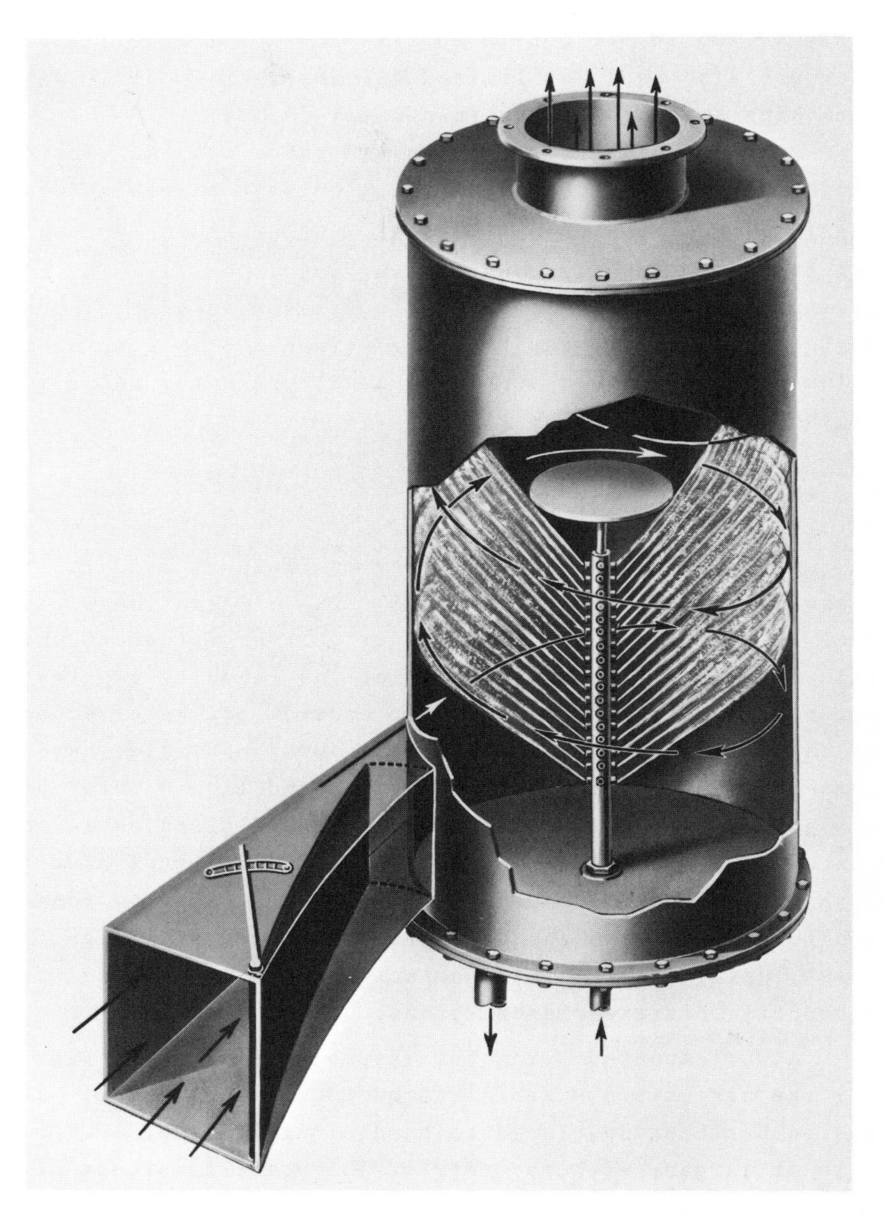

Fig. 6. Chemico P - A cyclonic scrubber (Chemical Construction Corp.).

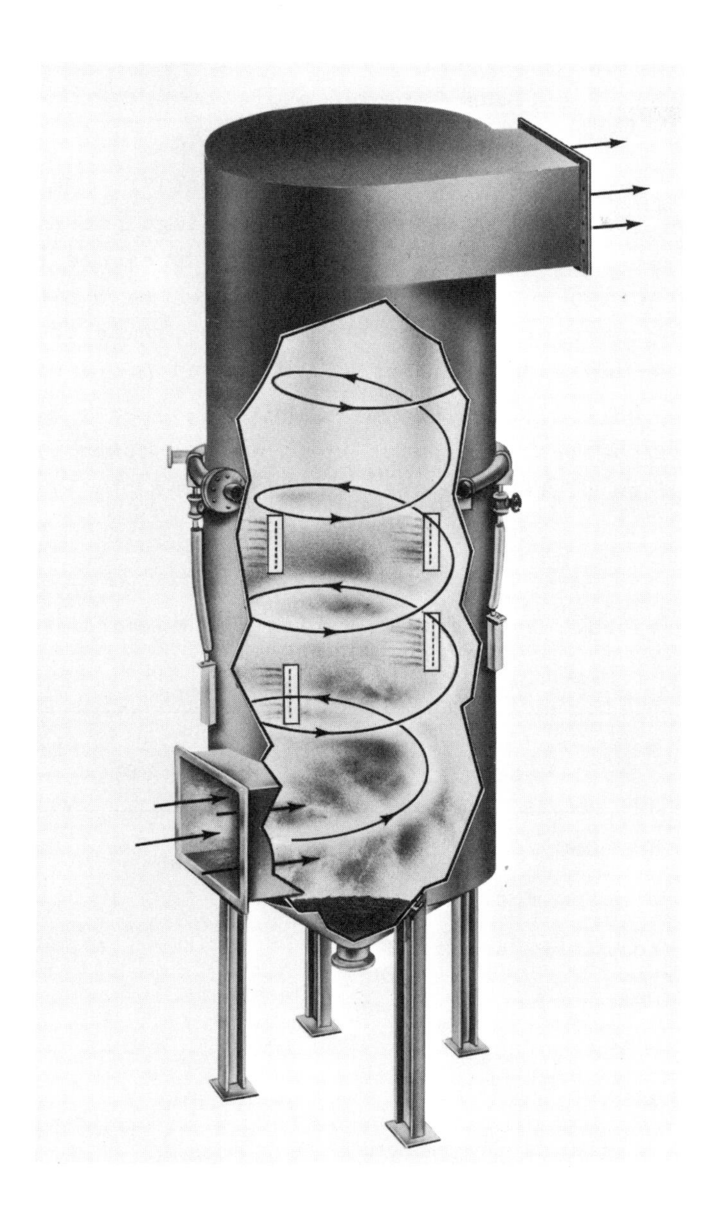

Fig. 7. Series 500 Mikro/Airetron Cyclonic Scrubber (Mikropul Ltd.).

line solution with recirculation is used. Its use for the conversion of silicon fluoride as follows:

$$3\ SiF_4 + 2H_2O \longrightarrow SiO_2 + 2H_2SiF_6$$

has been reported as very satisfactory (27). Normal operating conditions are 1-4 in. of water gauge pressure drop, liquid rate 5 gal/1000 cfm, and liquid pressure to the spray nozzles 50-150 psig, which is controlled to the higher pressures for particles approaching 1 μm in diameter. Table 7 gives examples of its applications and performances.

The Holmes-Schneible Multi Wash Collector avoids the use of high-pressure sprays by providing tower internals in the form of two impingement stages and one entrainment

TABLE 7

Performance Data on the Series 500 Mikro/Airetron Cyclonic Scrubber (28)

Application	Outlet (grains/scf)	Efficiency (%)
Lime kilns, CaO	0.05	99.0
Asphalt plants, rock dust	0.25	95.0
Incinerators, fly ash from		
(a) refuse,	0.034	98.0
(b) sludge	0.30	96.0
Alum plants $Al_2(SO_4)_3$	0.002	99.0
Cement kilns	0.30	85.0
Fertilizer plants SiF_4 from		
(a) continuous den	0.0004	99.0
(b) batch den	0.00015	98.5
(c) reaction belt (2 stages)	0.000033	99.3
Process exhaust	0.03 (vol %)	96.0

separator (see Fig. 8). The gases are introduced at the bottom, just above the cone. The cone serves as a wet cyclone, collecting the heavier materials and creating a vortex that sends the gases upwards in a spiraling motion. The washing agent is introduced just below the entrainment separator and cascades evenly over the impingement stage elements, carrying the collected material to the slurry outlet. The velocity of the carrier gases induces a turbulent action into the washing agent, which breaks down into finely divided spray patterns and, through intimate contact and impingement, washes out or absorbs the contaminants.

The spray is formed by the action of the gases on the liquid as it flows over the impingement stages; the scrubbing liquid is therefore supplied at no head pressure through an open pipe, as shown in Fig. 8. This arrangement, being virtually nonclogging, allows for recirculation of the scrubbing liquid if desired. Additional impingement stages may be added or removed to increase or reduce the severity of the operation as required.

Figure 9 shows a centrifugal gas scrubber that features a turbulent bubbling bed. This scrubber is primarily designed for the 3-8 μm range with medium pressure drops and a 99% + collection efficiency (29), but it can be operated for the efficient collection of particles in the 1-2 μm range at pressure drops up to 15 in. of water gauge. Operational control is provided for in the flow of scrubbing liquid to the unit, which, as in the previous design (Fig. 8), is delivered through low-pressure nozzles or simply an open pipe arrangement when liquid recirculation is employed. Design liquid flow rates are 2.0 gal/1000 cfm for pressure drops up to 10 in. water gauge and 4.0 gal/1000 cfm for higher pressure drops. Humidification nozzles, located beneath the scrubbing vanes, are usually installed for gas inlet temperatures over 500°F. Adjustment of the

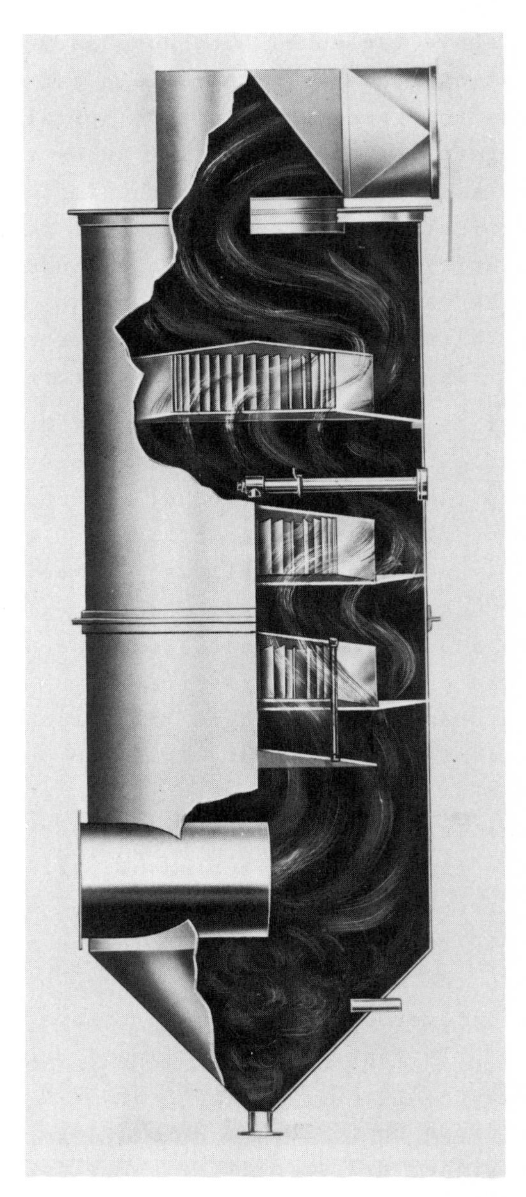

Fig. 8. Holmes-Schneible multiwash collector (W. C. Holmes & Co. Ltd.).

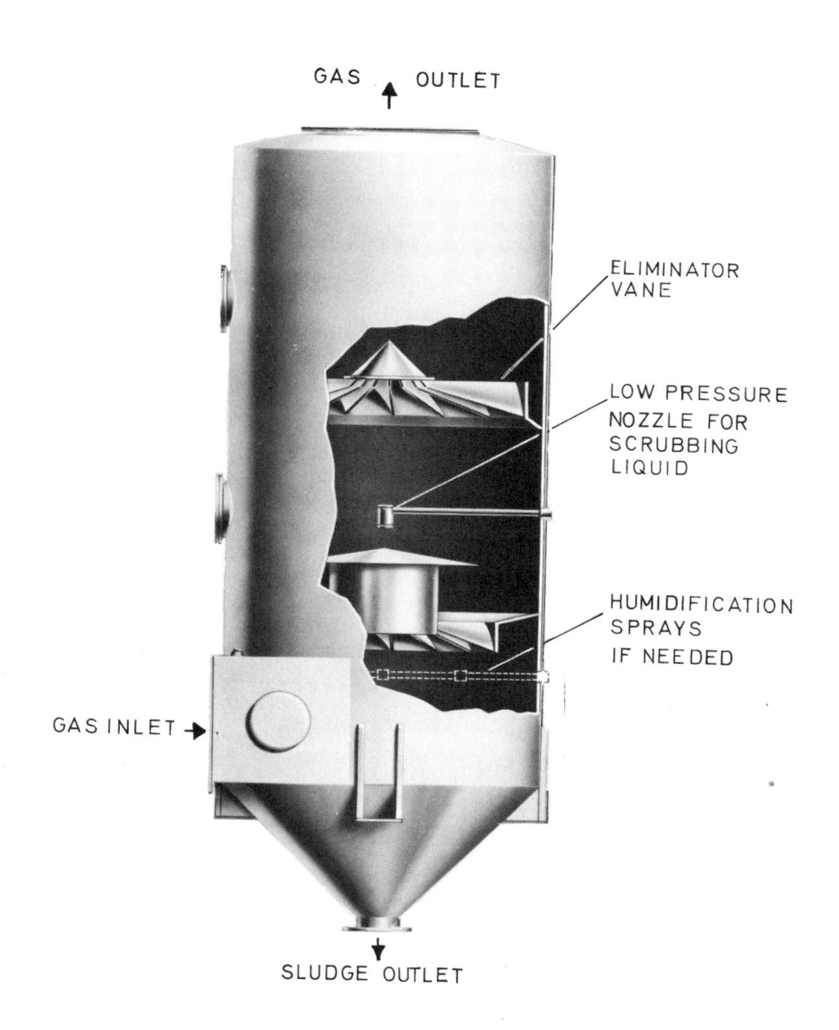

Fig. 9. Ducon multivane scrubber (Ducon International).

lower vane is provided to accommodate variation in the gas rate.

V. SELF-INDUCED SPRAY SCRUBBERS

The essential action of all wet scrubbers is the removal of pollutant, whether gaseous or particulate, from a gas stream by intimate contact with a scrubbing liquid. For a given quantity of liquid, the greater the surface area exposed to the gas flow, the more intimate the contact. In most wet scrubbers this surface area is provided by the formation of spray, and the surface area generated is inversely proportional to the spray droplet diameter; in short, the finer the spray, the more surface area exposed.

The formation of spray from a continuous liquid requires energy expenditure. In wet scrubbers, this energy may be provided by the scrubbing liquid in dissipating pressure head energy through spray nozzles, or by the gas using its kinetic energy by direct impingement onto a liquid surface. Scrubbers using this latter principle are referred to as "self-induced spray" scrubbers. They normally incorporate a Venturi or similar arrangement for gas acceleration during or prior to liquid impingement.

The formation of spray by the self-induced method has been studied by Nukiyama and Tanasawa (11, 30, 31). For gas impingement on a liquid wall in a Venturi, which in essence is the action by the self-induced spray, the following empirical equations were derived for the average droplet diameter, d_p, and the specific surface area, A:

$$d_p = \frac{585\sqrt{\sigma}}{u\sqrt{\rho_L}} + 597\left(\frac{\mu_L}{\sqrt{\sigma\rho_L}}\right)^{0.45}\left(\frac{1000Q_L}{Q}\right)^{1.5} \tag{15}$$

where d_p is the average droplet diameter (μm); u is the

relative velocity between air and liquid flows (m/sec); Q_L/Q is the rate of liquid volume flow to gas volume flow at the Venturi throat; ρ_L is the liquid density (g/cm^3); σ is the liquid surface tension (dynes/cm); and μ_L is the liquid viscosity (poise).

For the air-water system, the above equation reduces to:

$$d_p = \frac{16050}{u} + 1.41\ L_1^{1.5} \tag{16}$$

where L_1 is the number of gallons of water/1000 ft^3 air.

The specific surface of the droplets is given by:

$$A = 244L_1/d_p \tag{17}$$

where A is in ft^2/ft^3 of gas.

The spray generation by the self-induced mechanism is not as effective as with the spray nozzles. Their principal advantage lies in their ability to handle high gas rates with large dust concentrations. Water consumption is low, being normally less than 1 gal/1000 ft^3, with pressure drops within the range of 2-15 in. water gauge. They find extensive use in the metallurgical industry for dust and metal buffings removal. The average grade efficiency curve for the self-induced spray scrubber is shown in Fig. 10.

A. INDUSTRIAL DESIGNS

The Roto-Clone Type N is shown in Fig. 11. The air flows at a high velocity over the water surface through a stationary impeller, carrying the water with it in a turbulent sheet. The centrifugal force exerted by the rapid changes in flow direction causes efficient water/particle contact. The clean air flows upwards over a de-entrainment baffle to the suction fan. The water level falls or rises

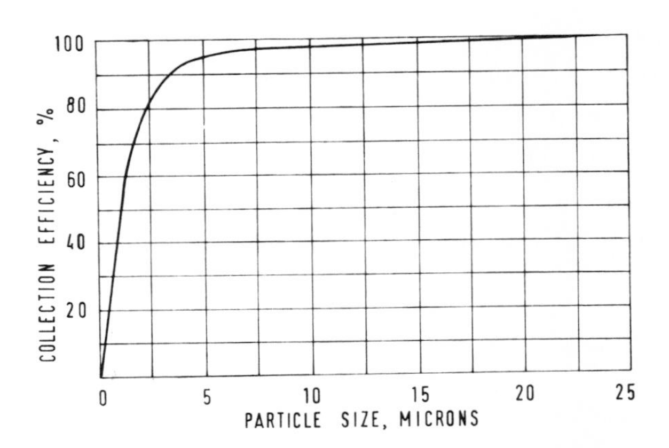

Fig. 10. Grade efficiency curve for self-induced spray scrubber; efficiency at 5 μm = 93%.

with an increase or decrease in air volume flow to maintain a constant velocity through the curved impeller. The unit pressure drop is 6 in. water gauge, which varies only slightly with air volume fluctuations. This design can handle volumes up to 48,000 cfm. The net consumption of water is low, about 1 gal/1000 ft^3 excluding evaporation. The water flow through the impeller is approximately 20 gal/100 ft^3 and is continuously recycled. This type of scrubber is used in abrasive cleaning and tumbling mill dust control foundry sand systems, and for many dryer, cooler, kiln, and materials handling operations in the chemical, mining, and rock product industries.

A different principle is used in the Turbulaire gas scrubber, which is shown schematically in Fig. 12. The flow of air is directed through a reduced cross-section of flow area onto the liquid surface, causing a turbulent bed effect. Because of the high turbulence in the scrubbing bath, the slurry can be highly concentrated: up to 50% solids. Water consumption as with most designs of this type is less than 1 gal/1000 ft^3. This type of unit has a wide

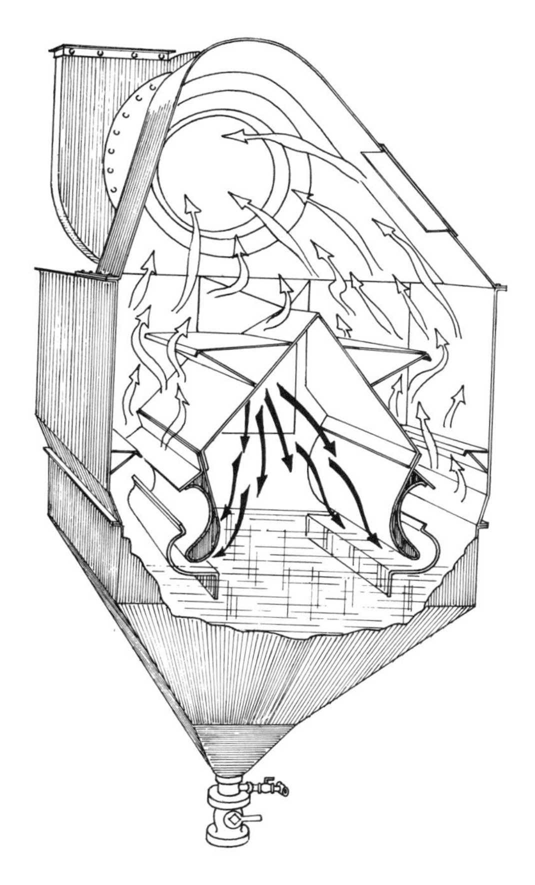

Fig. 11. Roto-clone Type N scrubber (American Air Filter Co., Inc.).

range of applications in the chemical, metallurgical, rock products, and waste disposal operations. Small capacity models are used for domestic incinerator smoke cleaning duties.

The operating principle of the Medusa Gas Scrubber is shown in Fig. 13. The design pressure drop over this unit may be varied with a direct effect on the collection efficiency by adjustment of the liquid level setting. The

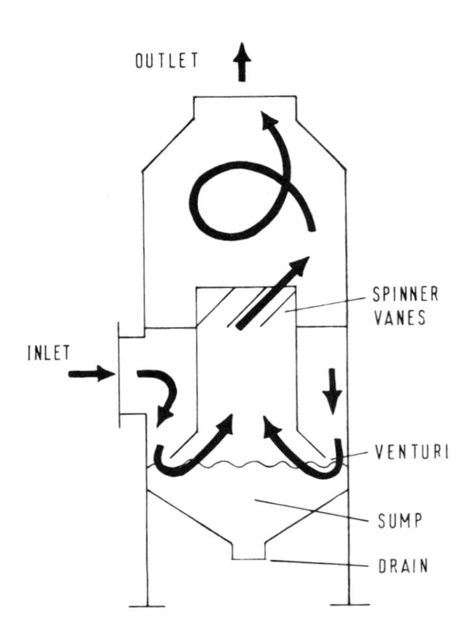

Fig. 12. Turbulaire Type D gas scrubber (Joy Manufacturing Co.).

scrubbing action takes place in three stages:

1. Impingement of the gas on the liquid surface.
2. Gas flow through the self-induced spray in the natural Venturi formed by the wave crest.
3. Gas flow through the vertical louvres or rods subjected to the spray of the atomized liquid.

This scrubber has general application for gas cleaning duties in the chemical and metallurgical industries.

VI. PLATE SCRUBBERS

In the simple plate scrubber, a cylindrical tower is divided into a number of stages or plates. The scrubbing liquid is fed to the top plate and flows horizontally across

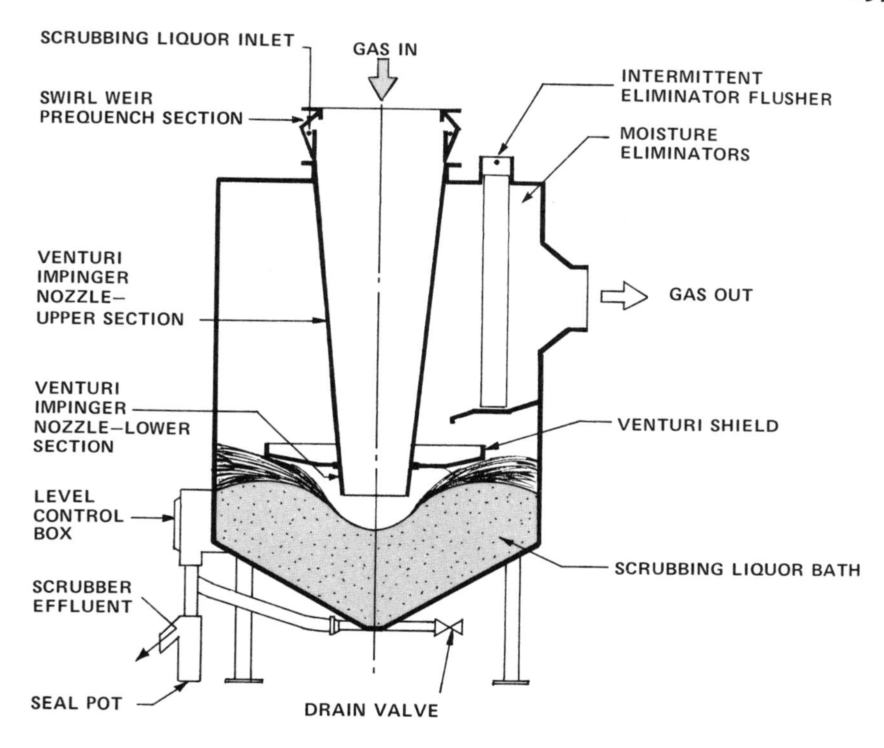

Fig. 13. Medusa gas scrubber (Krebs Engineers).

the plate to a down-comer pipe leading to the next plate; the water thus progresses from plate to plate to the outlet at the tower bottom. The gas flows up the tower, bubbling through openings in the plates into the liquid flow where gas absorption or particulate removal takes place. The gas is "cleaned" as it rises up the tower and the liquid becomes more contaminated with pollutant matter as it progresses down the tower.

A. LIQUID FLOW

With most wet scrubber applications, recycling of the scrubbing liquid is possible. With plate towers, the liquid flow across each plate must be sufficiently turbulent to prevent settling out of particulate solids. In some processes, a settling tank is provided in the recycle circuit with continuous sludge removal. In gas absorption processes, selected scrubbing liquids with high solubilities for the chemical pollutant are used and total recycling is not possible. The degree of recycle and makeup liquor required depends in each instance on the particular scrubbing duty.

Liquid height on the plates is determined directly by the weir height, which provides the overflow to the downcover. In general, the deeper the liquid on the plate, the greater is the gas-liquid contact, and thus the tray is more efficient. However, since the pressure drop over the tower as a whole is the sum of the liquid heights on the tray, a limit is set on the liquid depth. Pressure drop of about 3 in. water gauge per plate is normal.

Liquid flow rate is limited by the hydraulics of the tower. Excessive liquid flow, where the flow in is greater than flow out by gravity through the downcovers provided will cause a liquid buildup in the tower and make operation impossible. On the other hand, the flow must be sufficient to maintain a liquid seal in each liquid downcover, otherwise the gas would flow up the downcovers, bypassing the plates,and would result in inefficient operation. Typical values are in the range of 3-5 gal/1000 ft^3.

B. GAS FLOW

Superficial velocity is the total gas flow divided by the tower over-all cross-section area. The upper limit on this figure is set by the possibility of entrainment of liquid over the top; normal values are in the range 1-2 ft/sec. De-entrainment devices are usually provided at the tower top.

Velocity through plate openings for particulate-removal

processes is set much higher than in a gas absorption process. Velocities of the order of 15-20 ft/sec are normal, though 75 ft/sec has been quoted (2). The gas flow through the orifices resembles a high-velocity jet and causes turbulence in the liquid on the plates. The orifice openings add up to about 10% of the tower cross-section. Impingement plates are positioned over each hole, thus causing violent agitation at the liquid/gas interface. This action enhances the rate of mass transfer and also serves to keep the trays free from plugging up problems.

C. PLATE DESIGN

In addition to the weir height discussed above, the design of plates involves the selection of the type of opening to be provided for the gas flow. In wet scrubber applications, sieve plates, with impingement baffles placed above the perforations, are usual.

Plate towers are sometimes used for the removal of gaseous pollutants when the process is then a gas absorption one. As such, the operation is measured by calculating the individual tray and the over-all tower efficiencies. For solutes such as sulfur dioxide (SO_2) or ammonia (NH_3), which would be absorbed from a gas stream by a sulfuric acid (H_2SO_4) solution, a solubility curve exists that relates equilibrium concentrations of the solute between the gas and liquid phases. The gas and liquid flowing onto each plate are not in equilibrium, and transport of the solute occurs in the direction of equilibrium. The efficiency of a stage is a measure of how far the streams have approached equilibrium values; it is given by the equation

$$\eta_N(g) = \frac{Y_{N-1} - Y_N}{Y_{N-1} - Y_N^*} \tag{18}$$

where $\eta_N(g)$ is the efficiency of tray N calculated from the gas composition change over it; Y_{N-1} and Y_N are the mole fractions of the solute into and out of tray N, respectively; and Y_N^* is the equilibrium mole fraction of the gas corresponding to the liquid composition flowing from tray N, as read off the equilibrium solubility curve.

In designing gas absorption plate towers, it is possible to predict accurately from published data the efficiencies of various designs of trays at given liquid and gas loadings. These would normally be in the 40-60% range. It is therefore a relatively simple design problem to estimate the number of plates necessary to reduce the solute concentration of the gas stream to a desired value. The over-all tower efficiency is defined as the ratio of the number of 100% efficient trays to the actual number of trays required.

Plate scrubbers are often used for the dual operation of removing both particulate solids and gaseous pollutants from a gas stream (33). Also, in common with other wet scrubbers, plate scrubbers are used as gas coolers. While water is the most usual scrubbing liquid used, specialized acid or alkaline solutions are used for gas absorption duties when required. For the absorption of SO_2, solutions of calcium carbonate, ammonium sulfite, ammonium sulfate, dimethylamine, and sodium borate-boric acid have been used. For H_2S absorption, sodium phenolate, monoethanolamine (MEA), diethanolamine (DEA), triethanolamine (TEA), tripotassiumphosphate, and sodium carbonate have been used (34).

D. INDUSTRIAL DESIGNS

Figure 14 is the average grade efficiency curve for impingement plate gas scrubbers used for particulate collection duty.

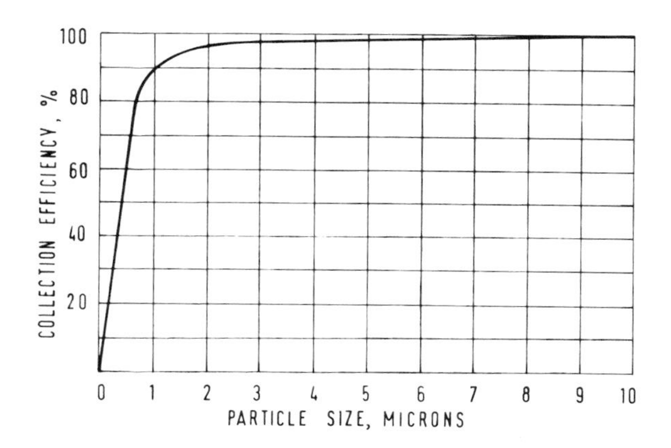

Fig. 14. Grade efficiency curve for impingement plate gas scrubbers; efficiency at 5 μm = 97%.

Figure 15(a) shows a two-plate gas scrubber. The plate, detailed in Fig. 15(b), has from 6500 to 30,000 orifices/m^2 (600-3000/ft^2). In the lower half of the tower, the hot and particle-laden gas rises through humidifying sprays, where it is quenched, saturated, and stripped of coarse particles. After passing through the plates, the gas passes through an entrainment separator. Slurry with concentration up to 65% has been recirculated with such scrubbers. The liquid consumption is small, being that lost to evaporation plus that required to maintain a free-flowing slurry. In addition to the impingement baffle plate shown in Fig. 15(b), many other designs are commercially available for collecting submicron particles or fumes. Pressure drop range is 1.5 - 8 in. water gauge depending on the number of stages, each having 1.5-2.5 in. water gauge pressure drop. For submicron particle removal, the impingement-baffle and venturi-slot plates are both used (as shown) and pressure drops up to 40 in. water gauge are sometimes realized.

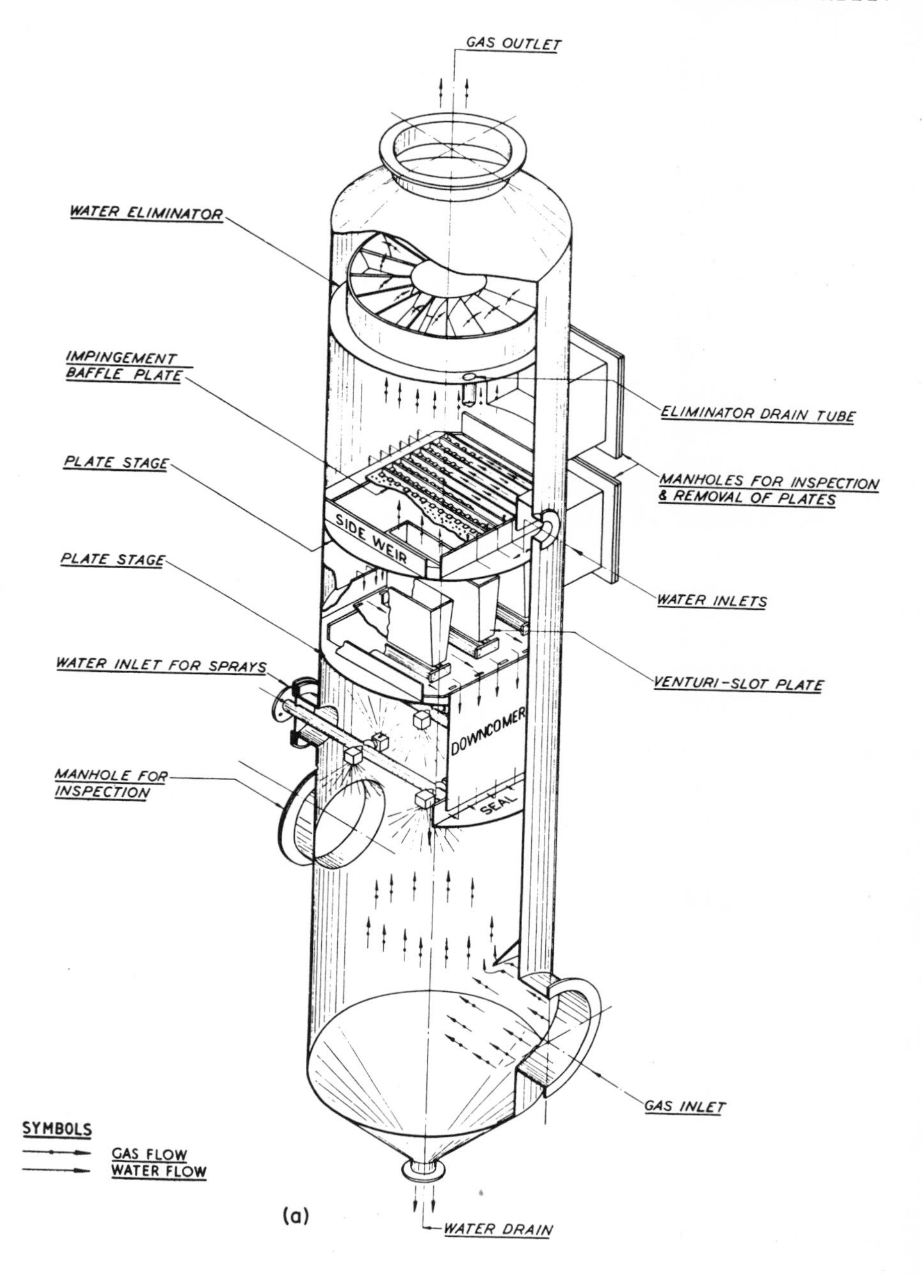

Fig. 15(a). Peabody gas scrubber (Peabody Ltd.).

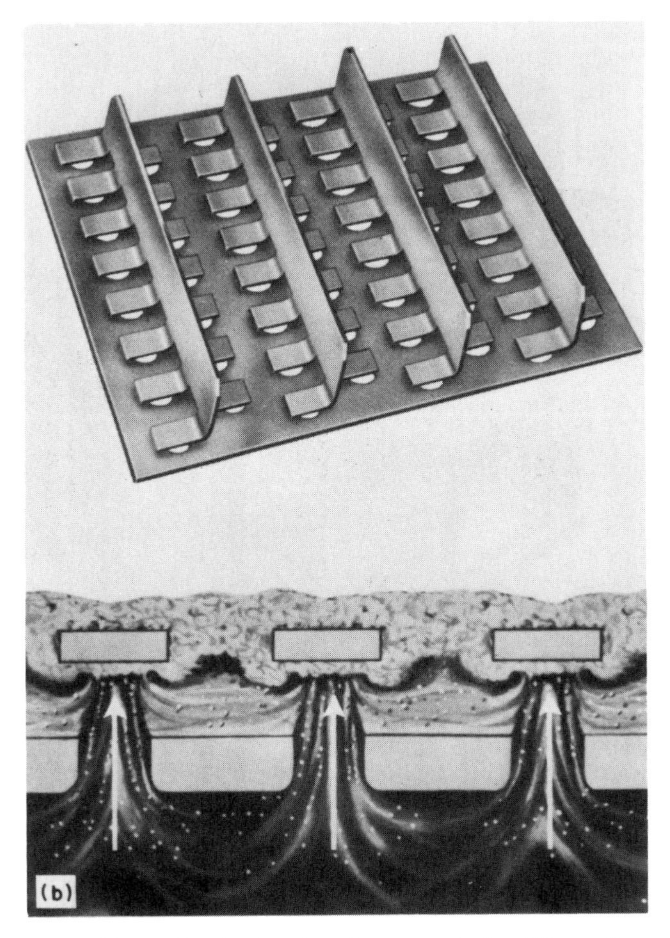

Fig. 15(b). Peabody impingement baffle plate stage (Peabody Ltd.).

The one-stage rectangular section plate scrubber shown in Fig. 16(a) uses the same principles. The effect of increasing the collection efficiency by adding additional stages is shown in Fig. 16(b). This type of unit is capable of handling high gas flows up to 500,000 cfm.

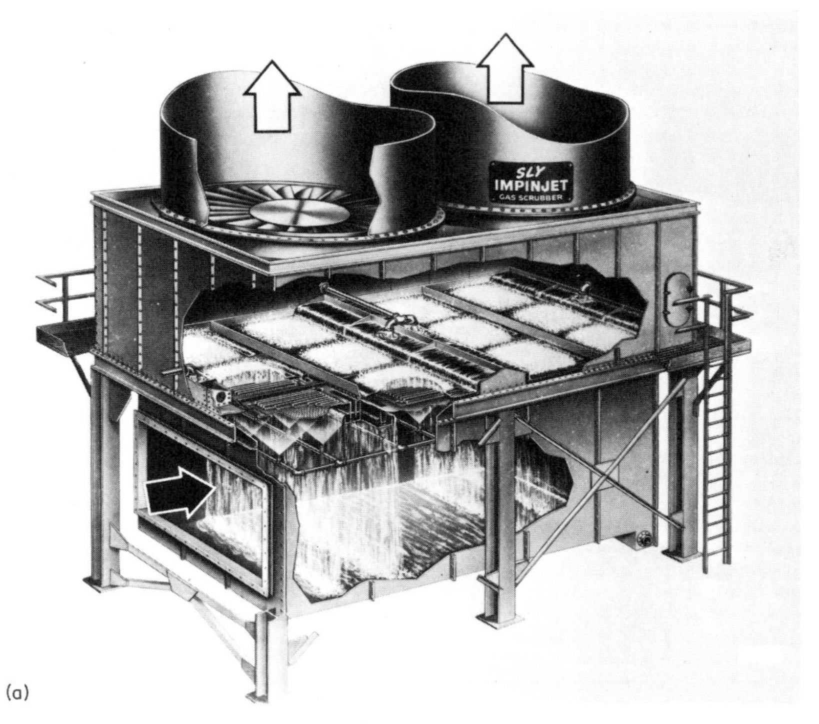

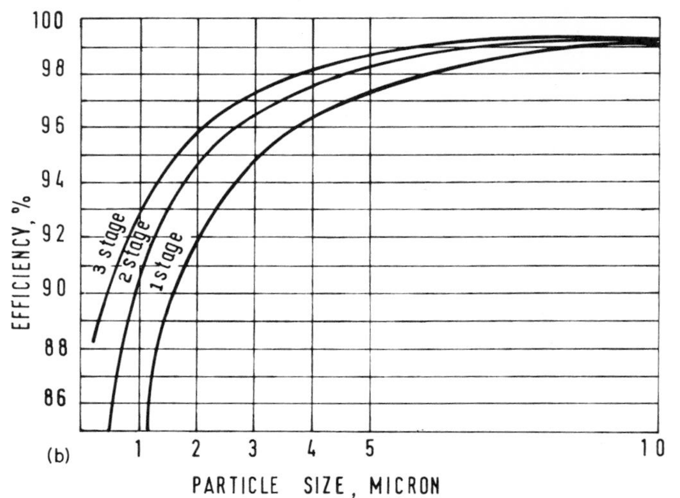

Fig. 16. (a) Single stage plate gas scrubber. (b) Effect on collection efficiency of additional stages. (W. W. Sly Manufacturing Co.)

VII. PACKED BED SCRUBBERS

Packed bed scrubbers are generally used for the removal from gas streams of gaseous pollutants rather than particulate matter. The scrubbing liquid is fed to the top of the tower. A distributor plate with a pattern of overflow pipes provides an even distribution of liquid to the top of the bed. The liquid flows down through the bed, wetting the packing and thus provides interfacial surface area for mass transfer of the pollutant chemical with the gas, which flows up the bed countercurrent to the liquid.

This type of tower is used extensively in the chemical industry for gas absorption, or other such processes which require mass transfer or reaction between a liquid and a gas stream. In gas cleaning duties, many variations of this simple design have been developed by equipment manufacturers to meet specific problems. The principles, however, remain the same as those for gas absorption processes. These principles are described fully in many standard texts (35) and are reviewed only summarily here. Mention has already been made in Sec. VI of gas absorption in plate towers. The main difference in the theoretical analysis arises from the continuous contact that is maintained between the liquid and gas streams in packed towers, whereas in plate towers, the contact is "stagewise."

A basic decision to be made in the design of a gas absorption tower is the selection of a suitable solvent that has a high and, if required, selective solubility for the particular gaseous constituent to be removed from the gas phase. Where the pollutant is particulate dust, water is invariably used as the scrubbing liquid. In gas absorption scrubbing, the gases to be cleaned are generally very dilute where the pollutant is of the order of 1% by volume or less; this is because such gas streams are the vent streams from a recovery unit and the pollutant chemical is thus the fraction that escaped "recovery," as dictated by the efficiency of the recovery process. In these instances, the concentration is often too small to justify a further recovery process, yet too large to be simply vented into the atmosphere.

The material balance and mass transfer rate equations for the packed bed gas scrubber may be combined in the form

$$Z = \frac{G_s}{K_y a} \int_{C_2}^{C_1} \frac{dC}{C - C^*} \quad (19)$$

where Z is the depth of packing required to reduce the gas concentration from C_1 to C_2; G_s is the total gas flow rate per unit cross section of tower; K_y is the over-all gas mass transfer coefficient; and "a" is the interfacial surface per cubic volume of packing; C* is the equilibrium concentration of the solute in the gas phase.

Equation (19) is usually abbreviated to

$$Z = H_t \times N_t \quad (20)$$

where H_t is the <u>height of a transfer unit</u>, has the dimensions of length, and

$$= \frac{G_s}{K_y a} \quad (20a)$$

N_t is the <u>number of transfer units</u>, is dimensionless, and

$$= \int_{C_2}^{C_1} \frac{dC}{C - C^*} \quad (20b)$$

The height of a transfer unit is a characteristic of the particular system (packing, gas, gas rate, and composition) and is often taken as a constant over fixed ranges of operation. Values for particular systems have been published in the literature for many of the more commonly used processes (36).

The number of transfer units must be calculated for the collection required, that is C_1 to C_2, and the equilibrium solubility curve for the gaseous pollutant must be known over this range of concentrations. Graphical integration of Eq. (20b) is often used to evaluate N_t. In industrial practice for gas scrubber design purposes, the approximate expression

$$N_t = \ell n \left(\frac{C_1}{C_2}\right) \tag{21}$$

is sometimes used [see Eq. (6)]. Equation (21) would be the exact integral of $\frac{dC}{C - C^*}$, where $C^*=0$, that is, the equilibrium line is taken as coinciding with the x axis of the equilibrium solubility curve. The error resulting from this approximation is negligible with very dilute solutions or where the gaseous pollutant is very soluble, as is the case, for example, in the absorption of ammonia (NH_3) out of air with a sulfuric acid (H_2SO_4) solution, or for particulate removal.

Another feature of the design of packed bed scrubbers is the type of packing used. Some manufacturers of packed bed equipment have developed their own patented packings for which they have specific performance claims for H_t and pressure drop. Packings are in the form of rings, saddles, wire gauze, and various other special designs that improve performance.

For a given gas rate, it is not operationally possible to increase the liquid rate beyond a fixed value or tower flooding will occur. The alternate case for a fixed liquid rate also applies. Tower flooding occurs when the flow of liquid out of the tower is less than the flow in, and the liquid builds up within and on top of the packed bed; at the same time, the pressure drop increases rapidly and the unit ceases to operate. The experimental curve of Lobo et al. (37) is useful in getting an approximate value of the flooding liquid and gas rates (see Fig. 17). The manufacturers of tower packings will normally provide more accurate flooding curves for their own packings. Towers are usually operated at 60% or less of the flooding point rate. Where the gas and liquid flows are in the same direction through the scrubber tower (cocurrent) or at right angles to each other (crosscurrent), the limitations set by flooding no longer apply.

A. INDUSTRIAL DESIGNS

A shallow bed of marbles is used in the scrubber shown in Fig. 18. The scrubbing mechanism is such that dust-laden gases (up to 40 grains/ft^3) can be efficiently cleaned. The

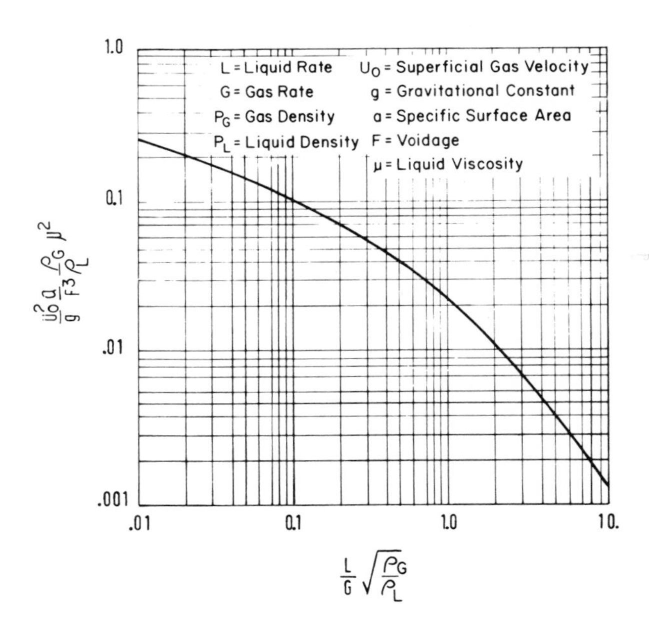

Fig. 17. Generalized correlation for flooding rates in packed towers (37).

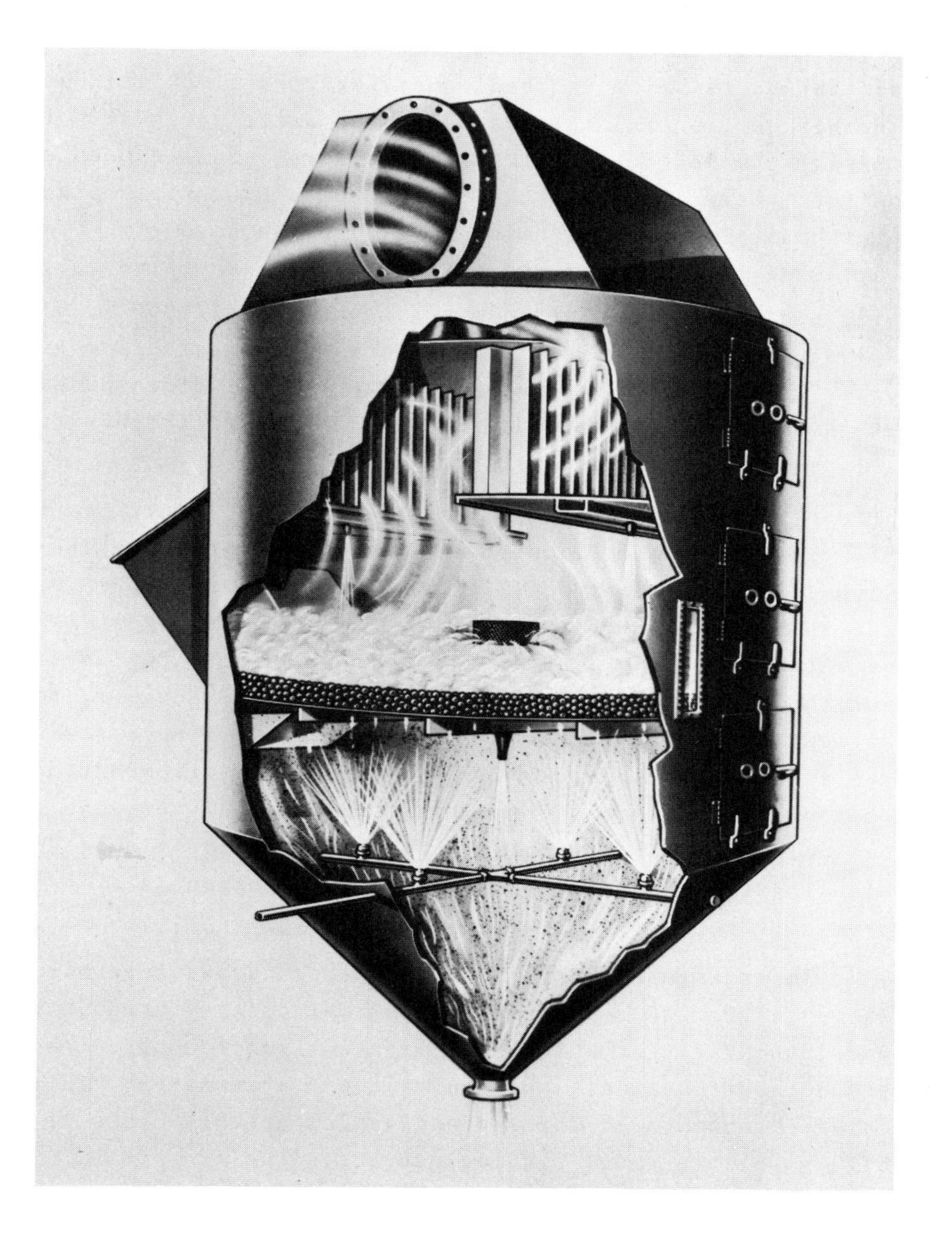

Fig. 18. Hydrofilter (Environeering Inc.).

sprays at the bottom section of the scrubber are directed to the underside of the bed. Heavier particles fall out in this section and the gas pulls the water spray with it through the bed to form a turbulent layer of fine bubbles on top. Overflow from this layer, containing the captured particles, flows down the central duct to the dirty water outlet at the drum bottom. Water consumption rates vary, with the inlet dust load varying from 2.5 gal/1000 ft^3 for 5 grains/scf to 8.0 gal/1000 ft^3 for 15 grains/scf or greater. The pressure drop range is from 4-6 in. of water gauge. The Hydrofilter is rated at 99% collection efficiency for all particles of 2 μm and over.

A cocurrent design is shown in Fig. 19. This has a bed of flexirings. Cocurrent scrubbers may be operated at high liquid and gas rates, since the pressure drop is low and there is no flooding limitation; thus the tower diameter of a cocurrent spray tower for a given duty is smaller than for the corresponding countercurrent tower. However, for efficient operation, large driving forces are necessary for gas absorption; these are not obtained with cocurrent operation. Cocurrent designs find application where the gaseous pollutant is very soluble in the scrubbing liquid and also in situations where the plant floor space is at a premium.

An arrangement for a crossflow packed bed scrubber is shown in Fig. 20. The air passes horizontally through two beds, the first of which is irrigated from the top. The second bed functions as an entrainment eliminator. This unit is recommended for gas containing soluble mists or highly soluble gases without particles.

A combination of scrubbing mechanisms is used in the Hexadyne Wet Scrubber, which finds application in the collection of particulate pollutants (see Fig. 21). The dust-

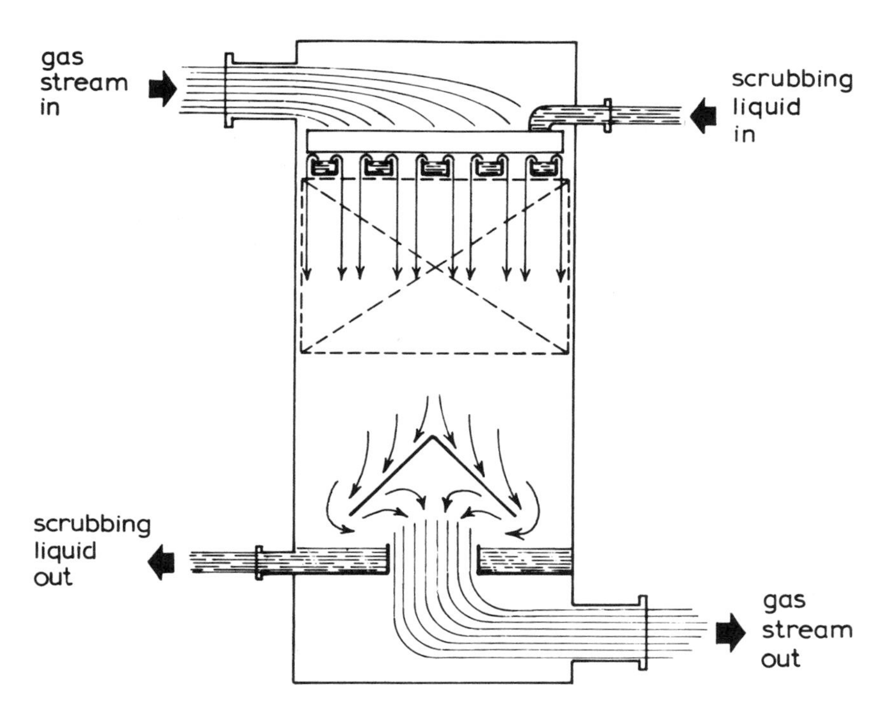

Fig. 19. Cocurrent packed tower with flexirings (Koch Engineering Co. Inc.).

laden gases enter from the top and are forced to make a 180° turn through the open passages of the inverted cone inertial separator. The inside walls of this separator are wetted from a low-pressure spray above; the larger particles are captured here and drawn to the bottom outlet down the central duct. The gas then flows downward through the packed bed section. This bed, which is flushed from above by distributed low-pressure sprays, consists of 120 layers, each layer 1/4 in. thick, of a "honeycomb" type of packing media. This packing is normally made of impregnated wood fiber, though other materials are used for specific scrubbing jobs. The water is atomized by the gas and packing and the dust par-

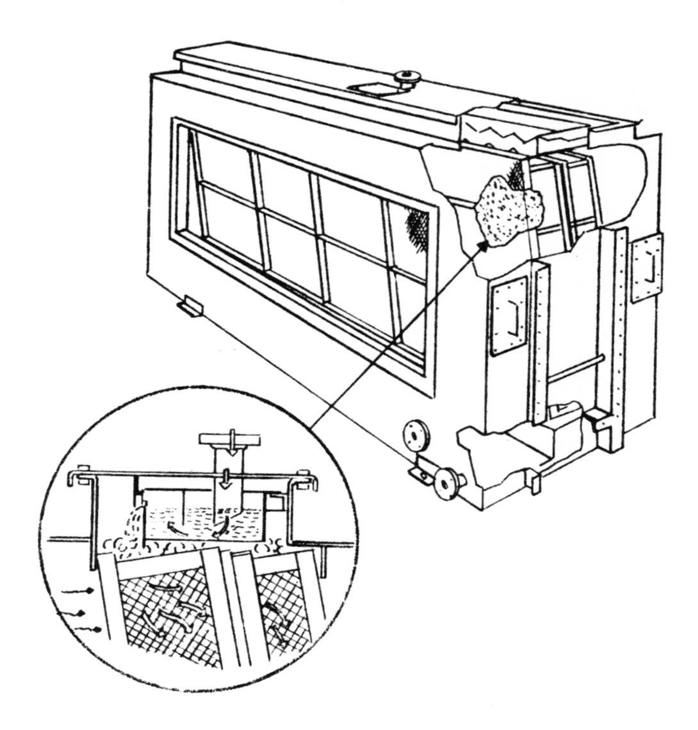

Fig. 20. Cross flow packed bed scrubber (Arco, Division of Envirotech Corp.).

ticles are captured by inertial impaction with the water spray. The water droplets containing the fine dust agglomerate into larger droplets and fall down to the outlet hopper. The gases finally pass through a moisture eliminator before exiting.

VIII. VENTURI SCRUBBERS

When a gas flows through a simple venturi (see Fig. 22), the pressure energy of the inlet gas is converted down

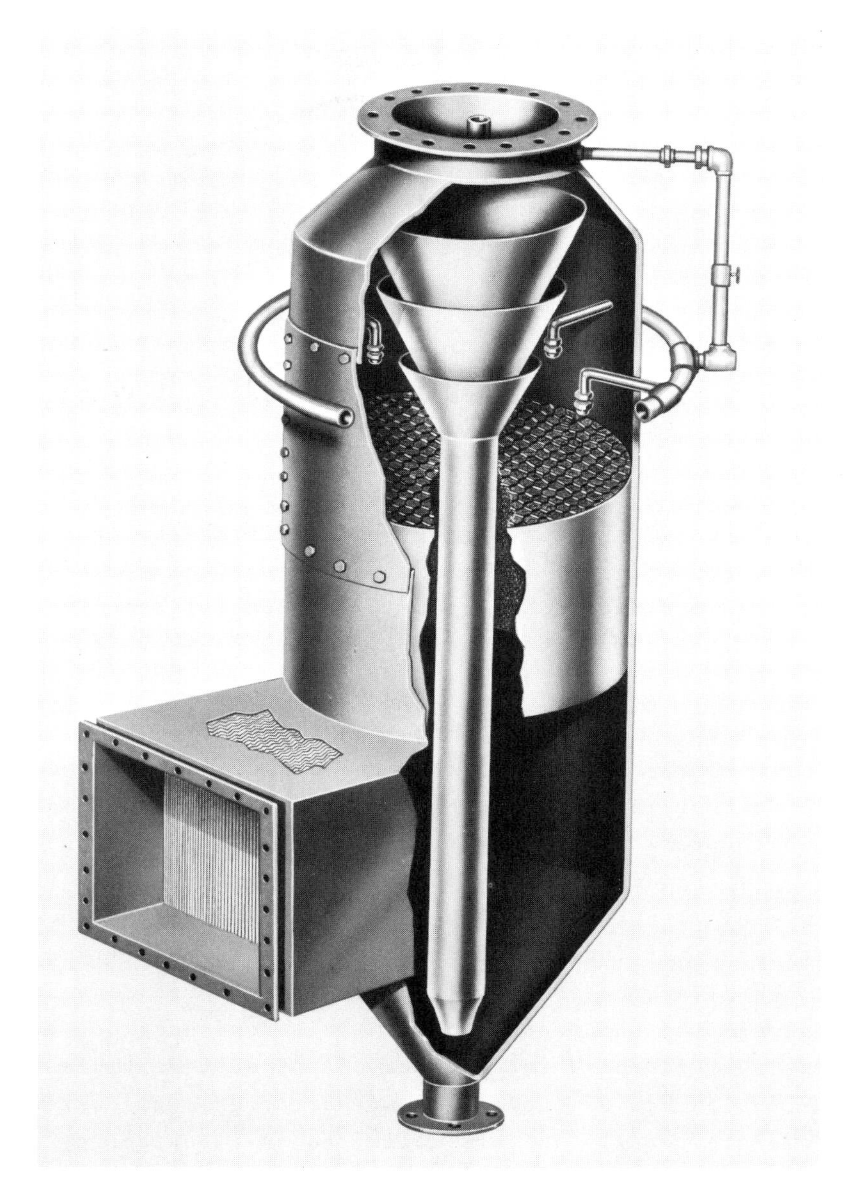

Fig. 21. Hexadyne Wet Scrubber (American Standard).

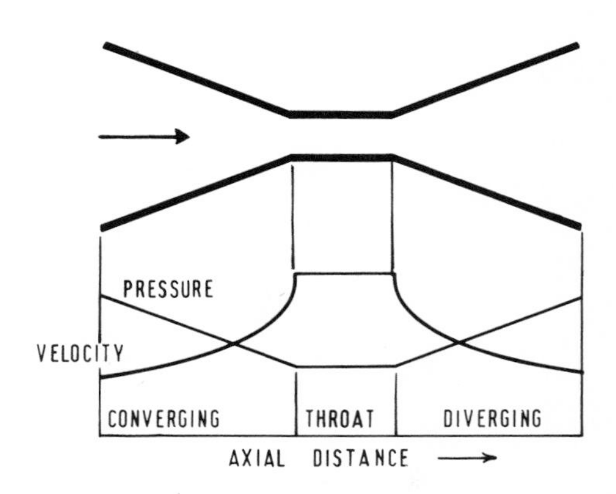

Fig. 22. Pressure-velocity effects in a Venturi scrubber.

the length of the converging section into kinetic energy. The reverse occurs in the diverging section. Frictional and irreversible losses cause a net loss of total energy in the gas stream that shows up as the pressure drop over the venturi.

In the venturi scrubber, the water is fed to the gas stream either at the inlet to the converging section or at the venturi throat. The high degree of turbulence at the throat causes the gas to become totally saturated with water The gas then flows to the diverging section of the Venturi, carrying with it as a very fine spray the excess (over-saturation) water that was fed to it. As the velocity decreases in the diverging section, the pressure increases and condensation of the vapor occurs. This condensation, occurring very rapidly, takes place on the dust particles in the gas, which act as nuclei. Dust particles, thus wetted, agglomerate readily and are collected in a low-energy centrifugal or spray collector.

Venturi scrubbers have high collection efficiencies (see Fig. 23) and correspondingly high pressure drops.

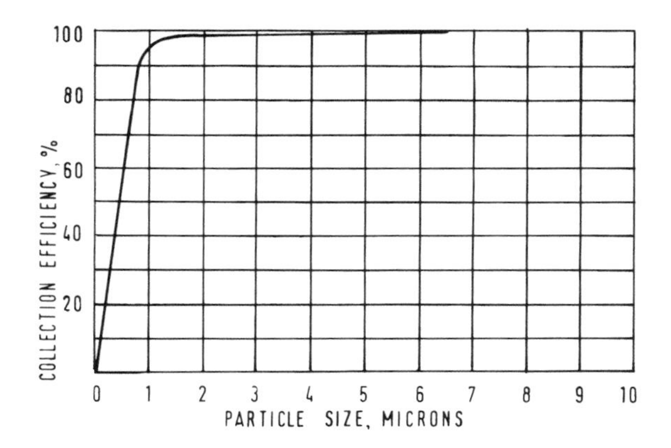

Fig. 23. Grade efficiency curve for a Venturi scrubber; efficiency at 5 μm = 99.6%.

While venturi scrubbers are of simple and compact design, thus having low initial and maintenance costs, they are expensive to operate at the high efficiency/high pressure drop range. Normal operational ranges are as follows: water pressure to venturi throat at 5-25 lb/in^2 and at rates between 2-10 gal/1000 ft^3. Pressure drops fall within the range 5-35 in. water gauge, though values of up to 70 in. water gauge have been quoted for specially designed duties (9, 10). The effects of gas and liquid flows may be seen from the curves of Fig. 24 (38). Throat velocities range from 300-400 ft/sec, though higher velocities are sometimes used for the collection of submicron particles.

The mechanism of particle collection in the venturi scrubber is predominantly by inertial impaction (39, 40). However, for the submicron particle, Brownian movement, and possibly electrostatic forces, contribute to the collection mechanism; it is not really known what is going on at this level.

The work of Nukiyama and Tanasawa (30, 41) on the prediction of droplet diameters [Eqs. (15-17)] has provided

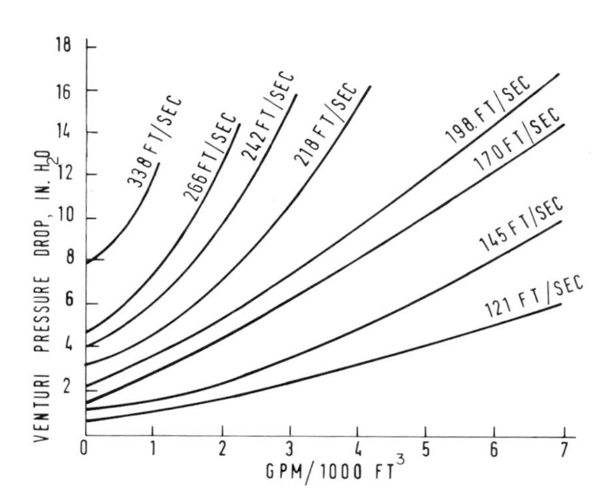

Fig. 24. Venturi pressure drop vs. water rate with throat gas velocity as a parameter (38).

the starting point for theoretical and experimental studies on the mechanism of particle collection in venturi scrubbers (40, 42, 43). The results of several investigators were compared by Boll (39) with the predictions from these equations. It was found that the mean drop sizes ranged from 25-200% of these predictions and thus provided a rough confirmation of these equations.

The existence of an optimum droplet size, following the lines of the discussion in Sec. III, would appear to be likely, and it has been suggested (44) that the spray droplet sizes should match the sizes of the particles to be collected; elsewhere (2), it has been suggested that a droplet size of 10 μm might be a more reasonable optimum. Experimental evidence in support of these or any other optimum has yet to be provided; nevertheless, it is evident that a much smaller droplet size than in spray towers is optimum for the venturi scrubber. Correlations of venturi scrubber efficiencies with specific surface area of spray droplets have been made by Johnstone and Roberts (42).

The prediction of pressure drops in venturis has been studied by Boll (39), who derived a model based on the standard drag correlations and on the Nukiyama and Tanasawa equations for mean droplet size. Agreement with experimental results was good, more so when a uniform liquid distribution was achieved. Laboratory studies on a venturi scrubber were carried out by Lapple and Kamack (45). They observed that the position of the water injection was of secondary importance to good water distribution, which can be achieved with feed points at or upstream of the throat. Moreover, it was noted from an extensive range of experiments on a variety of wet cyclone and venturi scrubber types, that the over-all collection efficiency was primarily dependent on the unit pressure drop and was influenced only marginally by the scrubber design itself.

A. INDUSTRIAL DESIGNS

Two variations of the venturi scrubber are shown in Figs. 25(a) and 26(a). Both designs show a centrifugal scrubber downstream of the venturi contactor. In the flooded wall venturi, shown in Fig. 25(a), the scrubbing liquid is introduced to the inlet of the Venturi converging section and provides a wetted wall flow to the throat, where atomization takes place. Pressure drop control, and hence control of the severity of operation, is provided for in the adjustable damper at the venturi throat. This allows a wide range of operation where the outlet concentration can be controlled by the setting of this damper. It is noted from Fig. 25(b) that the outlet dust concentration, in gr/scf, may be reduced from 0.06 to 0.03 by increasing the pressure drop from 15 to 20 in. of water gauge; increase in the pressure drop further from 20 to 100 in. water gauge is needed to remove the remaining 0.03 gr/scf.

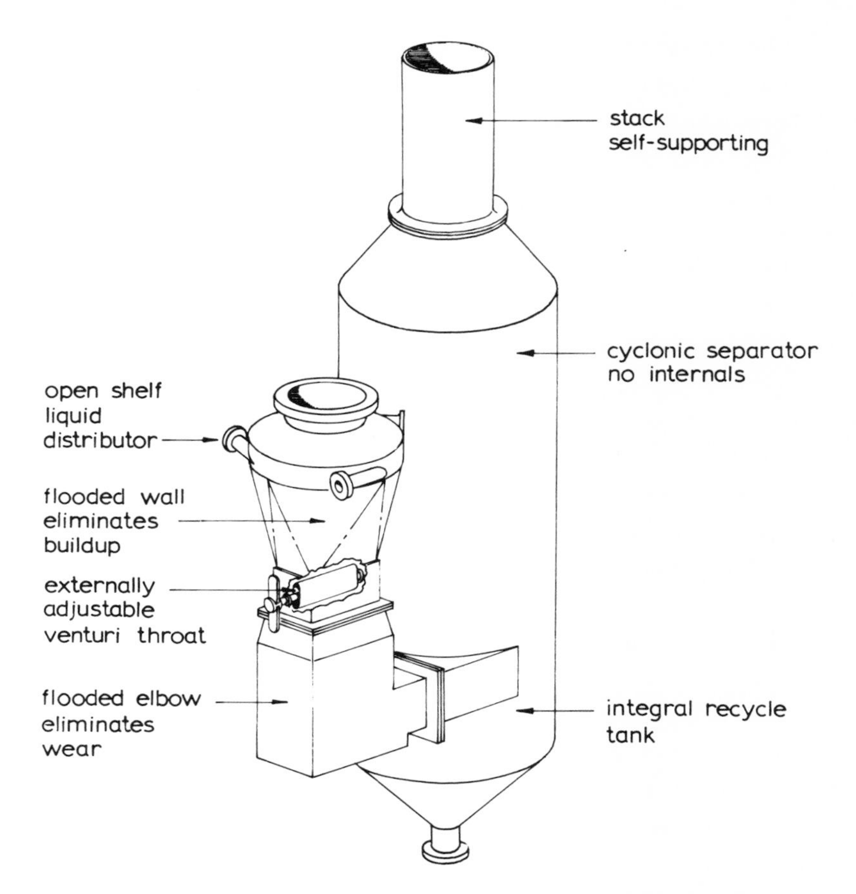

Fig. 25(a). Flooded wall Venturi scrubber (Air Pollution Industries).

It is for this reason that high-severity venturi operation is expensive.

In the PA venturi, shown in Fig. 26(a), the scrubbing liquid is introduced directly through jets at the venturi throat, where it is atomized by the high-velocity gas. Variations in either the gas flow or water injection rate will cause changes in the pressure drop. The pressure drop re-

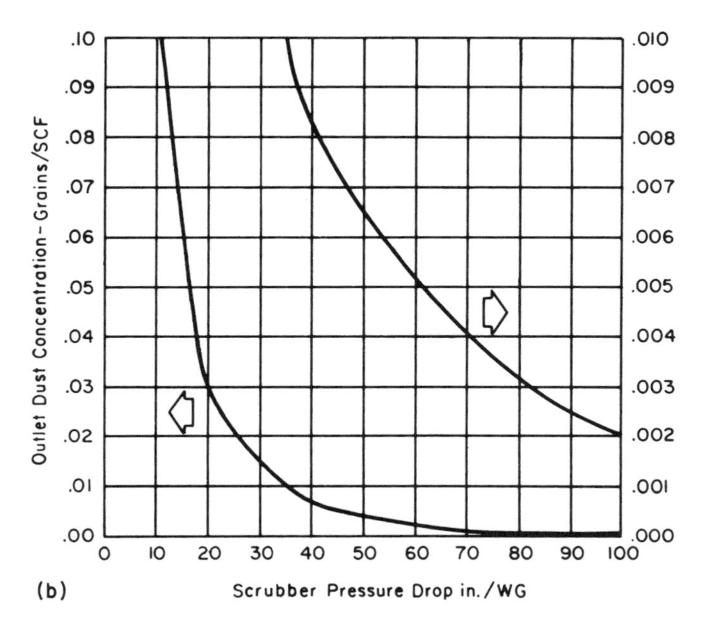

Fig. 25(b). Performance curve for flooded wall Venturi or blast furnace gas (Air Pollution Industries).

quired for efficient scrubbing will vary with the particle size of the specific dust, fume, or mist to be removed. Some gas cleaning problems require considerably less pressure drop than others, depending on the pollutant source and desired efficiency. This effect is shown in Fig. 26(b). The venturi scrubber principle is used in many other forms in commercial designs; some of these are shown in Figs. 27-29.

In Fig. 27, a mobile packed bed of spheres, irrigated by sprays from above, provides for the collection of wetted particles from the outlet of the venturi. An adjustable venturi throat may be automatically controlled to respond to changes in gas or liquid volumes and to maintain a constant pressure drop and collection efficiency.

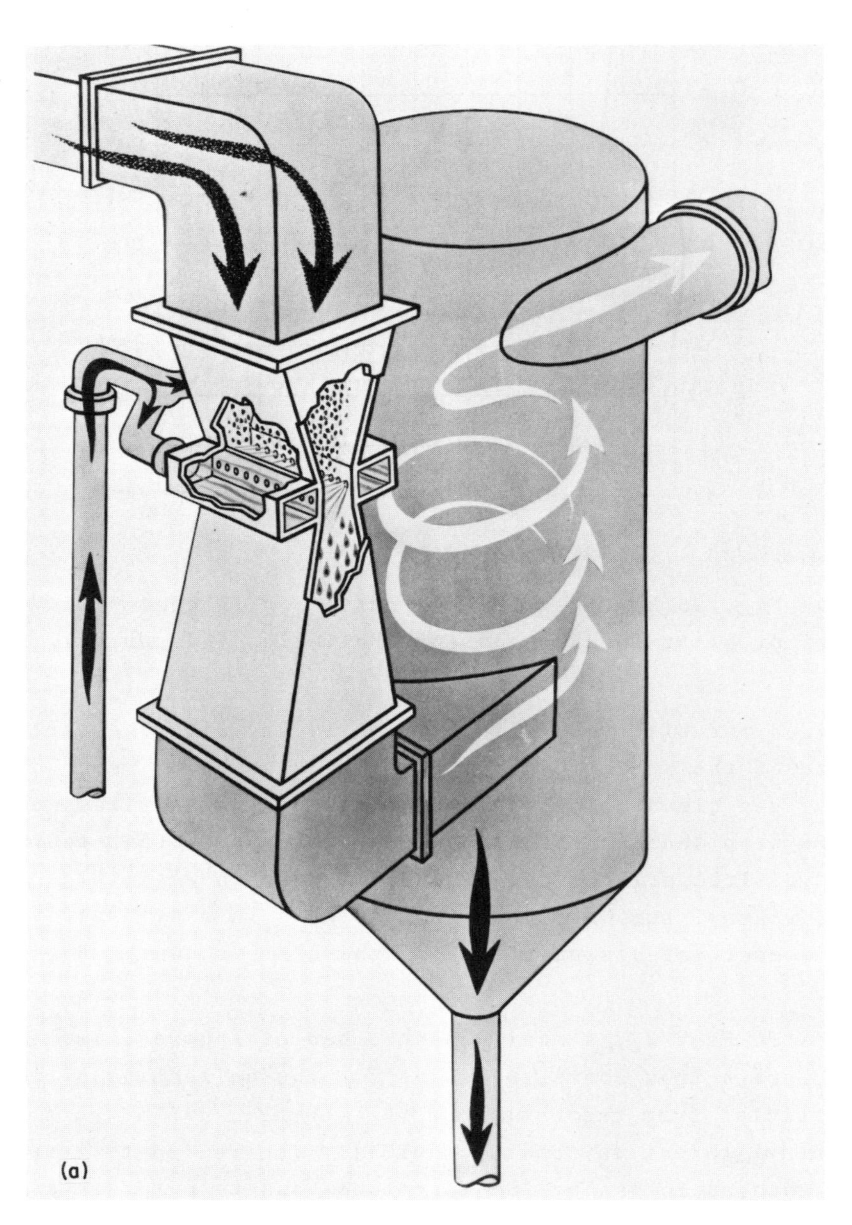

Fig. 26(a). P. A. Venturi scrubber (Chemical Construction Corp.).

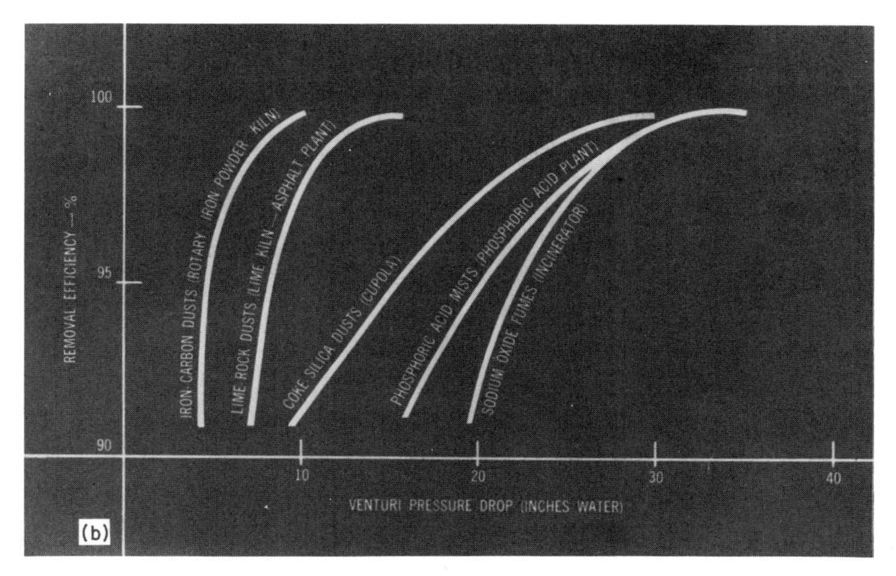

Fig. 26(b). Efficiency pressure drop curves for various dusts, fumes, and mists (Chemical Construction Corp.).

The Venturi principle is used somewhat differently in the design shown in Fig. 28. The adjustable conical baffle permits adjustment to the pressure drop during operation to achieve optimum performance. The Venturi is formed in the annular opening between the conical baffle and the sloped walls of the vessel. Both surfaces are wetted, the cone by a central open pipe, the outer walls by overflow over a serrated weir.

A multipass Venturi scrubber is shown in Fig. 29. Each pass contains a number of Venturi openings through which the gases must pass. Each opening is provided with its own spray and the "captured" particles fall to the scrubber bottom, where a liquid level provides a seal between the passes. The number of passes is determined by the efficiency required. A demister zone follows the last pass.

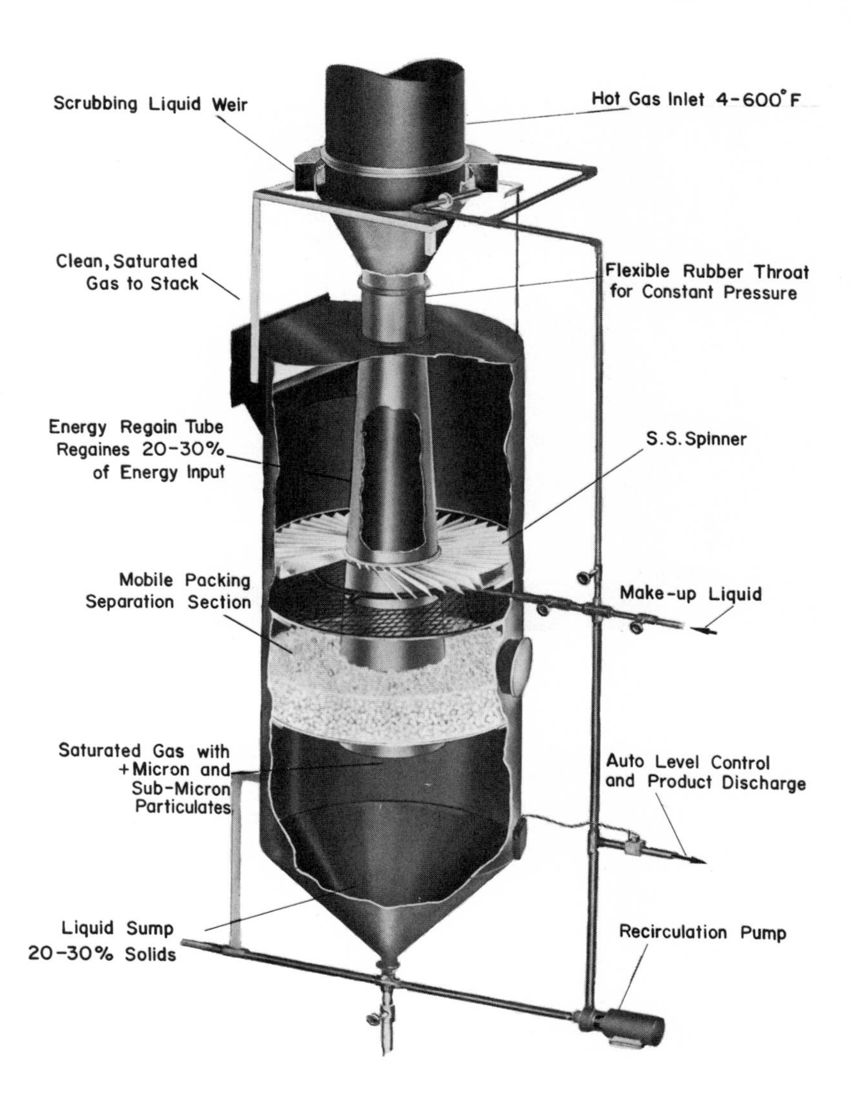

Fig. 27. Ventri-sphere wet scrubber (Air Correction Division of U. O. P.).

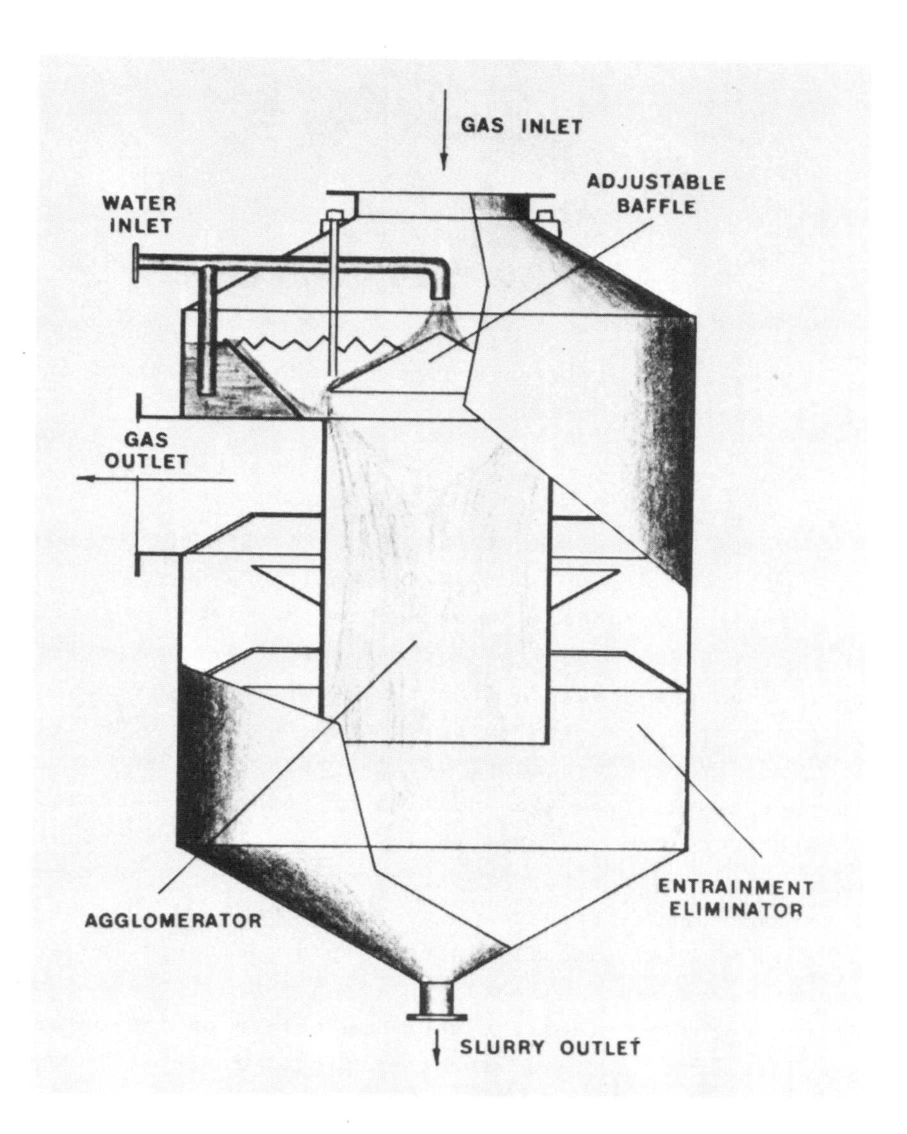

Fig. 28. Stansteel Model D scrubber (Stansteel Corporation).

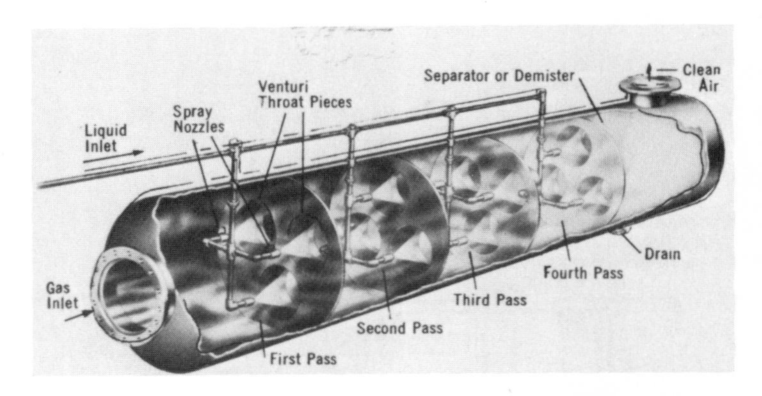

Fig. 29. Multi-Venturi gas scrubber (Koertrol Environmental Control Systems).

IX. MECHANICALLY INDUCED SPRAY SCRUBBERS

For the scrubbers described in the previous sections, the total energy required for the particulate mist or fume collection was obtained from the gas stream in terms of pressure drop over the unit, or from the scrubbing liquid in terms of pressure drop over the spray nozzles, or indeed, in most cases, from the addition of both these sources. Ultimately, of course, this energy is supplied to the blower or pump in the form of electrical energy or as heat energy to steam turbine drives.

In the mechanically induced spray scrubber, additional energy is provided directly to the scrubber, where a motor driven rotor provides the shearing action on the water to generate spray and also the centrifugal force to throw it outward to the container walls for collection. This type of scrubber is relatively small in size, requires very little water, and has performance characteristics comparable to the high-energy Venturi scrubber. Power consumption is high, however, varying from 3-10 hp/1000 cfm, and high maintenance costs due to corrosion are a possibility, depending on the particular duty.

In Fig. 30, detail is given of one of this type of scrubber. It combines a fan, water pump, air distributor, high kinetic energy water spray generator, moisture eliminators, and water solution reservoir. Power is supplied by a single motor mounted on the side of the fan housing and which drives the fan, pump, and spray generator from a central shaft. The gas is drawn into the unit through an inclined rectangular ducting. Baffles direct the gas downward, across the surface of the water solution, and then

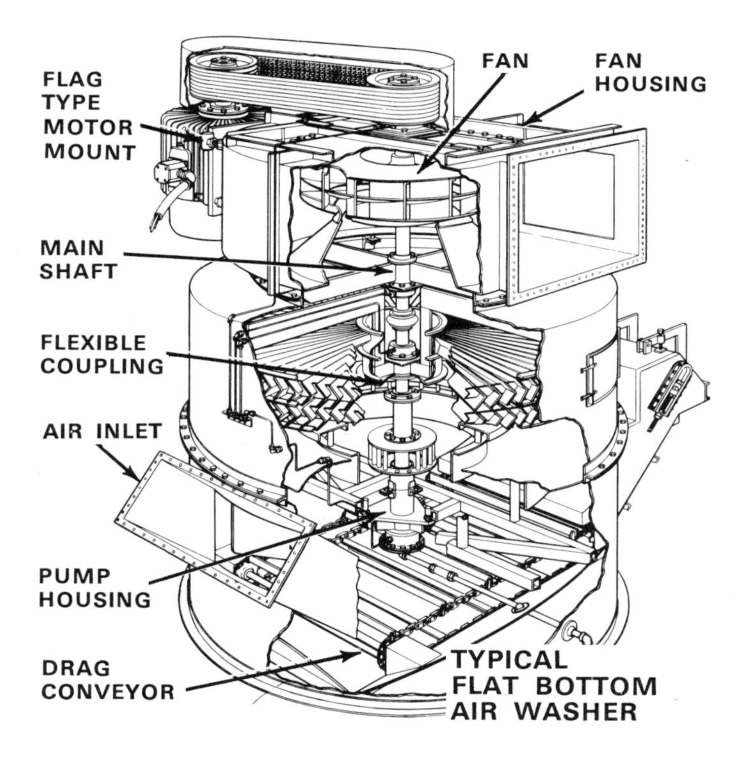

Fig. 30. Centrifugally generated water spray air washer (Centrispray Corp.).

upward into the wash chamber, where it is subjected to a uniform 360° high kinetic energy water spray. Particle collection is by inertial impaction; the captured particles drain back to the water sump below.

The wet scrubber shown in Fig. 31 may be conveniently fitted along the process pipe line to provide "on line" gas cleaning. It is made up of a mixer section, an eliminator section, and a booster fan section. The gas enters the mixer section under suction from the booster fan. Water sprays supply a boundary layer at the impingement element through which the gas passes, effecting the capture of dust particles by the water droplets. In the eliminator section, a helical motion is imparted to the gas, throwing the droplets to the surface of blind louvres, each of which is a separate airtight elimination chamber. The slurry of dust and water drains from these louvres to the sump below, where it flows by gravity to the slurry handling system. The clean air exits via the fan.

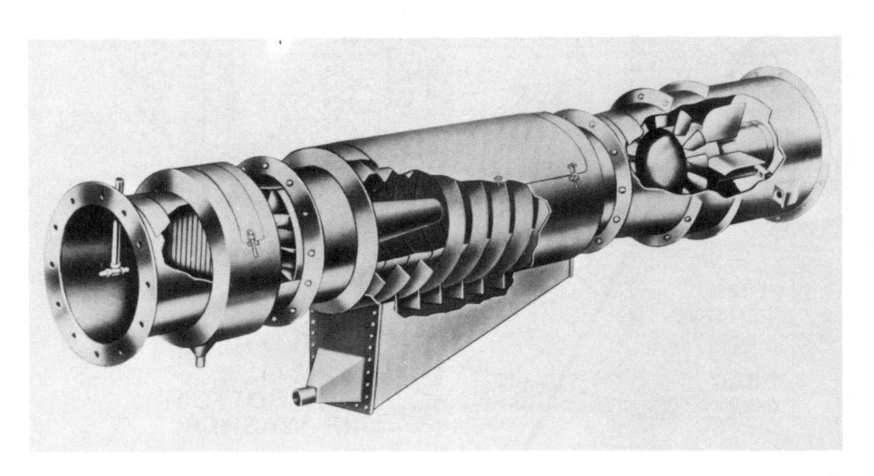

Fig. 31. Joy Microdyne scrubber (Joy Manufacturing Co.).

ACKNOWLEDGMENT

Finally, it should be said that though it was necessary in this chapter to select as examples a very small fraction of the wet gas scrubbers that are commercially available all over the world, it is in no way suggested that these few are the "best" examples of that particular type, nor is the performance data exclusive to that scrubber. The "best" will, of course, vary from district to district and from country to country, since it depends on so many local factors, such as the legal emission limits, the costs of power and water, the location and space available, and even, perhaps, on the direction of the prevailing winds! Anti-pollution equipment is probably the fastest growing technology today and it is just not possible to present an up-to-the-minute comprehensive survey of the equipment available with fair comment on the relative merits of each type. Nonetheless, a survey of the basic designs of modern wet gas scrubber types and their performance characteristics has been attempted in this chapter. The information was provided by the technical and trade publications and by the manufacturers themselves. Sincere thanks and due acknowledgment are here offered to the large number of equipment manufacturers who were most helpful in providing this information.

REFERENCES

1. C. J. Stairmand, Trans. Inst. Chem. Engrs. (London), 29, 356 (1951).
2. C. J. Stairmand, J. Inst. Fuel, 29, 181 (1956).
3. C. J. Stairmand, Chem. Eng. (London), 194, 310 (1956).
4. C. J. Stairmand, Filt. Sep., 7 (Jan. - Feb., 1970).

5. C. J. Stairmand, Report of the Working Party of the European Federation of Chemical Engineering and Directory of Establishments in Europe engaged on Air Pollution Research, 1968.

6. N. F. Imperato, Chem. Eng. (Oct. 14, 1968).

7. K. T. Semrau, J. Air Pollution Control Assoc., 13, 12 (1963).

8. K. E. Lunde and C. E. Lapple, Chem. Eng. Progr., 53, 8 (1957).

9. E. B. Hanf, Environ. Sci. Technol., 4, 2 (1970).

10. Control Techniques for Particulate Air Pollutants, National Air Pollution Control Administration Publication No. AP - 51, 1970.

11. W. Strauss, Industrial Gas Cleaning, Pergamon Press, New York, 1966.

12. K. T. Semrau, J. Air Pollution Control Assoc., 10, 3 (1960).

13. S. Calvert, in Air Pollution (A. C. Stern, ed.), Vol. 3, Academic Press, New York, 1968.

14. W. R. Marshall, in Chem. Eng. Progr. Monograph Ser., 50, 2 (1954).

15. C. N. Davies, in "Symposium on Particle Size Analysis," Trans. Inst. Chem. Engrs. and Soc. Chem. Ind. (Feb., 1947).

16. L. Schiller and A. Naumann, Zeit. Ver. Dent. Ing., 77, 318 (1933).

17. J. M. Coulson and J. F. Richardson, Chemical Engineering, 2nd ed., Vol. 2, Pergamon Press, New York, 1968.

18. C. J. Stairmand, Heat. Vent. Engr., 26 (1953).

19. Air Pollution Manual, Part 2, Control Equipment, Amer. Ind. Hyg., Assoc., 1966, Chap. 6.

20. A. T. Canton, Proc. Eng. Plant Control (March, 1970).

21. A. J. Teller, Chem. Eng. Progr., 63, 3 (1967).

22. I. S. Shah, Chem. Eng. Progr., 64, 9 (1968).

23. T. R. Coulter and R. L. Reveley, J. Technol. Assoc. Pulp Paper Ind., 54, 4 (1971).

24. H. F. Johnstone and R. V. Kleinschmidt, Trans. Am. Inst. Chem. Engrs., 34 (1938).

25. R. V. Kleinschmidt and A. W. Anthony, Jr., in Air Pollution (Louis McCabe, ed.), McGraw Hill, New York, 1952, Chap. 28.

26. C. J. Stairmand, Chem. Eng., 254, 375 (1971).

27. R. C. Specht and R. R. Calaceto, Chem. Eng. Progr., 63, 5 (1967).

28. Bulletin 767-1, Mikropul Ltd., Division of Slick Corp. New York, 1971.

29. H. L. Storch, Chem. Eng. Progr., 62, 4 (1966).

30. S. Nukiyama and Y. Tanasawa, Trans. Soc. Mech. Engrs. (Japan), 4, 14 (1938).

31. H. C. Lewis, D. G. Edwards, M. J. Goglia, R. I. Rice, and L. W. Smith, Ind. Eng. Chem., 40, 67 (1948).

32. J. C. Walling, Pulp Paper,36, 122 (June, 1971).

33. R. Kopita and T. G. Gleason, Chem. Eng. Progr., 64, 1 (1968).

34. J. Louis York, in Air Pollution (Louis McCabe, ed.), McGraw Hill, New York, 1952, Chap. 37.

35. R. E. Treybal, Mass Transfer Operations, 2nd ed., McGraw Hill, New York, 1968.

36. J. H. Perry, Chemical Engineers Handbook, 5th ed., McGraw Hill, New York, 1973.

37. W. E. Lobo, L. Friend, F. Hashmall, and F. A. Zenz, Trans. Am. Inst. Chem. Engrs., 41, 693 (1945).

38. J. F. Byrd and E. L. Dewey, Chem. Eng. Progr., 53, 9 (1957).

39. R. H. Boll, Particle Collection and Pressure Drop in Venturi Scrubbers, paper presented to the 69th National A.I.Ch.E. Meeting, May 1971.

40. F. O. Ekman and H. F. Johnstone, Ind. Eng. Chem., 43, 6 (1951).

41. S. Nukiyama and Y. Tanasawa, Trans. Soc. Mech. Engrs. (Japan), 5, 18 (1939).

42. H. F. Johnstone and M. H. Roberts, Ind. Eng. Chem., 41, 11 (1949).

43. H. F. Johnstone, R. B. Field, and M. C. Tassler, Ind. Eng. Chem., 46, 8 (1954).

44. R. M. G. Boucher, Chaleur Ind., 33 (1952).

45. C. E. Lapple and H. J. Kamack, Chem. Eng. Progr., 51, 3 (1955).

Chapter 6

CONTROL PRACTICE

Larry J. Shannon
A. Eugene Vandegrift
and Paul G. Gorman

Midwest Research Institute
Kansas City, Missouri

I. INTRODUCTION

The only feasible way of alleviating or averting the problems created by the release of particulate pollutants into the atmosphere is to control the pollutants at their source. Control or reduction of particulate pollutant emissions can be achieved by modifications to process equipment and operating procedures or by the installation of control devices on the pollution sources. Factors such as firing rates, raw materials, equipment design and operation, and charging methods can contribute to the nature and quantity of particulate emissions. The degree of reduction in particulate emissions that can be achieved by optimization of process equipment and operating methods is limited, and control devices are generally required in order to comply with pollution codes.

Air pollution codes generally specify allowable particulate emissions from a given source on the basis of mass emission rate (lb/hr), concentration (grains/scf), or plume opacity, or a combination of all three. In those cases where regulations utilize all three criteria, control equipment must be able to meet the most stringent regulation--generally the opacity regulation. Specification of the allowable outlet conditions sets the efficiency for which a control device must be designed, based on given inlet conditions. Currently, control equipment is rated primarily by one parameter--overall-mass efficiency. Efficiency based on over-all mass collection is useful when pollution regulations are based on mass emission rate or concentration. However, visibility reduction and plume opacity are influenced primarily by the 0.1-1.0-micron (μ) radius particles. Efficiency based on particle size collected is preferable in this case. Although, at present, data are too limited to be of value in establishing the efficiency of control devices based on particle size, efficiencies based on particle size will become more important in the future as stricter pollution control regulations are adopted.

In this chapter, air pollution control practices are discussed. The selection of the control device best suited for a specific application depends on the characteristics of the source and effluent, the efficiency required, and the economics of the situation. It is therefore useful to briefly survey control principles with particular emphasis on those control techniques that are not discussed elsewhere in the book. This survey is presented in the next section. Cost considerations are discussed in Sect. III and in the Appendix.

With the increasing emphasis on controlling micron and submicron particulates, a comparison of alternative collection techniques on a mass efficiency basis alone is not sufficient. Fractional efficiency information on each control device is needed to determine performance in the fine particle region. A comparison of control technique performance based on particle size distribution is presented in Sec. IV.

Finally, in Sec. VI, a discussion of actual industrial practice is presented that describes the effectiveness of specific control techniques for selected industrial processes. The processes have been arranged into categories that have similar operations and effluent properties. This arrangement allows extrapolation of control practice to operations and processes not discussed.

II. CONTROL PRINCIPLES

Nearly all industries require some form of air pollution control. Metallurgical operations, the production of cement, lime, fertilizer, and paper, power plants, acid production, and mineral beneficiation represent a few of the major manufacturing operations that require air pollution control equipment. Furthermore, each manufacturing operation has different air pollution problems, and the problems can be solved in several ways. In order to select the best method of reducing pollutant emissions, each solution should be thoroughly evaluated prior to implementation. As noted previously, steps such as substitution of fuels and raw materials and modification or replacement of processes should not be overlooked as possible solutions. If such steps are not feasible, gas-cleaning equipment will be required.

The proper choice of gas-cleaning equipment for any single problem depends on a number of variables. These can be grouped into four general areas: (a) degree of reduction of emissions required to meet emission standards; (b) process and efflent characteristics; (c) equipment capacities and limitations; and (d) capital investment and operating costs. The degree of reduction in emissions required will define the collection efficiency. Important gas stream and particle characteristics of the process include: volume, temperature, moisture content, particle size, density, and explosiveness. High gas temperatures without cooling preclude the use of fabric filters; explosive gas streams prohibit the use of electrostatic precipitators; and submicron particles cannot generally be efficiently removed with mechanical collectors. One of the major process considerations is to determine whether or not product recovery is involved. This is an important factor in deciding whether dry or wet collection should be selected. However, if the pollutants are present in the form of both dusts and gases, then a wet scrubber may be the only logical choice. If a wet scrubber is used, and product recovery is not a factor, then disposal facilities must be made available for the scrubber liquid effluent.

The choice of one suitable control device over another is principally determined by the degree of emission reduction required. Table 1 illustrates the usual ranges of collection efficiency for various equipment alternatives. Each alternative will have a specific cost associated with it, and the components of this cost should be carefully examined. Those alternatives that meet the efficiency, process, and plant facility requirements can then be evaluated in terms of cost. On this basis, the gas-cleaning system may be selected.

A brief description of each of the general types of control devices is presented below to acquaint the reader with each and to point out those industries and processes

TABLE 1

Air Pollution Control Equipment Collection Efficiencies

Equipment type	Efficiency range (on a total weight basis) (%)
Electrostatic precipitator[a]	80-99.5+
Fabric filters[b]	95-99.9
Mechanical collector	50-95
Wet collector	75-99+
Afterburner:	
Catalytic[c]	50-80
Direct flame	95-99

[a] Most electrostatic precipitators sold today are designed for 98-99.5% collection efficiency.

[b] Fabric filter collection efficiency is normally above 99.5%.

[c] Not normally applied in particulate control; has limited use because most particulates poison or desensitize the catalyst.

that utilize each type. A more complete description of each of the devices and treatment of theoretical considerations may be found in the previous chapters, as well as in many of the references given in the bibliographies following each chapter.

A. SETTLING CHAMBERS

The simplest method of removing particles from a moving gas stream is to allow them to settle out under the force of gravity. Large particles will often do so on the floor of a horizontal duct, which functions in this instance as a simple settling chamber, while specially designed chambers will act as more efficient collectors of larger particles. For these particles, particularly if they are abrasive, simple settling chambers are a preferred means of collection because of the low energy requirements as well as the long maintenance-free periods obtained with this type of unit.

For the collection of large particles the simple settling chambers have a number of advantages and disadvantages. The advantages are:

1. Simple construction.
2. Low initial cost and maintenance.
3. Low pressure losses and no temperature and pressure limitations, except those of materials of construction.
4. Dry disposal of collected materials.
5. No problems with abrasive materials.

The major disadvantage for the horizontal flow unit is the very large space requirement of the chamber.

B. CYCLONES

Cyclonic collectors are generally round, conically shaped vessels in which the gas stream enters tangentially and follows a spiral path to the outlet. The spiral motion produces the centrifugal forces that cause the particulate matter to move toward the periphery of the vessel, collect on the walls, and then fall to the bottom of the vessel.

Centrifugal force is the major force causing separation of the particulate in a cyclone separator. This force (F_c) is equal to the product of the particulate mass (M_p) and centrifugal acceleration (V_p^2/R), where V_p is the particle velocity and R is the radius of motion (curvature).

$$F_c = (M_p) \frac{V_p^2}{R} \tag{1}$$

The centrifugal forces cause the particles to move outward toward the wall of the cyclone. However, this movement of the particle through the gas stream is opposed by frictional drag on the particle caused by the relative motion of the particle and gas.

The centrifugal and frictional forces, plus the force of gravity, combine to determine the collection efficiency. This collection efficiency increases with:

1. Dust particle size.
2. Particle density.
3. Gas velocity.
4. Cyclone body length.
5. Smoothness of cyclone wall.

Although efficiency increases with increasing gas velocity, it does so at a lower rate than that at which the pressure drop increases. For a given cyclone and dust combination, an optimum velocity exists, beyond which turbulence increases more rapidly than separation efficiency, and efficiency decreases (1).

The cyclonic collectors are generally of two types: the large-diameter, lower-efficiency cyclones, and the smaller-diameter, multitube high-efficiency units. The larger cyclones have lower efficiencies, especially on particles less than about 5 μ. However, they have low initial cost and usually operate at pressure drops of 1-3 in. of water. The multitube cyclones are capable of efficiencies exceeding 90%, but the cost is higher and their pressure drop is usually 3-5 in. of water. They are also more susceptible to plugging and erosion.

C. WET SCRUBBERS

Wet collectors use water "sprays" to collect and remove particulate matter. There are many variations of wet collectors but they may generally be classified as low- or high-energy scrubbers. Low-energy scrubbers of 1-6 in. of pressure drop may consist of simple spray towers, packed towers, or impingement plate towers. Water requirements may run 3-6 gal/1000 ft^3 of gas and collection efficiencies can exceed 90-95%.

The high-energy scrubber, or Venturi, imparts high velocity to the gas stream by means of a converging-diverging duct section, and contacts the gas stream with injected water. The high velocities provide increased collection efficiency, up to 99.5% but the pressure drop may range from 10-60 in. of water, requiring a draft fan with high-power input.

Another wet scrubbing device is the disintegration scrubber, in which the atomization of the liquid is caused by a rapidly moving rotor. The water is injected axially and separated into fine droplets by the rotor, while the relative velocity of the gas is maintained at 200-300 ft/sec through the system. It is normal practice to preclean the gases to a concentration below 1 grain/ft^3 before passing them to the disintegrator to minimize erosion of the rotor vanes. Power consumption is high--of the order of 16-20 hp/1000 ft^3/min of gas treated. The scrubber is very effective, removing 90% of the 1-μ particles present and 70% of the 0.5-μ particles (2).

Wet scrubbers can provide high collection efficiency but their use may involve treatment of liquid wastes with settling ponds. They also saturate the gas stream and may produce a visible steam plume.

D. ELECTROSTATIC PRECIPITATORS

The operating principle of electrostatic precipitation requires three basic steps:

1. Electrical charging of the suspended particulate matter.
2. Collection of the charged particulate matter on a grounded surface.
3. Removal of the particulate matter form the collecting surfaces by mechanical vibration (rapping) or flushing with liquids.

The electrical charging is accomplished by passing the suspended particles through a high-voltage, direct-current corona. Gas velocities range from 3-15 ft/sec. This low linear velocity promotes deposition and minimizes re-entrainment. However, this also means that the precipitators will be large in size, or cross-sectional area, to achieve the low gas velocities. Uniform flow distribution is also an important factor that must be considered in design of ductwork.

The collection efficiency of electrostatic precipitators is expressed by the Deutsch equation as:

$$E = 1 - e^{-wA/Q} \tag{2}$$

where E is the weight fraction of dust collected; w is the migration velocity of dust particle toward the collecting electrode, ft/sec; A is the area of collecting electrode, ft^2; and Q is the gas flow rate, acf/sec.

This equation shows the exponential relationship of efficiency to the area of the collecting electrode. Thus moderate increases in collector efficiency for an existing unit may require a rather large increase in collecting surfaces.

The proper operation of an electrostatic precipitator is dependent on the electrical resistivity of the particles. Preconditioning with water sprays may be required to impart beneficial resistivity character to the particles. Proper control of operating voltages must also be provided if efficient particulate removal is to be maintained. The precipitator generally has high initial cost but it is capable of high collection efficiency, exceeding 99%, at a pressure drop less than 0.5 in. of water.

E. FABRIC FILTERS

Fabric filter systems, i.e., baghouses, usually consist of tubular bags made of woven synthetic fabric or fiberglass, in which the dirty gases pass through the fabric while the particles are collected on the upstream side by the filtering action of the fabric. The dust retained on the bags is periodically shaken or blown off and falls into a collecting hopper for removal.

Fabric filters usually provide very high collection efficiencies, exceeding 99.5%, at pressure drops usually ranging from 4-6 in. of water. The maximum operating temperature for a baghouse is 550°F using fiberglass bags. However, there may also be a minimum temperature limitation so as to maintain the gas temperature at 50-75°F above the dew point. Inlet dust loadings may range from 0.1-10.0 grains/ft^3 of gas. Higher concentrations in some industries are removed by a precleaning device, such as a low-efficiency cyclone.

Baghouses do provide high collection efficiency at moderate pressure drop, but initial cost is relatively high, especially when precooling systems are required.

Baghouses also may be large and take up considerable space, and they frequently entail high maintenance costs for bag replacement. However, replacement of bags need not impair baghouse operation if the unit is compartmented so that one section can be taken out of service for maintenance while the others continue to operate.

F. MIST ELIMINATORS

One of the most commonly used types of mist eliminators is the mesh filter, which consists of an evenly spaced knitted wire or plastic mesh, usually mounted in a horizontal bed. Rising mist droplets strike the wire surface, flow down the wire to a wire junction, coalesce, and flow to the bottom surface of the bed, where the liquid disengages in the form of large droplets and returns by gravity to the process equipment (1).

Operating pressure drop is usually less than 1 in. of water with gas velocities of 10-15 ft/sec. Advantages in the use of this type of collector are low initial cost, low maintenance, high removal efficiency, and recovery of valuable products without dilution. However, these units should not be used in services where the material can cause plugging of the mesh unless provision is made for flushing out accumulated solids.

Another type of mist eliminator consists of packed beds of fibers. These may operate at velocities ranging from 5-90 ft/sec and, therefore, have correspondingly higher pressure drops of 5-15 in. of water. Collection efficiencies may be in excess of 99% on liquid particles. However, this type of collector may not be suitable for solid particles unless the grain loading is very low. One such application is the recovery of platinum in nitric acid plants.

Other types of mist eliminators are impingement baffle mist eliminators and packed bed mist eliminators, which may not achieve efficiencies as high as those discussed above, but do prevent loss or carryover of larger droplets.

G. GRANULAR FILTERS

A granular gas filter is a stationary device consisting of a bed of separate, relatively close-packed granules that make up the collection surfaces for the filter. Fixed bed or close-packed moving bed devices are included in this definition, but fluid or dispersed beds, where the granular particles are kept in motion by the gas being treated, are excluded. In order to prevent the collected particulate matter from plugging the interstices formed between the granules and thus causing excessive pressure drops, granular filters embody some means for either periodic or continuous removal of particulates from the collecting surfaces.

Among the earliest of granular filter tests were those carried out in 1948-1949 at the General Electric Company, Hanford Works (3, 4). Sand bed filters were used to remove radioactive particles from ventilation system air. On the basis of preliminary tests, large-scale sand filters were designed and set up at Hanford with a capacity of 30,000 cfm each. The design face velocity used was 6 ft/min, which means that the beds covered an area of 5,000 ft^2. The filtering layer consisted of a 24 in. depth of 20-40 mesh Hanford sand. Seven layers of coarser sand were also used in the bed to aid in flow distribution. Over-all pressure drops were in the range of 4-7 in. of water, and an average collection efficiency for these large units was determined to be about 99.7%. The potential lifetime of these sand beds was estimated at about 5 years, based on estimated accumulation of particulates.

Although granular filters appear to have merit as gas-cleaning devices, they have never come into general use as a preferred mode of industrial gas cleaning. The majority of such devices that have been proposed are found in the patent literature and, with some exceptions, little or no work has been directed toward development or implementation of the patented designs on an industrial scale.

H. AFTERBURNERS

Afterburners are gas-cleaning devices that use a furnace for the combustion of gaseous and particulate matter. Combustion is accomplished either by direct-flame incineration or by catalytic combustion.

The disposal of particulate matter by combustion is limited to residue-free vapors, mists, and particulate matter that are readily combustible, as well as to particle sizes that require short furnace retention time and small furnace size. Afterburners are normally used to dispose of fumes, vapors, and odors when relatively small volumes of gases and low concentrations of particulate matter are involved.

Advantages of the direct-flame incineration afterburner include: (a) high removal efficiency of submicron odor-causing particulate matter; (b) simultaneous disposal of combustible gaseous and particulate matter; (c) compatibility with existing combustion equipment; (d) relatively small space requirements; (e) simple construction; and (f) low maintenance.

Disadvantages include: (a) high operational costs, including fuel and instrumentation; (b) fire hazards; and (c) excessive weight.

Advantages of the catalytic afterburner include: (a) reduced fuel requirements; and (b) reduced temperature, insulation requirements, and fire hazards.

Disadvantages of catalytic afterburners include: (a) high initial cost; (b) sensitivity to catalytic poisoning; (c) inorganic particles must be removed and organic droplets must be vaporized before combustion to prevent damage and plugging of the catalyst; and (d) catalysts may require frequent reactivation.

III. CONTROL EQUIPMENT COSTS

The actual cost of installing and operating air pollution control equipment is a function of many direct and indirect factors. The principal costs of concern to the control equipment user are those directly associated with the capital investment, installation, and operation of control devices.

Capital investment costs can include the cost of: (a) land acquisition; (b) engineering design; (c) control equipment; (d) auxiliary equipment and replacement parts; (e) structure modification; (f) installation; and (g) startup. Items a, b, e, f, and g are highly variable, depending upon geographical location, company practice, and the specific plant in which equipment is installed. For example, companies with a central engineering staff may do their own system design and installation and purchase only the basic control equipment hardware. Other companies may option for a complete turn-key contract.

Table 2 illustrates the typical ranges of purchase costs for various types of control devices. These purchase costs include built-in instrumentation and pumps. The purchase cost usually varies with the size and coll-

TABLE 2

Range Of Typical Purchase Costs For Air Pollution Control Equipment

Type	Cost ($/acfm)
Dry Centrifugal (Simple and Multiple)	0.15-0.80
Electrostatic Precipitator (High Efficiency)	0.80-1.25
Fabric Filter	1.00-1.50
Wet Scrubber	0.25-0.75
Afterburners:	
Direct flame	1.0-3.0
Catalytic	1.0-5.0
Mist eliminator	0.2-3.0

ection efficiency of the control device, and also must be brought to a common base year (see Appendix).

Installation costs depend upon factors such as: (a) plant age; (b) transportation; (c) space limitations; (d) degree of preassembly; (e) special equipment required for installation (e.g., cranes, helicopters); and (f) labor rates or union contract requirements. Table 3 presents the range of installation costs that have been reported in a recent EPA publication (1). Data obtained by the authors on recent EPA funded programs support the figures given in Table 3 (5, 6).

Maintenance and operating costs are difficult to define and assess because of the different accounting and maintenance procedures of control equipment users. Maintenance cost is the expenditure required to sustain the operation of a control device at its designed efficiency with a scheduled maintenance program and necessary replacement of any defective parts. Table 4 shows annual maintenance cost factors for several types of particulate control devices. Simple, low-efficiency control devices have low maintenance costs; complex, high-efficiency devices have high maintenance costs.

Annual operating cost is the expense of operating a control device at its designed collection efficiency. This cost depends on the following factors: (a) the gas volume cleaned; (b) the pressure drop across the system; (c) the operating time; (d) the consumption and cost of electricity; (e) the mechanical efficiency of the fan; and (f) the scrubbing-liquor consumption and costs (where applicable).

The effect of the variations in the above cost components is to make single point estimates of control equipment cost for a specific case very difficult. The appendix presents more detailed cost information for control equipment. The cost data are given for various types of equipment as a function of operating capacity. The cost elements and other parameters used in developing the cost data are discussed in detail.

TABLE 3

Installation Cost For Various Types Of Control Devices Expressed As A Percentage Of Purchase Costs (1)

Control device	Cost (%)			
	Low	Typical	High	Extreme high
Gravitational	33	67	100	--
Dry centrifugal	35	50	100	400
Electrostatic precipitators	40	70	100	400
Fabric filters	50	75	100	400
Wet collector:				
Low, medium energy	50	100	200	400
High energy[a]	100	200	400	500
Afterburners	10	25	100	400

[a] High-energy wet collectors usually require more expensive fans and motors.

TABLE 4

Annual Maintenance Costs For All Generic Types Of Control Devices (1, 5, 6)

Control device	Dollars per ACFM		
	Low	Typical	High
Gravitational and dry centrifugal collectors	0.005	0.015	0.025
Electrostatic precipitators:			
High voltage	0.01	0.02	0.03
Low voltage[a]	0.005	0.014	0.02
Fabric filters	0.02	0.05	0.10
Wet collectors	0.02	0.04	0.06
Afterburners:			
Direct flame	0.03[b]	0.06[c]	0.10[c]
Catalytic	0.07	0.20	0.35

[a] Low-voltage precipitators are used for air conditioning or industrial applications involving collection of liquid particles that will drain from the collector plates.

[b] Metal liner with outside insulation.

[c] Refractory lined.

IV. COMPARISON OF CONTROL DEVICE PERFORMANCE

The collection efficiency of control devices is one of the most meaningful parameters to use in comparing the performance of air pollution control devices. Current control equipment performance is usually related to over-all mass efficiency. However, specification of control equipment efficiency in terms of penetration or fractional efficiency in specific particle size ranges is a more revealing method for rating control equipment performance. The fractional efficiency (on a mass basis) of a control device is defined as the efficiency of removal of specific sizes of particles. For example, if a stream contains 10 lb of 1 μm particles, and a control device removes 9 lb of this material, the fractional efficiency of the device at 1 μm is 90%. Penetration of a specific particle size pollutant through a control device is defined as

$$[P]_{d_1} = 1-[EFF]_{d_1} \tag{3}$$

where $[P]_{d_1}$ is the penetration at diameter d_1 and $[EFF]_{d_1}$ is the fractional efficiency of the control device at particle size of d_1.

As part of a broad-based study on particulate pollution from stationary sources, the authors compiled and analyzed available data on the fractional efficiency characteristics of the major types of control equipment (7). The fractional efficiency data for each type of control device varied over a wide range, since the over-all efficiency and design of the devices varied over a wide range. Also, available data on fractional efficiency of control devices apply to specific industrial sources in only a few cases. Many of the data reported in the literature were obtained from laboratory studies or pilot-scale units, and tests were often conducted using specific test dusts or monodisperse aerosols. In many cases the published data were only identified as "typical" for a certain type of control device.

Since a wide variation existed in the data for each type of device, the data were examined with the objective of drawing general curves that would represent low, medium, and high over-all efficiencies for each type of control device. The data for each type of control device and the information regarding specific designs, operating conditions, and testing procedures were carefully assessed to determine the general fractional efficiency curves for each type of control device that best represented low, medium, and high over-all efficiencies. The resulting general fractional efficiency curves are shown in Fig. 1.

As noted previously, the fractional efficiency of each class of control device varied over a wide range. However, as shown in Fig. 1, the data do reflect the expected efficiency characteristics for specific classes of devices. That is, multiclones exhibit a higher efficiency curve than cyclones, and Venturi scrubbers show a higher efficiency than low-pressure drop scrubbers. It is also readily apparent that high-efficiency electrostatic precipitators and fabric filters are superior collection devices, especially in the smaller particle size ranges.

V. INDUSTRIAL APPLICATIONS OF CONTROL DEVICES

Selected applications of the major categories of gas-cleaning devices are discussed in this section. To facilitate discussion of control device applications, stationary particulate pollution sources selected for discussion

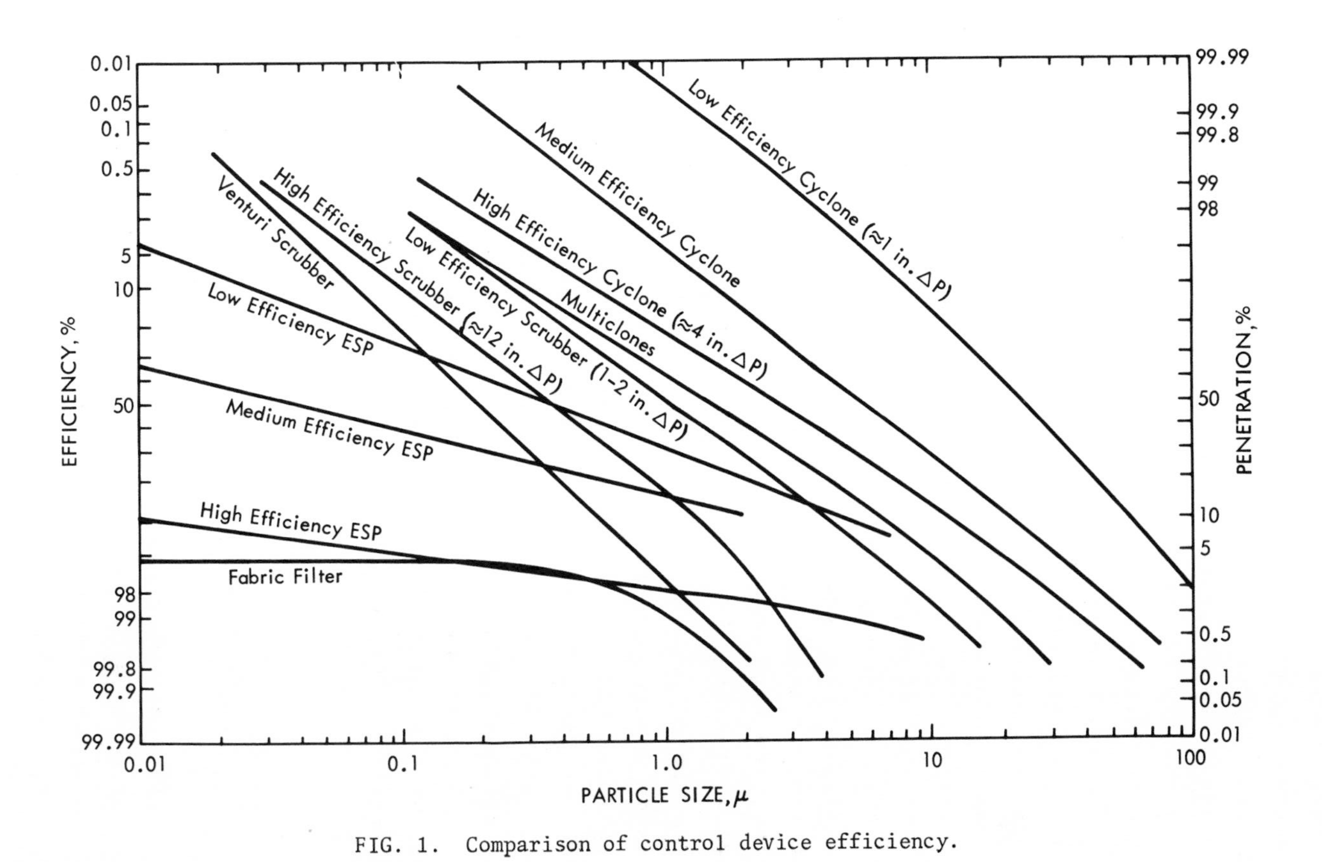

FIG. 1. Comparison of control device efficiency.

are divided into the broad categories of combustion, metallurgical, mineral refining, and chemical processes.

A. COMBUSTION PROCESSES

More than 29 million stationary combustion sources are currently in operation in the United States: (a) electric-utility power-generating plants; (b) industrial power plants and process heaters; and (c) space-heating units. Coal, oil, and gas are burned in a wide variety of equipment (1).

Particulate emissions vary widely from unit to unit because processes, practices, and fuels all affect emission levels. For each fuel, several different processes are used for atationary combustion. Coal-fired electric generating plants utilize pulverized, cyclone, and stoker-fired boilers. Burners, combustion chambers, draft systems, heat transfer characteristics, and combustion controls of industrial units may vary widely. Steam, hot water, and warm air furnaces are in common use for domestic heating.

Stationary combustion sources are divided into electric utility, industrial, and commercial and residential groups for a more detailed discussion in the following sections.

1. Electric Utilities

The production of power by the combustion of coal, fuel oil, and gas contributes large quantities of particulates, sulfur oxides, and nitrogen oxides to the atmosphere. Coal-fired units are the dominant source of particulate emissions. Particulate emissions from oil-fired power boilers are about 1% of the emissions from coal-fired equipment. Use of natural gas as a fuel nearly eliminates particulate emissions. Since emissions depend so strongly on the type of fuel burned, discussion of control practices is presented for each fuel type in succeeding paragraphs.

a. Coal-Fired: Electrostatic precipitators, cyclones, settling chambers, and wet scrubbers have been used to control particulate emissions from coal-fired power plants. A comparison of the cost of various types and combinations of control equipment is given in Table 5 (8).

Electrostatic precipitators are used extensively in power plants. A recent survey shows that in recent years high-efficiency electrostatic precipitators are being installed almost exclusively (9). Recent trends toward the use of low-sulfur coal may require increasing the capacity of the electrostatic precipitators. A reduction in the sulfur content of the coal is apt to increase fly-ash resistivity, thereby reducing the efficiency of existing electrostatic precipitators (8).

Multiple small-diameter cyclones are used on mechanical draft combustion units either as precleaners for electrostatic precipitators or as final cleaners. Efficiencies of well-designed units range from 90% for some stoker-fired units to 60% for coal-fired cyclone furnaces.

Wet scrubbers are used to a limited extent in the stacks of coal-fired units to control particulate emissions during soot blowing. The problems that limit the use of wet scrubbing include high corrosion rates, high or fluctuating pressure drops, adverse effects on stack gas dispersion, and waste disposal.

A turbulent contact absorber system has undergone testing for absorption of SO_2 in alkaline solution and simultaneous removal of fly ash for a coal-burning power plant. With this system, fly ash collection efficiencies of 98% and over-all SO_2 removal of 91% can be expected at wet scrubber pressure drops of about 4.5 in. water gauge.

TABLE 5

Air-Cleaning Equipment Installed Cost
Based on 1000 MW Unit (1968) (8)

Furnished, Delivered, and Erected
(Supports and Flues Not Included)

Unit Cost	Mechanical (cyclone) (75%)	Electro-static precip. (95%)	Electro-static precip. (99%)	Comb. mech. electro. (95%)	Comb. mech. electro. (99%)	Comb. electro-mech. (99%)	Bag filter
\$/MW at 300°F	730	3,330	5,200	3,830	4,470	6,160	7,650
\$/lb steam/hr at 300°F	0.11	0.48	0.76	0.56	0.69	0.90	1.12
\$/ton coal/hr at 300°F	2,300	10,400	16,200	12,000	14,800	19,300	23,800
\$/CFM flue gas at 300°F	0.25	1.00	1.50	1.15	1.40	1.80	2.25
\$/CFM flue gas at 700°F	0.20	0.85	1.35	1.00	1.15	1.55	--

For a generating capacity of 25 MW equivalent to 100,000 cfm of flue gases at 300°F, the investment and operating costs are approximately \$10/kW and \$1.17/ton of coal, respectively (10). More recent EPA data in the March 21, 1972 Federal Register indicates somewhat higher costs--\$18.7 to \$25.7/kW (1971 dollars), for new plants with calcium-base scrubbing systems. Retrofitting of existing plants shows greater costs, ranging from \$28.6 to \$61.8/kW.

Two limestone injection SO_2 removal systems, designed by Combustion Engineering, have been installed on operating boiler units. This design incorporates a marble-bed wet scrubber having <6 in. of pressure drop and is guaranteed to remove 83% of the SO_2 and 99% of the particulate matter (11).

Fabric filter systems have not been used on coal-fired power plants operated by electric utilities to date. However, fabric filter installations are currently being tested with hopes of future application. (12)

b. Oil-Fired: The only air pollution control devices that have found ready acceptance on oil-fired power plant boilers are dust collectors used to control particulates during soot blowing. This equipment serves principally to collect particulate matter larger than 10 μm. Small-diameter multiple cyclones are the most common soot-control devices installed. The emission from an oil-fired unit without special collection equipment is comparable to a coal-fired unit equipped with a control device of better than 99% collection efficiency (12).

2. Industrial Power Generation

Large industrial plants may generate their own electric power or process steam by the combustion of coal, fuel oil, and gas. Emissions from these combustion units parallel those from electric utilities but are lower in total quantity because of the smaller quantity of fuel burned.

a. Coal-Fired: Control equipment for coal-fired industrial boilers is very similar to that employed by electric utilities. Cyclones, multiclones and electrostatic precipitators are used in both groups, although the use of electrostatic precipitators has not been as prevalent for industrial boilers.

The increased emphasis on reducing air pollution has apparently prompted many of the industrial coal-fired boiler operators to switch from coal to oil or gas. This route to control may be a simpler matter for the industrial operator than for the electric utility boilers, which consume larger quantities of fuel.

b. Oil-Fired: The use of control devices on oil-fired units is usually limited to periods when soot-blowing operations are in progress or in areas where restrictive legislation requires low particulate loadings and low opacity of stack effluents. Multiple cyclones and electrostatic precipitators have been used for these purposes (13).

c. Gas-Fired: Control equipment has not been necessary on natural gas-fired units.

3. Commercial, Institutional, and Residential Furnaces

Commercial, institutional, and residential furnaces are primarily used for space heating. The general character of emissions from these sources parallels that of electric utility and industrial sources, with allowance made for differences in furnace types and operating procedures. In general, combustion efficiency will be lower in these units and outlet grain loadings higher as a result.

Control equipment is not generally used on these small furnaces. Reference 14 discusses the use of various combustion-improving devices to reduce emissions of smoke, carbon monoxide, nitrogen oxides, and total gaseous hydrocarbons from high-pressure atomizing-gun burners used in domestic oil-fired furnaces. The devices (primarily improved nozzles) were designed to improve combustion in older inefficient furnaces.

B. METALLURGICAL PROCESSES

Metallurgical processes contain some of the most difficult sources to control because effluent streams are characterized by metallic fumes and high volumetric flowrates. In many cases, the particulates are corrosive, sticky, and have high angles of repose. In addition to the high volumetric flowrates, the carrier-gases are at high temperatures and may be corrosive. Control practices for the iron and steel industry, iron foundries, and primary nonferrous metallurgy are reviewed in the following sections.

1. Iron and Steel Industry

The manufacture of iron and steel involves many different processes. Some of the processes produce large quantities of particulates and gaseous emissions, while other processes are relatively free of air pollution problems. The pyrometallurgical processes inherent in the iron and steel manufacturing processes emit a very fine particulate material.

The major sources of particulate air pollution are: sintering plants, coke-oven plants, blast-furnace operations, steel-making furnaces (especially those using large amounts of oxygen for steel making), and materials-handling operations.

The operation and types of furnaces associated with the iron and steel industry vary widely. Therefore the criteria for the type of control equipment are determined by the specific operation. A tabulation of the types of equipment used in the various processes is given in Table 6 (15).

The cost of the different types of control equipment for these applications has been reported in a NAPCA document, "A Cost Analysis of Air Pollution Controls in Integrated Iron and Steel Industry." Section V of that report presents the cost/effectiveness investigation and the cost models that were developed (16).

The applied control systems are more thoroughly discussed in a companion document, "A Systems Analysis Study of the Integrated Iron and Steel Industry" (15). Excerpts of Sec. VI of that report are included herein.

a. Sinter Machines. Sintering machines generally accept and process a wide variety of feeds and produce a considerable quantity of emissions of variable quantity and nature (15). Emissions from sinter plants vary widely in quantity and particle size and depend upon, among other things, raw mix composition, types of machines, and exhaust-fan characteristics.

Sinter machine emission sources can be divided into two categories: (a) windbox emissions, and (b) discharge-end emissions. Windbox emission dusts are mainly generated early in the sintering process, and again later, when the flame front has reached the bottom of the bed. Emissions from the discharge end parallel those of the windbox on the basis of pounds emitted per ton of sinter. However, grain loadings are higher at the discharge end.

Cyclones, electrostatic precipitators, Venturi scrubbers, and baghouses have been used in various combinations at the various points of emission. However, dry-type cleaners are best suited for this application because the sulfur content of the gas streams can lead to corrosion problems

TABLE 6

Representative Distribution Of The Type And Number Of Emission-Control Devices Installed In Emission-Control Applications In The Integrated Iron And Steel Industry (15)

Iron-or steel-making segment	Type of emission-control equipment			
	Mechanical	Scrubbers	Precipitators	Fabrics
Sinter plant	17	2	9	3
Blast furnace[a]	13[b]	51	108	0
Open-hearth furnace	0	6	93	0
Basic oxygen furnace	0	15	23	0
Electric furnace	0	5	1	29
Scarfing	4	4	3	2

[a] Final control equipment.

[b] Dust collectors followed by other equipment are not considered.

in wet systems. Because of the large particle size, cyclones applied to sintering plants usually operate at over 90% efficiency by weight. However, cyclone exit loadings range from 0.2-0.6 grain/ft^3.

Electrostatic precipitators have been installed in series with cyclones. One such installation is reported to operate at an efficiency of 95%, and the final discharge contains only 0.05 grain/scf (17). Another installation handled 457,000 cfm with an inlet loading of 2.5 grains/scf and yielded an output loading of 0.038 grain/scf, an efficiency of 98.5%. However, the materials charged to the sintering machine have changed from straight ore fines to ore, flue dusts, and lime. The characteristics of the ore used have also changed. These changes in materials have resulted in an increased output loading of 0.25 grain/scf, a decrease in collection efficiency to 90% (15). The use of self-fluxing sinter has also impaired the operation of electrostatic precipitators in this service at other plants.

Only two sinter plants are known to have wet scrubber installations. Operating problems occur because of erosion and imbalance of fan blades. Dust carried over to the exhaust fans is moist and has a tendency to accumulate on the blades.

There are only three known applications of baghouses at sinter plants (15).

b. Blast Furnace. Under normal conditions, the untreated gases from a blast furnace contain from 7-30 grains of dust/scf of gas. Most of the particles are larger than 50 μ. Blast-furnace gas-cleaning systems normally reduce particulate loading to less than 0.01 grain/scf to prevent fouling of the stoves in which the gas is burned. These systems are composed of settling chambers, cyclones, low-efficiency wet scrubbers, and high-efficiency wet scrubbers or electrostatic precipitators connected in series (1).

One of the main reasons for cleaning blast-furnace gas is to render it sufficiently clean for use as fuel. Recovered dust is returned to the iron-making process.

Blast-furnace gas is cleaned in three stages; the first two, at least, are used almost universally throughout the industry. The majority of furnaces have secondary cleaning facilities as well. The three stages and the equipment used in each, along with average outlet dust loading from these stages, are:

1. Preliminary cleaning: settling chamber or dry-type cyclone (3-6 grains/scf).
2. Primary cleaning: gas washer or wet scrubbers (0.05-0.07 grain/scf).
3. Secondary cleaning: electrostatic precipitator or high-energy scrubbers (0.004-0.08 grain/scf).

The Great Lakes Steel Corporation (Detroit) has three blast furnaces equipped with high-energy scrubbers. These units have made it possible to clean the gas at a low cost, although there have been troublesome problems in design and operation of equipment (18). High performance of a Venturi scrubber can be achieved only if the blast furnace is operating at a high enough top pressure to provide the required pressure drop. Lack of sufficient top pressure has usually required use of an electrostatic precipitator as the final gas-cleaning unit.

The use of electrostatic precipitators for cleaning blast-furnace gas has come about because of the requirements for cleaner gas for the hot-blast stoves. The operation of these units has been relatively trouble-free because the blast furnace is an almost continuous producer of gas and because of a high percentage of the particulate emissions removed by the wet scrubbing systems that have previously been used to clean up the gases. The wet scrubbers also serve to condition the gases, for both temperature and resistivity, prior to their entry into the electrostatic precipitator.

c. Open Hearth Furnace. The small size of the particulates emitted from these furnaces requires high-efficiency collection equipment, such as Venturi scrubbers, electrostatic precipitators, and baghouses.

A description of the application of Venturi scrubbers to control open-hearth stack emissions is given in Ref. 19. Reference 20 describes the application of a Venturi scrubber on a 200-ton open hearth. It is pointed out that the investment and operating costs of a Venturi scrubber system would compete very favorably with those of a precipitator-waste-heat boiler system for cleaning the gas from open-hearth shops that have an anticipated operating life of no more than a few years.

References 21 and 22 discuss using electrostatic precipitators to collect the fume from open-hearth furnaces. Another electrostatic precipitator system was put into service in 1965 by Weirton Steel Division, National Steel Corporation. This system is used for the control of fumes from 550-ton and 600-ton capacity oxygen-lanced furnaces.

The major problem with respect to actual efficiency of electrostatic precipitators on open hearths is the open-hearth process itself. The problem stems from the variation in the properties of emissions from the open-hearth furnace during a heat. During the period of a heat, the moisture content of the gases may drop from a normal value of 18% to 2%, with a resultant increase in resistivity and drop in precipitator efficiency. The situation may be corrected by steam injection.

A glass-fabric baghouse has been applied to the collection of fume from an oxygen-lanced open-hearth furnace by Bethlehem Steel Corporation in Sparrows Point, Maryland. This unit is described in a 1966 article (23). The authors state that the capital cost estimates favored the baghouses over an electrostatic precipitator, and the subsequent operating experience has shown the maintenance and operating cost to be approximately half that required on an electrostatic precipitator. This baghouse serves a 380-ton/heat furnace having a waste heat boiler and economizer that cool the gas to below 500°F (23).

d. Basic Oxygen Furnace. The basic oxygen furnace (BOF) creates more emissions than the open-hearth furnace, and the particles are smaller. All basic oxygen furnaces in the U.S. are equipped with high-efficiency electrostatic precipitators or Venturi scrubbers. Final effluent from these control devices will contain 0.03-0.12 grain/scf. Inlet loadings may vary from 2.0-5.0 grains/scf.

Problems associated with the use of wet scrubbers on basic oxygen furnaces include inadequate water treatment facilities or lack of sufficient water and the abrasive and corrosive nature of the dust-laden water. The dirty water from Venturi scrubbers is cleaned by a combination of liquid cyclones, clarifiers, and vacuum filters. The recovered dust may be sintered if the composition is suitable; otherwise it is hauled to disposal.

References 24 and 25 discuss the use of wet scrubbers on BOF furnaces. With a wet scrubber system, it is not essential to burn all the CO in the hood, since a wet system can tolerate a considerable amount of CO without danger of explosion.

Problems associated with applications of electrostatic precipitators to basic oxygen furnaces are basically those of variability in gas flow, in the moisture content, and in temperature of the entering gases, as well as maintenance. Also, collection is lower during the initial phase of oxygen lancing before the temperature and water sprays produce a properly conditioned gas stream for efficient collection .

The hoods over the BOF are a necessary part of the collection system and can result in operating problems. The gap between the BOF and the hood is usually dictated by the anticipated operating conditions and the buildup on the mouth of the furnace ("skull"). Excess buildup can restrict the flow of air required for combustion of the carbon monoxide, with the result that a significant amount of carbon monoxide may reach the electrostatic precipitator with possible disastrous results.

The hot gases leaving the BOF are cooled by heat exchange and water sprays to a preferred temperature of 450-500°F. The moisture content of gas going to the precipitators is quite important; it should be kept between 20% and 30% to ensure adequate conductivity of the dust layer.

No installations of fabric filter devices exist on BOF units in the U.S. However, fabric filters have been applied in Europe. The European system uses refractory surfaces in an accumulator to absorb heat from the gases and depends upon the cyclic nature of the BOF for its operation. The refractory accumulator absorbs the heat from the gases during the blowing cycle, cooling the gases to approximately 250°F before the baghouse. As soon as the blowing cycle is over, the accumulator is cooled by backblowing atmospheric air through it (26).

e. Electric Furnaces. The characteristically small particle size of electric-arc steel furnace fume precludes the use of dry centrifugal collectors, settling chambers, etc. High-efficiency scrubbing systems, electorstatic precipitators, and baghouses are used to control fumes from electric-arc steel furnaces.

High-energy scrubbers installed on one oxygen-lanced electric furnace producing a dust concentration of 3.2-6.4 grains/scf reduced the dust output to the range of 0.256-0.0512 grain/scf. Baghouses reduced it to the range of 0.004-0.0064 grain/scf (15). The only known installation of an electrostatic precipitator on an electric furnace plant is at Jones and Laughlin Steel Corporation in Cleveland. The precipitators are considered to be operating satisfactorily. Precipitators installed in 1955 at Bethlehem Steel Corporation in Los Angeles were replaced by baghouses in 1967.

Fabric filters have been successfully applied to the control of emissions from electric furnaces ranging up to 100-150 net-tons capacity, and for multiple-furnace shops, as well as one-furnace shops. The application and operating problems for a baghouse installed on a 150-ton electric furnace are discussed in a recent article by W. W. Blintzer and D. R. Kleintop of Lukens Steel Company (27).

Wherever fluorspar is employed, the fluorides in its off-gas attack glass-filter media. Hence, fiberglass in any form is not recommended on furnaces employing fluorspar as a fluxing agent. However, other synthetic fabrics work well.

2. Iron Foundries

Gray iron foundries range from primitive, unmechanized hand operations to highly mechanized plants in which operators are assisted by electrical, mechanical, and hydraulic equipment. Cupola, electric-arc, electric-induction, and reverberatory air furnaces are used to obtain molten metal for production of castings. Electric-arc and induction furnaces are used mostly to produce specially alloyed irons.

Electrostatic precipitators, high-energy scrubbers, and fabric filters are capable of removing the fine particles from cupola gases. Regardless of whether electrostatic precipitators or baghouses are used as the means of gas cleaning, it is necessary to maintain efficient secondary combustion in the cupola stacks (recuperative preheater). Otherwise, the operation of the gas-cleaning equipment is adversely affected. Maintaining a reducing atmosphere in the cupola stack will allow unburned oil vapor and tarry matter, as well as coke fines and other combustibles, to be carried over into the gas-cleaning equipment. The secondary combustion process does, however, cause an increase in the volume of the gases to be treated (28).

General dust collector selections and efficiencies for various foundry operations are given in Ref. 29, while relative cost comparisons of the various types of dust collectors applied to hot-blast cupola waste gases are presented in Ref. 30.

Electrostatic precipitators installed on cupolas have failed to attain consistently high collection efficiencies because of the wide variation of gas stream conditions. To satisfactorily apply electorstatic precipitation to foundry-cupola-fume collection, it is essential to determine the temperature at which the peak resistivity of the fume occurs and to design the gas-conditioning system to provide a gas temperature well away from that at which the resistivity reaches its maximum. Electrostatic precipitators have been employed in Europe to a greater degree for cupola operations and utilize both wet and dry electrode cleaning techniques. There have been several electrostatic precipitator installations to control particulates from the larger electric furnaces.

Fabric filters are being used increasingly for cleaning cupola gases whenever high collection efficiencies are desired. There are several installations employing completely automatic, tubular dust collectors using synthetic filter bags in conjunction with cooling of the hot gases from cupolas prior to filtration. The primary difficulty in their use arises from poor control of inlet gas temperature. The temperature of the gas stream from the top of a cupola may be as high as 2200°F. Therefore the gases must be cooled prior to entering the baghouse. Cooling can be effected by evaporative water cooling or other means of heat exchange or by dilution with ambient air. When the inlet temperature is too high, the bags burn out; when it is too low, the fabric blinds from condensation of water vapor. The fabric collector temperature is generally limited to the capability of the fabric media and, at present, glass is operable up to 550°F. Currently there are also about 50 installations employing fabric filters to control dust emitted from electric furnaces.

Various types of wet scrubbers are widely used for control of cupola emissions (30). The SO_2 content of the cupola waste gases must be considered in designing a wet-scrubber system. It is not sufficient to neutralize the scrubbing water to a pH value of 7 so that it enters the scrubber in a neutral state. An acidic reaction occurs after it contacts the cupola waste gases and a serious corrosion danger exists. Thus the wash water should be adjusted to a pH of 9 so that in the gas washer itself the acidic range is not reached.

Although the wet-scrubber systems absorb some of the SO_2 from the waste gases, the odor of SO_2 may be more perceptible than with other collection systems because wet scrubbing lowers the temperature of the gases, which causes a decrease in plume rise (30). The carbon content of the dust makes wetting more difficult and thus puts higher requirements on wet scrubbers.

Wet scrubbers pose the problem of sludge removal and potential water pollution, which must be taken into account in computing the cost of this type of collector.

Wet caps have been in common use on cupola gases, but these only remove the coarser particles and are usually not adequate for compliance with local codes.

3. Primary Nonferrous Metallurgy

The term primary nonferrous metallurgy, as used here, will include smelting and refining of copper, lead, zinc, and aluminum. The air pollution problems of the nonferrous metallurgical industry are extremely varied. However, one typical characteristic exists--in almost all the processes in the production of nonferrous metals, the particulates emitted are metallic fumes generally submicron in size. Adequate control of these emissions can be achieved only by the use of high-efficiency control equipment.

Electrostatic precipitators, usually preceded by mechanical collectors, are almost universally applied to the control of particulates from copper smelting operations. The precipitators are normally more massive and rugged than their counterparts in the power generation or other industries. Mild-steel construction of electorstatic precipitators is accomodated, by maintaining sufficient gas temperatures to preclude corrosion, with temperatures ranging from 300-650°F on converters and from 600-900°F on roasters. Actual collection efficiency usually is reported in the 98.5-99.5% range (31). Mechanical collectors, used as the primary cleaners, are typically of the large diameter (24 in. or more) multiclone type.

Fabric filters have also been used as secondary collectors on converter gases.

German practice in the control of dust emissions from primary copper smelting and refining is summarized in Table 7.

Lead smeltering equipment, such as sintering machines, blast furnaces, and reverberatory furnaces are also controlled by the use of settling chambers or centrifugal collectors in combination with electrostatic precipitators and baghouses. Dust and fume are recovered from the sinter machine gas stream by settling in large flues and electrostatic precipitators or baghouses (33). Collect-

TABLE 7

Operation Data For Dust Collectors Applied To Primary Copper Smelting And Refining (32)

Type of separator	Maximum efficiency (%)	Draft required (in. water)	Utilization
Dust chambers	30-60	< 0.2	Beyond sintering machine and shaft furnace
Cyclones ≦ 16 in. in diameter	85-95	3-4	For secondary purification beyond reverberatory furnaces
Electrostatic precipitators	96-99	0.2-0.6	For higher demands and fine dust beyond roasting, sintering, and shaft furnaces
Cloth filters	99+	2-6	For dry air beyond coolers for converter gas

ion efficiencies are up to 96% for precipitators and 99.5% for baghouses (33).

Table 8 summarizes German practice in the control of emissions from lead smelters.

Since practically all zinc ores, as mined, are too low in zinc content for direct reduction processes, they must be concentrated first. After concentration, the first step in the extractive metallurgy of zinc is generally roasting the concentrate.

For efficient recovery of zinc, the sulfur content of the concentrate is reduced to 2% by roasting. Multiple-hearth or Ropp raosting may be followed by sintering, or double-pass sintering may be used alone. Sintering in zinc smelters produces SO_2, which is converted into sulfuric acid by the contact process. The sinter-plant gases can be precleaned by electrostatic precipitators, baghouses, or wet scrubbers (35). The collection of sintering fume involves a large volume of gases in the range of 1,300,000 scf/ton zinc product. Zinc smelter sintering fume is difficult to collect efficiently by electrical means because of its inherently high electrical resistivity. This requires close control of gas temperature and moisture content to maintain efficient collection (36). Horizontal-flow, plate-type precipitators have been installed on most of the newer zinc sintering machines. Moisture and/or steam is normally added to improve dust resistivity for optimum precipitation. Mild-steel construction is common, and installed costs for base collectors of 50,000 cfm would be $3.50/cfm (31).

Fluid-bed roasters have been used to process agglomerated feed. Exhaust gases were cleaned in a waste-heat boiler, cyclones, and an electrostatic precipitator in series (37).

TABLE 8

Lead Smelter Control Equipment (34)

Process	Control device		Efficiency	
	Primary	Secondary	Primary	Secondary
Sinter machine	Centrifugal	Electrostatic precipitator, bag filter	80-90	95-99
Blast furnace	Centrifugal	Electrostatic precipitator	80-90	95-99
Reverberatory furnace	Waste heat Boilers Tubular coolers	Electrostatic precipitator, bag filter	70-80	95-99

Reference 38 describes the fume-recovery facilities at a Canadian smelter operation. These facilities used electrostatic precipitators for controlling the emissions from zinc-roasting and lead-sintering operations. The zinc-roaster gases entered the precipitator at 420°F without pretreatment. The lead-sintering gases required conditioning in a spray tower (39). The precipitators were very versatile, but the paramount problem was corrosion caused by condensed vapors. Wet scrubbers were also used on the refinery melting kettles.

Aluminum is produced by the electrolysis of alumina (Al_2O_3) in fused cryolite (AlF_3-3NaF). Essentially all the alumina used is extracted from bauxite. While all aluminum production employs the basic Hall-Heroult process, several variations in cell construction have evolved, based upon the method of anode manufacture. Differences in cell construction, because of the type of cell employed, affect the types and quantities of pollution generated. The types of cells are: (a) the prebake cell; (2) the horizontal stud Soderberg cell; and (c) the vertical stud Soderberg cell.

The vertical stud Soderberg cell captures cell effluents most effectively because placement of the anode studs in a vertical position allows a metal skirt to be fixed to the lower end of the steel anode jacket. The skirt reaches down to the encrusted bath and effectively encloses the fuming bath. Gases are therefore drawn off with little dilution by air and are concentrated enough to allow combustion of the tarry hydrocarbons. During burning, hydrocarbons are reportedly reduced from 3 to 0.1% by volume, and most fluoridated carbon compounds converted to hydrofluoric acid. The effective oxidation of tar is a great aid to subsequent collection because tar contamination and plugging of ducts are avoided. After burning, exhaust fumes from each pot are sent to a central header fan and control equipment. One piece of control equipment will commonly treat exhaust fumes from 15 pots.

Control devices used for vertical stud Soderberg cells have included multiclones and spray-type scrubbers. The scrubbers use high-pressure sprays to contact countercurrently the gases and particulate. Such a system can remove 95% of the fluorides entering the control system, but no figures for total particulate collection efficiencies are reported. Exhausts may be treated by bag filters coated with lime or alumina, or by electrostatic precipitators, but the residual tar creates a fouling problem in the collection system.

A recent addition to the family of collectors is the sieve-plate scrubber (40). The device has been put into operation in Norway. A three-plate tower has removed 97% of the hydrogen fluoride, 80% of the solid fluoride particles, and 70% of the total particulate in incoming gas streams. Higher absorption efficiencies are possible with the addition of a fourth plate. The design allows self-cleaning of the plates by the use of sprays directed at the underside of each plate, where heavy tar and particle deposition occur. High-velocity droplets, impelled by the air flowing through the restriction, are blown against the plate and are forced to the top by the air stream, thereby preventing plugging of the sieves. Particulate collection is by impingement on the plates and the water droplets. The tower produces hydrofluoric acid, used as recirculation liquor for the first plate.

Another collection system recently reported incorporates dry cyclones, an electrostatic precipitator, and two scrubbers (40). The system is reported to achieve 99.9% gaseous fluoride collection, but particulate efficiencies were not given.

Effective capture of all effluents for a horizontal stud Soderberg cell is much more of a problem, because open channels are required for replacement and readjustment of the studs. As a result, hooding is less complete; therefore larger volumes of air are entrained. Large volumes of exhaust air create a dilute mixture of hydrocarbons. Because burning is not possible, a tar-fouling problem occurs in ducts and control equipment. Cyclones or multiclones and baghouses have fouled too easily to be an effective answer to control with horizontal stud Soderberg cells. Where electrostatic precipitators are attempted, the plates require water flushing to prevent fouling by tars. Existing controls have consisted of scrubbers, either grid packings in a vertical scrubber or the high-velocity spray type. The lack of primary collectors makes these scrubbers especially susceptible to plugging by particulate matter.

To solve this problem, the Alcan Company recently developed the floating-bed scrubber (40). Collection efficiencies for the floating bed scrubber are detailed in Table 9 for a single-bed installation having pressure drops of 4-6 in. of water (40).

Hooding for prebake cells is similar to the vertical stud Soderberg, but ventilation rates are much greater. One exception to this generalization is the Pechiney process, which employs no local exhaust ventilation over cells at all, but instead relies upon roof-monitor emissions control to collect pollutants.

Particulate emissions from prebake pots contain none of the tar found in the horizontal stud Soderberg; instead, "dusting" of the carbon anode produces carbon particles in addition to alumina, etc. Current controls have consisted of dry-type cyclones or electrostatic precipitators followed by wet scrubbers. Scrubbers are 12-15 ft in diameter and 40-60 ft high, with internally mounted spray headers.

TABLE 9

Collection Efficiencies For The Floating Bed Scrubber Used On Horizontal Stud Soderberg Cell (40)

Substance	Collection efficiency (%)
Total fluorides	90+
Hydrogen fluorides	98+
Particulates	80-90[a]

[a]Anticipated efficiency at higher pot ventilation rates of 3600 scfm and when doors are closed.

The newest controls being developed have relied upon a dry absorption of fluoride gases (and perhaps particulates) on finely divided alumina powders. The alumina dust and any other particulate are then collected in a baghouse and the catch is sent to the cell as the feed. A solid-particle coating of alumina on the inside bags also aids in fluoride collection. Dry collection avoids all the plugging and high water costs of scrubbing towers, while allowing the dust to be returned directly to cells in a dry form. Performance data are extremely scarce on this unit, since it is still in the development stages, but total particulate collection of 99% and gaseous fluoride efficiencies of 95% have been estimated.

Table 10 summarizes efficiency data for current and prototype control devices for aluminum reduction cells (40). Existing particulate control efficiencies vary from 40-60%, except where an electrostatic precipitator is used in conjunction with other devices.

TABLE 10

Current And Newest Air Pollution Controls For Primary Aluminum Potline Air Pollution Controls (40)

Type of cell	Existing collectors	Est. removal efficiencies (%)		Latest collectors	Est. removal efficiencies (%)	
		Fluorides[a]	Particulates		Fluorides[a]	Particulates
H.S.[b] Soderberg	Spray scrubbers	80-90	40-50	1. Floating-bed scrubber[c]	90	80-90
				2. Wetted-plate electrostatic (with conditioning of flue gases)	90	99
Prebake	Multiclones	0	< 60	1. Fluidized alumina contacts cell exhausts, followed by collection in alumina-coated baghouse	99	96-98
	Dry electrostatic precipitators		90			
	Spray scrubbers	80-90	40-50[d] < 10[e]	2. Counterflow packed scrubber	90	95
V.S.[f] Solderberg	Multiclones	0	< 60	Sieve-plate scrubber	95	70
	Spray scrubbers	80-90	40-50			

[a] Gaseous and particulate fluorides.

[b] H.S. = horizontal stud.

[c] One section of bed employed.

[d] When used after multiclones.

[e] When used after e.s.p.

[f] V.S. = vertical stud.

C. MINERAL PROCESSING INDUSTRIES

For the purpose of the present discussion, the mineral processing industries will be taken as cement manufacture, lime manufacture, and hot-mix asphalt production.

1. Cement Manufacture

Portland cement is made by either the wet or dry process. The hot kiln gases are the main source of emission, and they present a major problem because gas volumes are large; they contain acid gases such as H_2S and SO_2, varying amounts of H_2O, and a temperature range usually above 500 or 600°F (41). A kiln producing 20 tons/hr of cement clinker will produce about 240,000 lb/hr of exit gases, or about 92,000 acfm (42).

Although a number of types of dust collectors are used in the cement industry, only the high-efficiency collectors, such as the electrostatic precipitator and fabric filter, sometimes used in series with inertial collectors, effectively collect fine dust. Multicyclones alone are not an acceptable means of reducing dust emission from the kiln to the atmosphere. Multicyclones, when preceding other equipment, can be expected to scalp off about 70 wt%, or all of the coarser particles. Table 11 gives ranges of dust emissions for various combinations of control devices (43).

The operation of electrostatic precipitators has not been entirely satisfactory in the past because of decreasing efficiency over extended periods caused by the effects of the cement dust on the high-voltage components. Also, when kilns have been shut down and then restarted, it may be necessary to bypass the electrostatic precipitator for periods up to 24 hr because of the danger of explosion from the presence of combustible gas or coal dust. In a

TABLE 11

Ranges Of Dust Emissions From Control Systems Serving Dry- And Wet-Type Cement Kilns (43)

Source	Type of dust collector	Range of dust emissions from collector	
		Grains/scf[a]	lb/ton of cement
Kiln, dry type	Multicyclones	1.55-3.06	26.2-68.6
	Electrical precipitators	0.04-0.15	1.7-5.7
	Multicyclone and electrical pre-cipitators	0.03-1.3	0.6-29.4
	Multicyclone and cloth filter	0.039	0.7
Kiln, wet type	Electrical pre-cipitators	0.03-0.73	0.52-9.9
	Multicyclone and electrical pre-cipitators	0.04-0.06	4.3-24.2
	Cloth filter	0.015	0.35

[a] Grains/scf = grains/standard cubic foot of gas corrected to 60°F and 1 atm pressure.

wet-process plant the performance of an electrostatic precipitator is greatly enhanced by the extra water vapor present in the exhaust gases from the slurry. Dry-process kilns do not have this water in the feed, and it is often necessary to add it as an aid to precipitator operation.

Fiberglass baghouse filters have had much success in controlling kiln emissions. Bag life averages 2 years or more (41). Moisture condensation in glass-fabric filters can present problems. However, dew point temperatures are normally avoided by proper application of insulation to ducting, etc., and by proper operation to avoid condensation.

The simplicity of design and operation of the fiberglass filter system, which lowers the cost, is balanced to some extent by increased fan power needed to overcome pressure drop across the baghouse. Many baghouses operate with a pressure drop of 3-7 in. of water.

2. Lime Manufacture

The major potential source of particulates in lime manufacture is the calcining kiln. Emissions vary with kiln type and the composition of limestone burned.

Vertical kilns do not produce as much dust as do rotary kilns because of the larger size of the limestone charged, the low gas velocities, and the smaller amount of attrition. Nevertheless, vertical kilns are apt to be considered dusty by modern air pollution standards.

Rotary kilns constitute the largest single source of particulate matter in the lime industry. Abrasion of limestone in the kiln produces dust. The stone becomes more friable as it approaches the decomposition temperature, and dusting increases. Simultaneously with dusting from attrition, the high-velocity gases from direct-fuel-fire combustion blow the dust from the kiln. This dust is difficult to control and collect. It is hot, dry, difficult to wet, and prone to be electrostatically charged. It is of mixed composition, varying all the way from raw limestone to final calcined product. It will also be mixed with fly ash, tars, and unburned carbon if pulverized coal is used as the fuel.

The gases leaving a rotary kiln are usually first passed through a settling chamber to settle out the coarse particles. In some cases, dry cyclones may also be used for this primary collection. From 65-85% of the particulate matter may be collected here (44). The major dust-control problem is the dust passing the primary collector.

Plants in some areas have installed high-efficiency cyclonic secondary collectors on their kiln operations. However, to meet emission requirements, plants are increasingly turning to wet scrubbers and glass bag collectors. Electrostatic precipitators have also been found satisfactory.

A number of installations are reported making use of glass fiber bag collectors handling gas flows as high as 150,000 acfm at temperatures in the range of 350-550°F, with average particle sizes of 25 μ after precleaning with dust settlers. For the larger gas volumes, the baghouse is compartmented so that only one section at a time is cleaned. A 12-compartment baghouse for a 500 ton/day kiln has been reported. The cleaning cycle (shaking is not employed with glass bags) depends on dust loading, but is usually a 10-15 min. cycle. Design air-to-cloth ratio with one compartment out for cleaning is in the range of 1.95:1 to 2.2:1. Since kiln gases are frequently discharged when they are still too hot to be handled directly by the bags, it is usual practice to cool the gases by water spray, air dilution, or by a combination of the two. Insulation of the baghouse is not usually required, unless the moisture content of the gases is quite high, as might be the case with wet feed. Bag life up to 2 years is reported.

While the use of electrostatic precipitators tends to be costly for the lime industry, one installation has been reported using a single-stage precipitator as a secondary collector at a capital cost of $1.25/cfm. It handles 160,000 cfm at 450-500°F inlet conditions in which 90-95% of the inlet dust is $\leq$10 μm. It is designed with a gas velocity of 3.3 ft/sec and a residence time of 5.2 sec, and has an on-stream efficiency of 95% (44).

One of the advantages of a wet scrubber is that it can include a prehumidification section and eliminate the need for precooling the gases. A typical installation for a 180-200 ton/day kiln with 40-50,000 acfm at 900-1400°F and a 5-10 grains/ft^3 dust loading, would require a 9-10 ft diameter scrubber, 32 ft tall. Scrubbing water requirement is 4 gal/1000 ft^3 of gas processed. Pressure drop is 8 in. of water. For the 200-ton/day kiln installation, fan brake horsepower would be 150. Collection efficiency is stated to be 99.7%. Scrubber cost is reported to be $0.50/cfm of cooled saturated exhaust gas for 304 stainless steel construction and $0.25/cfm for carbon steel (44).

A dust removal efficiency of 96-97% has been reported for a 335-ton/day kiln using a combination Venturi scrubber and cyclonic separator. A pressure drop of 7-11 in. water was used. Inlet gas volume was 60-62,000 cfm at 350°F. Water supplied to the Venturi throat was 1500 gpm at 50 psig pressure. Cleaned exhaust gases were discharged at 160-165°F and nearly saturated with water vapor (44).

3. Hot-Mix Asphalt Paving Plants

Generally, a hot-mix asphalt plant consists of a rotary dryer, screening and classifying equipment, an aggregate weighing system, a mixer, storage bins, and conveying equipment.

The major source of dust is the rotary dryer. However, while dust from the rotary dryer is the greatest source, dust collected from the vibrating screens, the bucket elevator, storage bins, and weigh hopper is also significant. In some plants, the dryer dust problem is handled separately from the other sources. However, the trend is to combine both the dryer and ventline, or fugitive, sources together with a single collector fan system. Total ventilation requirements for the rotary dryer and the secondary dust sources vary according to the size of plant. For a 6000 lb/batch plant, 22,000 scfm is typical, of which ~ 3000 scfm is required for the secondary sources.

A typical asphalt plant dryer effluent will contain 20-30 grains/ft^3 (45). However, the loading can vary widely.

The collection of dusts from the dryer is usually by means of cyclones for the primary dust collection, followed by a higher-efficiency type of secondary collector. One recently installed control system consisting of cyclones followed by low-pressure drop wet scrubbers is reported to attain 99.89% over-all collection efficiency (46). Measurements of cyclone-scrubber systems on 14 asphalt plant dryers showed > 98.0% efficiency (47).

Electrostatic precipitators do not find much application to asphalt plants because they have a rather high first cost (48).

Fabric filters are subject to operating difficulties when installed in an intermittent process, such as asphalt dryers (45). Precautions must be taken to prevent overheating and to prevent condensation during shutdowns.

The air-to-cloth ratios for fabric filters will vary from 3.5:1 to 6.25:1. Reverse air supply is normally heated, and the collector housing insulated, to prevent condensation (49).

Wet collectors are in wide use in the asphalt paving industry and generally give good results without serious maintenance problems. The scrubbing liquid from these collectors requires a settling pond of adequate size. This pond should be at least 6 ft deep and hold at least 2 hr discharge of the scrubber. The sludge collected in the pond must be pumped or dredged out and removed to an appropriate disposal area, avoiding any chances for stream pollution.

D. CHEMICAL PROCESSES

Acid manufacturing processes will be discussed as an example of chemical processes that require the use of particulate air pollution control devices. Acid manufacturing processes generally include those for sulfuric, phosphoric, nitric, and hydrochloric. However, only sulfuric and phosphoric acid manufacture discharge significant particulate emissions.

1. Sulfuric Acid Manufacture

All sulfuric acid is made by either the chamber or the contact process. Elemental sulfur, or any sulfur-bearing material, is a potential raw material for both these processes. Contact-process establishments account for about 97% of the U.S. production.

Electorstatic precipitators, packed-bed separators, mesh-type mist eliminators, ceramic filters, and sonic agglomerators have been used to reduce the emission of acid spray and mist from contact acid plants (49). Most modern plants are equipped with high-efficiency electrostatic precipitators or mesh-type eliminators in which 99%

of the mist is recovered (1). Recovery equipment is rarely employed in chamber plants (49).

The electrostatic precipitator and mesh-type eliminator can provide removal efficiency of up to 99.9%. However, when oleum is produced, the portion of acid mist particles smaller than 3 μm in diameter is higher. This size seriously affects the low-pressure drop mesh-type eliminators; efficiency decreases sharply, and may be less than 40% (49).

The wire mesh mist eliminator has the lowest first cost for effective removal of particles larger than about 3 μm in diameter. However, corrosion possibilities may require frequent replacement of this type of control device; therefore costs will rise, and these increased costs must be considered (1).

The wire mesh eliminator is commonly constructed with two beds in series and operates with pressure drops of 1-3 in. of water (49). Typical gas velocities for these units range from 11-18 ft/sec (1).

The high-efficiency glass-fiber mist eliminator is capable of operating with collection efficiencies of over 99%. Pressure drop is usually 5-10 in. watergauge. The glass-fiber mist eliminator is also capable of maintaining high efficiency at varying tail-gas flow rates (49).

A recent development in this area involves the use of a teflon fiber mist pad. This device is reported to be 98% efficient, even when inlet loading is as low as 10 mg/ft^3 (50).

Electrostatic precipitators are highly effective for removal of acid-mist particles. Electrostatic precipitators may be either the wet or dry type. The dry type, which is suitable only for concentrated acid, is much less expensive, but more susceptible to corrosion. Wet-type precipitators are suitable for use only with dilute acid;

this necessitates prior humidification of stack gases. This also permits removal of SO_3 by converting it to acid mist. However, the humidification step appreciably increases the cost of a wet-type installation (49).

The lead-constructed electrostatic precipitator, for low-strength acid mist emissions, is used throughout the industry as the primary means of emission control. The use of mild-steel electorstatic precipitators for oleum stack cleanup has been reported as successful with considerable savings in cost (51).

Venturi scrubbers are capable of high efficiency, but at the expense of high pressure drop. They have not been used on contact acid plants, but have been used on sulfuric acid concentrators to give outlet (i.e., grain loading) mist loadings of 0.5-3.0 mg/ft^3 (49).

2. Phosphoric Acid Manufacture

Phosphoric acid is manufactured by two processes: (a) the thermal process; and (b) the wet process. The thermal process proceeds by burning elemental phosphorous to the pentoxide, followed by a hydration step. The wet-acid process involves treatment of phosphate rock with sulfuric acid. About 75% of the phosphoric acid produced comes from the wet-acid process, and about 92% of the wet-process phosphoric acid goes into fertilizer production. About 19% of the thermal-process phosphoric acid goes into fertilizer production, with the rest of the acid from both processes going into the production of other industrial chemicals. Only the thermal process is discussed here.

In the manufacture of phosphoric acid from elemental phosphorus, three steps are involved: (a) burning of the phosphorous; (b) hydration of the resulting phosphorous pentoxide; and (c) the collection of the mists formed. The principal atmospheric emission from the manufacture of phosphoric acid by the thermal process is acid mist in the absorber discharge gas.

Venturi scrubbers, packed scrubbers, glass-fiber mist eliminators, wire-mesh mist eliminators, and electorstatic precipitators are used as abatement equipment at phosphoric acid plants (52).

Packed and open-tower scrubbers have been used widely to collect phosphoric acid mist. Scrubbing is inexpensive and simple, but high collection efficiency is not usually obtained. Some plants have improved efficiency by installing wire-mesh mist eliminators after the scrubber.

Venturi scrubbers are capable of operating at high collection efficiencies on phosphoric acid mist (52). The extremely small size of the mist particles usually requires pressure differentials in excess of 40 in. of water. It is reported that Venturi scrubbers can reduce emissions to 0.10 mg/scf (53). Venturi scrubbers are not used solely for abatement, since they can actually hydrate part of the phosphorous pentoxide vapor, agglomerate the mist particles, and cool the stack gases; they may then be followed by cyclonic separators. This combination can recover up to 99.9% of the acid mist at pressure drops of 35-60 in. water gauge.

Cyclonic-type collectors are used in some plants, but because of the small particle size of the emitted acid mist, other devices usually supplement these collectors. Supplemental collectors are typically wire-mesh mist eliminator pads of low pressure differential.

Glass-fiber mist eliminators are capable of high collection efficiency in removing phosphoric acid mist from absorber effluent gas streams. When the mist eliminators are operating at a superficial vapor velocity of less than 1 fps and at pressure differentials of about 20 in. of water, collection efficiencies of 99.9% are attainable (54).

The high-energy wire-mesh contactor is a recently developed device that is reported to give collection efficiencies that exceed 99.9% at pressure differentials ranging from 35-41 in. of water (55).

TABLE 12

Operating Characteristics Of Phosphoric Acid Mist Electrostatic Precipitators (52)

Installation	Gas flow rate[a] (scfm)	Inlet temperature (°F)	Inlet mist conc. as noted	Outlet mist conc. as noted	Collector efficiency (%)	Rated capacity (%)
1	3,160	227	7.45[b]	0.08[b]	98.9	147
2	14,100	292	14.21[b]	0.415[b]	97.1	119
3	3,540	173	3,468[c]	0.10[c]	99.9+	75
4	3,900	192	3,650[c]	0.16[c]	99.9+	114
5	3,570	195	4,060[c]	0.24[c]	99.9+	104
6	7,300	234	278[d]	10.34[d]	96.3	101

[a]29.92 in. Hg and 32°F.

[b]Grains/scf dry gas as P_2O_5.

[c]Milligrams 80% H_3PO_4/cf dry gas at temperature.

[d]Milligrams mist/scf dry gas at 60°F as P_2O_5.

Electrostatic precipitators are also used in phosphoric acid plants. Table 12 illustrates electrostatic precipitator data for six installations (52).

APPENDIX

COST RELATIONSHIPS FOR AIR POLLUTION CONTROL EQUIPMENT

A. INTRODUCTION

The discussion in Sec. III, Chap. 6, pointed out the fact that variations in cost factors make it difficult to estimate costs for a specific application. The most uniform and complete discussion of the economic factors in air pollution control is presented in a U.S. Department of Health, Education, and Welfare Report, "Control Techniques for Particulate Air Pollutants" (1). Information presented in this appendix has been abstracted from that report. Readers" interest in specific details underlying the computation of the cost data are referred to this publication.

Cost information for control devices is given in Fig. A-1 to A-20 where, for various types of equipment, operating capacity is plotted against cost. The estimated purchase cost curves show the dollar amounts charged by manufacturers for basic control equipment, exclusive of transportation charges to the installation site. This basic control equipment includes built-in auxiliary parts of the control unit, such as instrumentation and solution pumps. The installed cost curves include the purchase costs, auxiliary equipment costs, and costs for field installation. Annualized cost curves include utilities, labor, supplies, and materials (i.e., maintenance and operations costs), as well as taxes, insurance, and interest (i.e., capital charges).

The upper and lower curves in these figures indicate the expected range of costs, with the expected average cost falling approximately in the middle. Although quantitative values for collection efficiency and gas volume capacity are not listed, higher collection efficiency, which involves more intricate engineering design, results in higher costs. Control equipment is designed for a nominal gas volume capacity, but under actual operating conditions the volume may vary. Similarly, the efficiency of control equipment will vary from application to application as particle characteristics, such as wettability, density, shape, and size distribution, differ. For example, a control device designed to operate on 50,000 acfm of gas with a nominal collection efficiency of 95% may have an effective operating range of from 45,000-55,000 acfm, and its collection efficiency may range from 90-97%.

To make the cost estimation problem manageable, nominal high, medium, and low collection efficiencies have been selected for each type of control equipment, except fabric filters. For fabric filters, the nominal high, medium, and low curves reflect construction variations. The purchase, installation, and total annualized costs of operation are plotted for each of the three efficiency levels over the gas volume range indicated. Purchase, installation, and total annualized costs for fabric filters are plotted for variations in filter construction and cleaning methods.

Generalized categories of control equipment are discussed rather than specific designs because of uncertainties in size, efficiency, and cost. If required, more detailed information on the cost of various engineering innovations (e.g., packed towers of specific design to accommodate a corrosive gas stream) could be requested from the manufacturers of the specific equipment.

The adjustment of cost data from the past several

years to a given year can be made using the Marshall and Stevens Equipment Cost Index, which is published in the periodical Chemical Engineering. The M and S Equipment Cost Index is based on a value of 100 in the year of 1926. The 1968 base value is used for equipment costs in this appendix, except as noted.

Adjustments in costs from one year to another are accomplished by multiplying the known cost by the ratio of the index values at the two years involved to obtain the desired cost. The choice of the M and S Index over several other indexes that are available was based on the fact that the industries included in the M and S Index cover a large segment of the control equipment market and the general observation that all the indexes have been moving at about the same rate in recent years.

Use of these indexes over more than a few years leads to considerable uncertainty and is not recommended. It is hoped that users of the book will continually update the data from the best sources available and not resort to changing the cost index alone. Equipment is continually being developed and many specific items do not follow the average trends.

There is a wide variation in costs at any given time. This is caused, in part, by the inventory and backlog situation of the individual manufacturers and suppliers. The best cost information is a written quotation for a specific item to be purchased during a given period. In a number of instances, quotes on equipment are available and can be used in adjusting the literature data. Essentially the same thing can be done, through discussions with experienced persons, for typical control equipment.

B. SETTLING CHAMBERS

The purchase costs of settling chambers are shown for three different efficiencies in Fig. A-1. The efficiencies are based on the assumption of essentially complete removal of 87-μm, 50-μm, and 25-μm particles, and are designated as low, medium, and high efficiencies, respectively. The low and medium-efficiency collectors are simple expansion chambers, and the high-efficiency collector is a multiple-tray settling chamber, commonly called a Howard separator. The total installed cost for each efficiency is shown in Fig. A-2. The total installed cost is the sum of the purchase and installation costs. The installation costs were assumed to range from 33-100% of the purchase cost; this range results in a cost band for each efficiency, as shown in Fig. A-2. No annualized cost curves are presented for these collectors because operation and maintenance costs, other than for removal

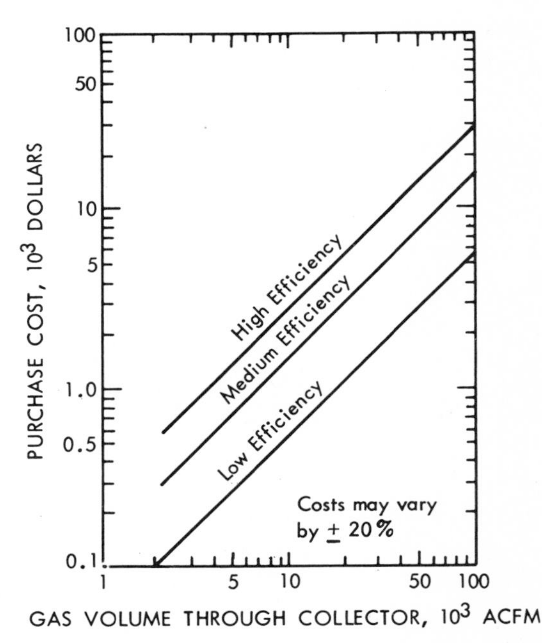

FIG. A-1. Purchase cost of gravitational collectors.

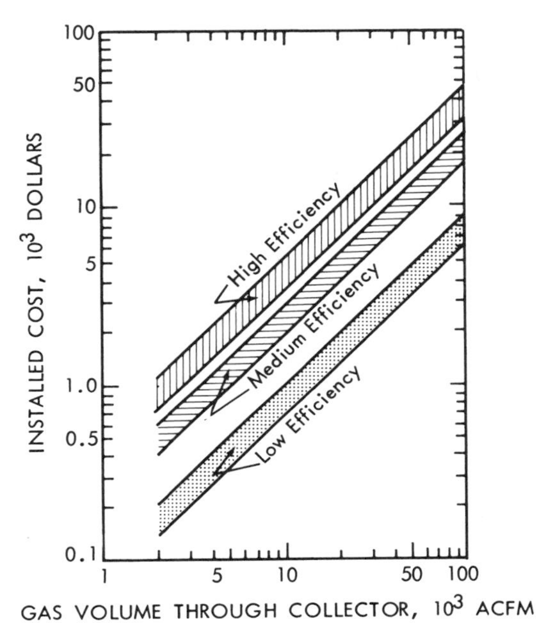

FIG. A-2. Installed cost of gravitational collectors.

and disposal of collected materials, usually are negligible, except where corrosion may be a problem.

C. DRY CENTRIFUGAL COLLECTORS

The costs of purchasing, installing, and operating mechanical centrifugal collectors are given in Fig. A-3, A-4, and A-5, respectively. The curves in these figures show costs for collectors that operate at nominal efficiencies of 50%, 70%, and 95%. Costs are plotted for equipment sizes ranging from 10,000 to 1,000,000 acfm. The assumptions used in calculating annual operation and maintenance costs for dry centrifugal collectors are as follows:

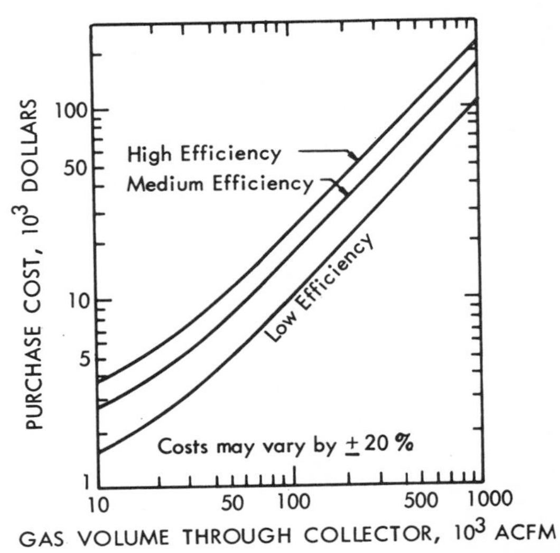

FIG. A-3. Purchase cost of dry centrifugal collectors.

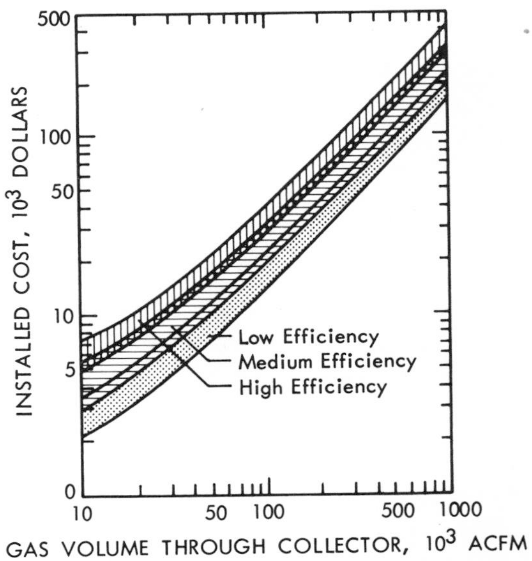

FIG. A-4. Installed cost of dry centrifugal collectors.

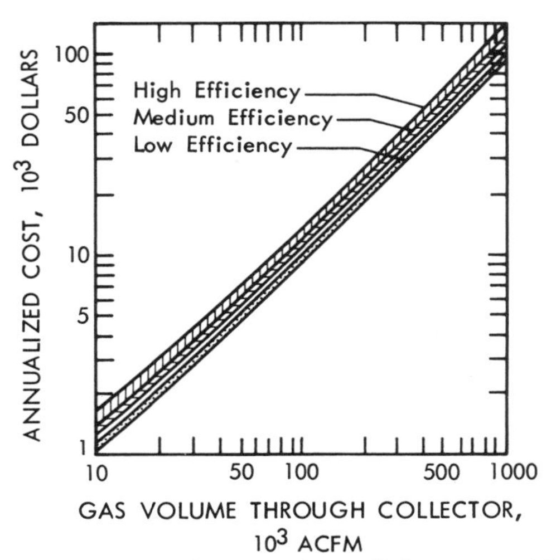

FIG. A-5. Annualized cost of operation of dry centrifugal collectors.

1. Annual operating time is 8760 hr.
2. Collector pressure drop is 3 in. of water.
3. Power cost is $0.011/kWh.
4. Maintenance cost is $0.015/acfm.

D. WET COLLECTORS

The costs of purchasing, installing, and operating wet collectors are given in Fig. A-6, A-7, and A-8, respectively, as a function of equipment size. The curves in these figures show costs for collectors that operate at nominal efficiencies of 75%, 90%, and 99%. The basic hardware costs for medium- and high-efficiency collection equipment are reported by manufacturers to lie in the same cost range and both appear on the same curve in Fig. A-6. The higher installed cost of a high collection efficiency system in Fig. A-7 results from the need for

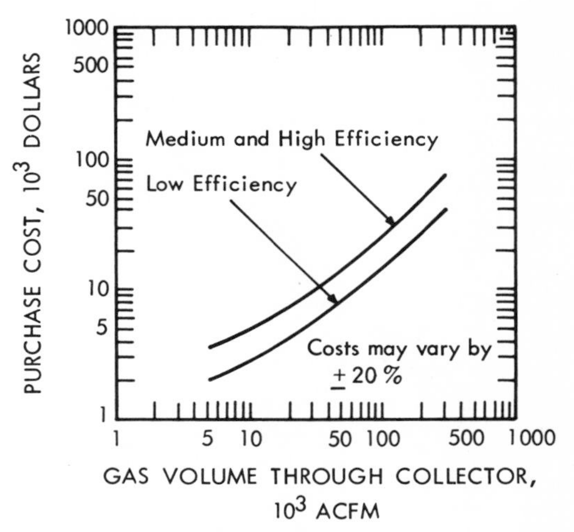

FIG. A-6. Purchase cost of wet collectors.

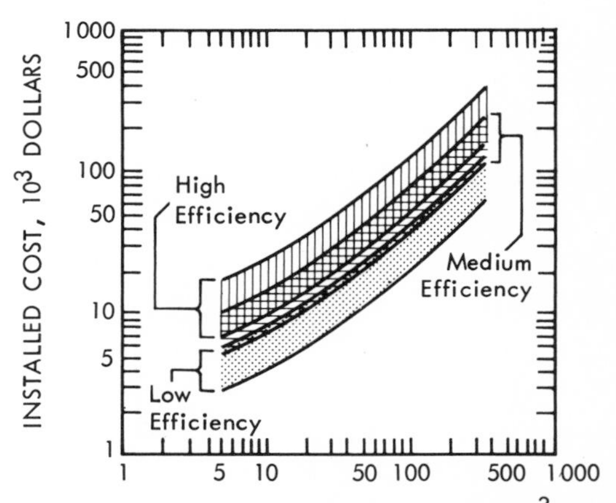

FIG. A-7. Installed cost of wet collectors.

larger, more expensive auxiliary equipment. The assumptions used in calculating annual operating and maintenance costs for wet collectors are as follows:

1. Annual operating time is 8760 hr.

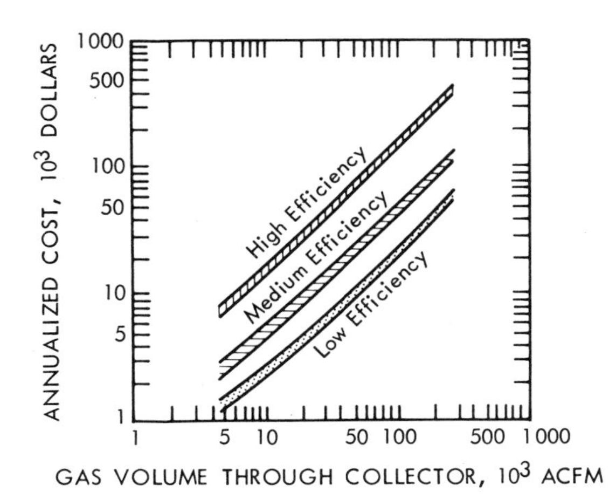

FIG. A-8. Annualized cost of operation of wet collectors.

2. Contact power requirements:
   0.0013 horsepower/acfm for 75% efficiency
   0.0035 horsepower/acfm for 90% efficiency
   0.015 horsepower/acfm for 99% efficiency.
3. Power cost is \$0.011/kWh.
4. Maintenance cost is \$0.04/acfm.
5. Head required for liquor circulation in collection system is 30 ft.
6. Liquor circulation is 0.008 gal/acfm.
7. Liquor consumption is 0.0005 gal/hr-acfm.
8. Liquor cost is 0.0005/gal.

E. ELECTORSTATIC PRECIPITATORS

The costs of purchasing, installing, and operating high-voltage electrostatic precipitators are given in Figs. A-9, A-10, and A-11, respectively. The curves in these figures show costs for collectors that operate at nominal

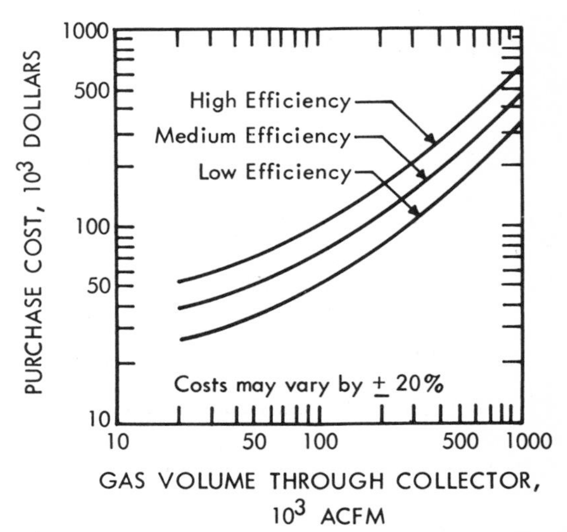

FIG. A-9. Purchase cost of high-voltage electrostatic precipitators.

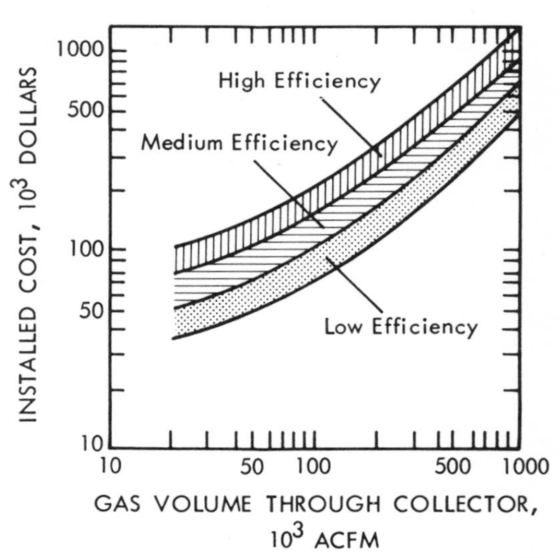

FIG. A-10. Installed cost of high-voltage electrostatic precipitators.

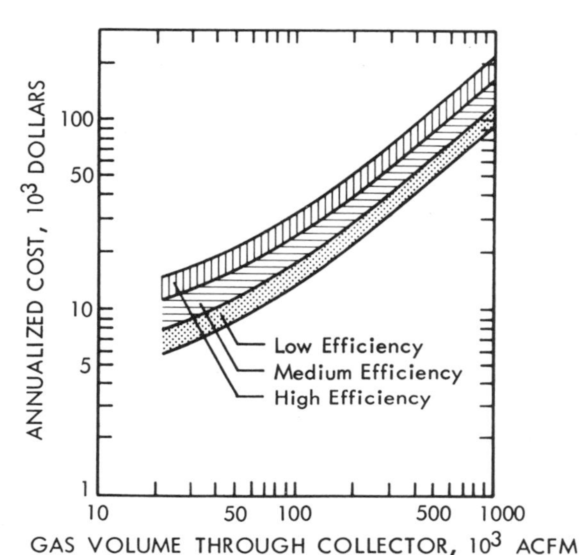

FIG. A-11. Annualized cost of operation of high-voltage electrostatic precipitators.

efficiencies of 90%, 95%, and 99.5%. These costs are plotted for equipment sizes ranging from 20,000-1,000,000 acfm. The assumptions used in calculating annual operation and maintenance costs for high-voltage electrostatic precipitators are as follows:

1. Annual operating time is 8760 hr.
2. Electrical power requirements:
   0.00019 kW/acfm for low efficiency
   0.00026 kW/acfm for medium efficiency
   0.00034 kW/acfm for high efficiency.
3. Power cost is $0.011/kWh.
4. Maintenance cost is $0.02/acfm.

The curves in Figs. A-12, A-13, and A-14 indicate purchase cost, installed cost, and operation cost of low-voltage electrostatic precipitators for low and high collection efficiencies based on design gas velocities of 150

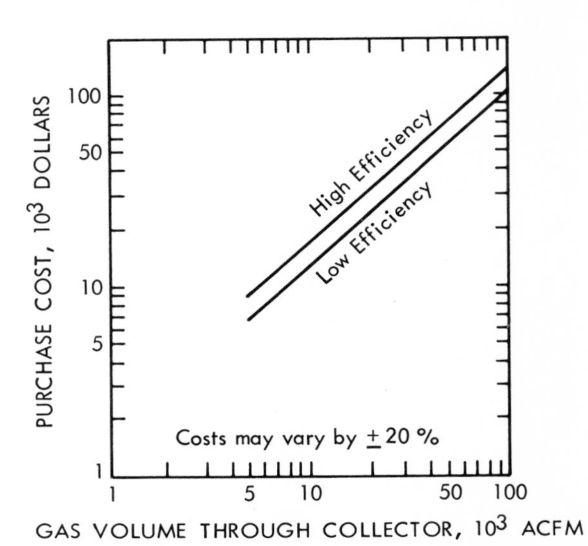

FIG. A-12. Purchase cost of low-voltage electrostatic precipitators.

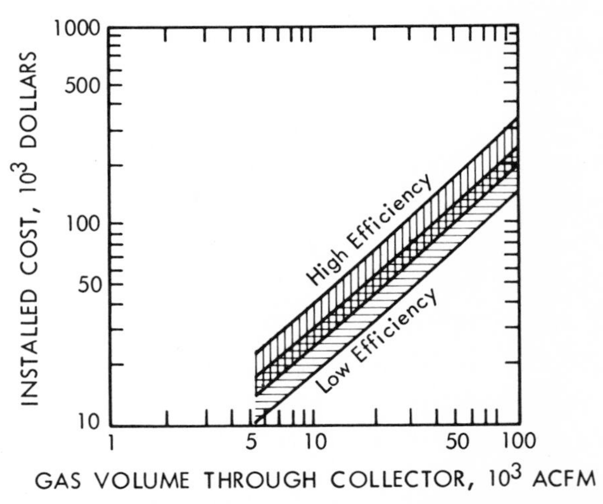

FIG. A-13. Installed cost of low-voltage electrostatic precipitators.

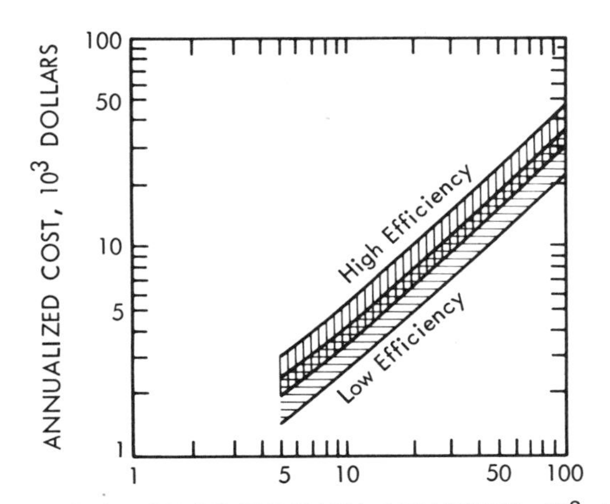

FIG. A-14. Annualized cost of operation of low-voltage electrostatic precipitators.

and 125 ft/min, respectively. Packaged modular low-voltage precipitators with flow rates of less than 1500 acfm are used to collect oil mist from machining operations. Purchase cost of such a unit usually is less than $1200. The assumptions used in calculating annual operation and maintenance costs for low-voltage electrostatic precipitators are as follows:

1. Annual operating time is 8760 hr.
2. Electrical power requirements:
   0.000015 kW/acfm for low efficiency
   0.000040 kW/acfm for high efficiency.
3. Power cost is $0.011/kWh.
4. Maintenance cost is $0.02/acfm.

F. FABRIC FILTERS

Figures A-15, A-16, and A-17 show purchase cost, installed cost, and annualized cost of control for three

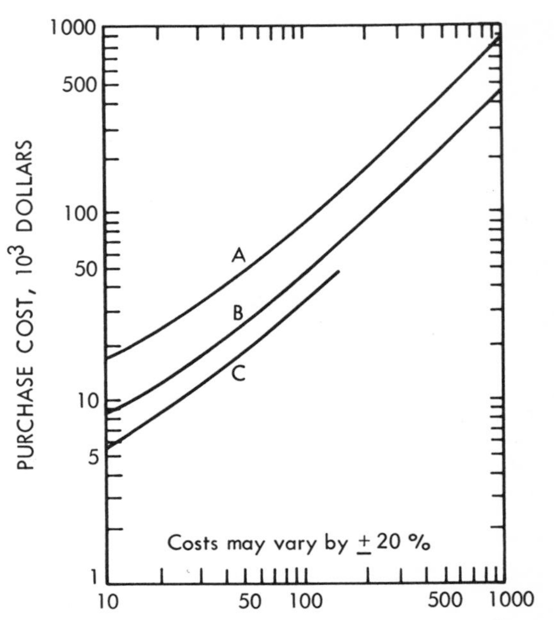

FIG. A-15. Purchase cost of fabric filters.

different types of filters. Each of the three filters is designed with about the same efficiency--99.9%. Costs are plotted for equipment sizes ranging from 10,000 to 1,000,000 acfm.

The control cost curves represent the following different types of filter installations:

1. Curve A represents a fabric filter installation with high-temperature synthetic woven fibers (including fiberglass) and felted fibers cleaned continuously and automatically.

2. Curve B represents an installation using medium-temperature synthetic woven and felted fibers, such as Orlon or Dacron, cleaned continuously and automatically.

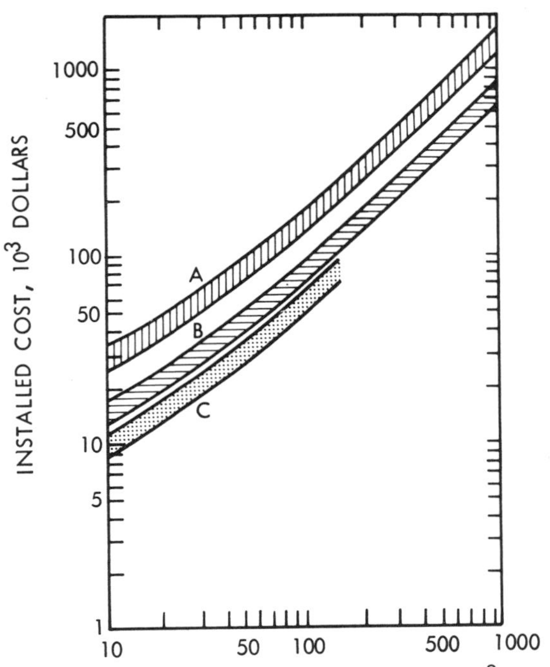

FIG. A-16. Installed cost of fabric filters.

3. Curve C is the least expensive installation. Woven natural fibers are used in a single compartment. Filters are intermittently cleaned. This equipment is rarely designed for processes handling over 150,000 acfm.

These control cost curves do not include data for furnace hoods, ventilation ductwork, and precoolers that may appear only in certain installations. The assumptions for calculating operating and maintenance costs are as follows:

1. Annual operating time is 8760 hr.
2. Pressure drop of the gas through the three types of fabric filters is 4 in. of water.
3. Power cost is \$0.05/acfm.
4. Maintenance cost is \$0.05/acfm.

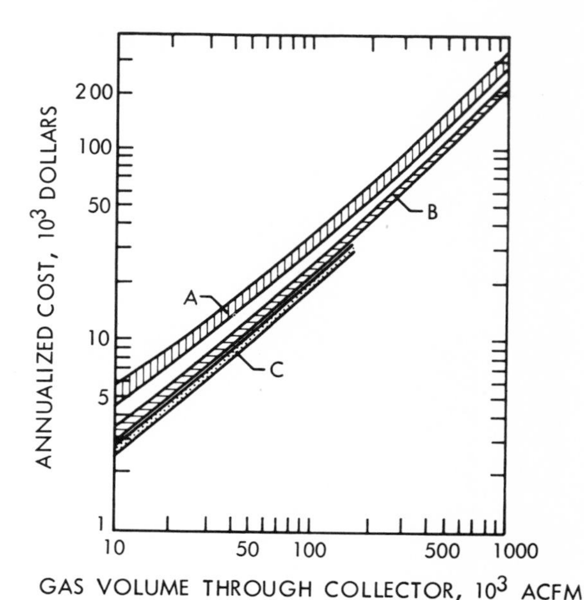

FIG. A-17. Annualized cost of operation of fabric filters.

G. AFTERBURNERS

Afterburners are separated into four categories: (a) direct flame; (b) catalytic; (3) direct flame with heat recovery; and (d) catalytic with heat recovery. Figure A-18 shows that purchase costs of direct-flame afterburners have a wider range than those of catalytic afterburners.

Figure A-19 shows the installation costs for afterburners. Heat exchangers are considered accessory equipment and appear as part of the installation cost. Installation costs may range from 10-100% of the purchase costs, although in some situations they may be as high as 400%. Differences in installation costs are due to the differences in structural supports, ductwork, and foundations.

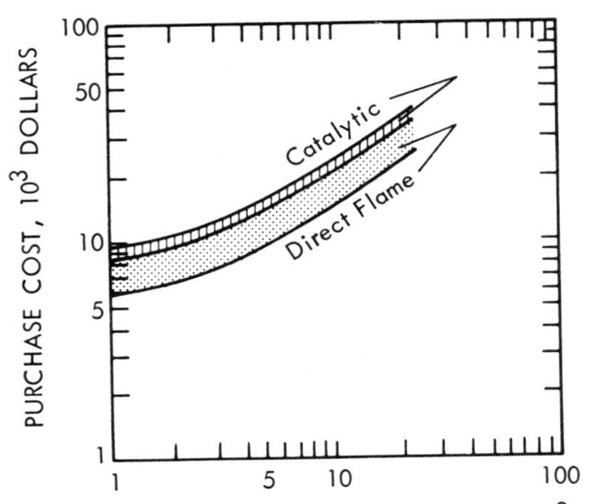

FIG. A-18. Purchase cost of afterburners.

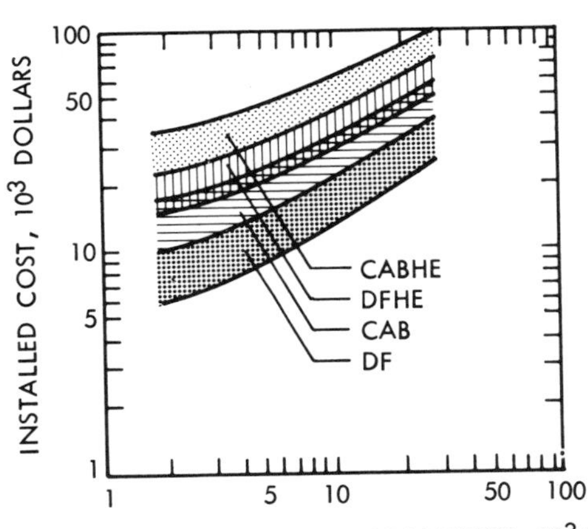

FIG. A-19. Installed cost of afterburners.

Installation costs for the addition of equipment to existing plant facilities will be higher than similar costs for new plants. Other factors accounting for different installation fees are the degree of instrumentation required, engineering fees in manufacturers' bids, startup tests and adjustments, heat exchangers, auxiliary fans, and utilities. The assumptions for calculating operation and maintenance costs are as follows:

1. Annual operating time is 8760 hr.
2. Fuel cost:
   $0.57/1000 acfm-hr for direct-flame afterburner with no heat recovery
   $0.23/1000 acfm-hr for direct-flame afterburner with heat recovery
   $0.28/1000 acfm-hr for catalytic afterburner with no heat recovery
   $0.14/1000 acfm-hr for catalytic afterburner with heat recovery.
3. Maintenance cost:
   $0.06/acfm for direct-flame afterburner
   $0.20/acfm for catalytic afterburner.
4. Pressure drop through all afterburner types is 1 in. of water.
5. Power cost is $0.011/kWh.

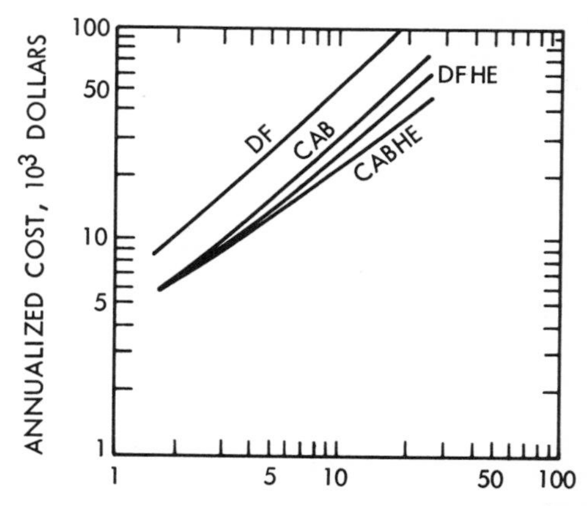

FIG. A-20. Annualized cost of operation of afterburners.

Cost comparisons presented in Fig. A-20 show that the direct-flame afterburner without a heat exchanger is the most expensive. The lower curve in Fig. A-20 shows that the annualized cost of a direct-flame afterburner with heat recovery is lower than the cost of a catalytic afterburner without heat recovery.

REFERENCES

1. Control Techniques for Particulate Air Pollutants, U.S. Department of Health, Education and Welfare, Washington, D.C., 1969.
2. Person, R. A., Control of Emissions from Ferroalloy Furnace Processing, Union Carbide Corporation, Niagara Falls, New York, 1969.
3. Lapple, C. E., "Interim Report: Stack Contamination-200 Areas," HDC-611, August 6, 1948.
4. Work, J. B., "Decontamination of Separation Plant Ventilation Air," HW-11529, November 10, 1948.
5. Vandegrift, A. E. and L. J. Shannon, "Particulate Pollutant System Study," Volume III, Handbook of Emission Properties, Midwest Research Institute, EPA Contract No. CPA 22-69-104, May 1, 1971.
6. Shannon, L. J., "Engineering and Cost Study of Emissions Eontrol in the Grain and Feed Industry," Monthly Report 5, Midwest Research Institute, EPA Contract No. 68-02-0213, December 6, 1971.
7. Shannon, L. J. and P. G. Gorman, "Particulate Pollutant System Study," Volume II, Fine Particle Emissions, Midwest Research Institute, EPA Contract No.22-69-104, 1 August 1971.

8. Air Pollution and the Regulated Electric Power and and Natural Gas Industries, Federal Power Commission Staff Report, Washington, D.C., September 1968.
9. O'Connor, J. R., and J. F. Citarella, An Air Pollution Control Cost Study of the Steam-Electric Power-Generating Industry, APCA Annual Meeting in New York, June 1969.
10. Pollock, W. A., H. P. Tomany, and G. Frieling, "Flue-Gas Scrubber," Mech.Eng., 21-25, (August 1967).
11. Miller, D. M., and J. Jonakin, "Kansas Power and Light to Trap Sulfur with Flue Gas Scrubber," Electrical World, March 4, 1968.
12. Report on Sulfur Dioxide and Fly Ash Emissions from Electric Utility Boilers, Public Service Electric and Gas Company (New Jersey), February 1967.
13. Smith, W. S., "Atmospheric Emissions from Fuel Oil Combustion--An Inventory Guide," PHS Publication No. 999-AP-2, November 1962.
14. Howekamp, D. P., and M. K. Hooper, Effects of Combustion Improving Devices on Air Pollutant Emissions from Residential Oil-Fired Furnaces, presented at 63rd Annual Meeting of the Air Pollution Control Association, St. Louis, Missouri, June 14-18, 1970.
15. Lownie, H. W., and J. Varga, A Systems Analysis Study of the Integrated Iron and Steel Industry, Battelle Memorial Institute, Contract No. PH 22-68-65, May 15, 1969.
16. Barnes, T. M., and H. W. Lownie, A Cost Analysis of Air-Pollution Controls in the Integrated Iron and Steel Industry, Battelle Memorial Institute, Columbus, Ohio, May 1968.
17. Devitt, T. W., The Integrated Iron and Steel Industry Air Pollution Problem, NAPCA, Cincinnati, Ohio, December 1968.

18. Hipp, N. E., and J. R. Westerholm, "Developments in Gas Cleaning--Great Lakes Steel Corporation," Iron Steel Eng., August 1967.

19. Broman, C. U., and R. R. Iseli, "Control of Open-Hearth Stack Emissions with Venturi Type Scrubber," Blast Furnace and Steel Plant, February 1968.

20. Bishop, C. A. et al., "Successful Cleaning of Open-Hearth Exhaust Gas with a High-Energy Scrubber," J. Air Pollution Control Assoc., 11 (2), 83-87(1961).

21. Schneider, R. L., "Engineering Operation and Maintenance of Electrostatic Precipitators on Open-Hearth Furnaces," J. Air Pollution Control Assoc., 13(8), 348-53 (1963).

22. Elliott, A. C., and A. J. Lafreniere, "Metallurgical Dust Collection in the Open Hearth and the Sinter Plant," Trans. Canadian Institute Mining and Metals, 1962.

23. Herrick, R. A., J. W. Olsen, and F. A. Ray, "Oxygen-Lanced Open-Hearth Furnace Fume Cleaning with a Glass Fabric Baghouse," J. Air Pollution Control Assoc., January 1966.

24. Willet, H. P., and D. E. Pike, "The Venturi Scrubber for Cleaning Oxygen," Iron Steel Eng., July 1961.

25. Wilkinson, F. M., "Wet Washing of BOF Gases--Lackawanna, Iron Steel Eng., September 1967.

26. Finney, J. A., and J. DeCoster, "A Cloth Filter Gas Cleaning System for Oxygen Converters," Iron Steel Eng., March 1965.

27. Blintzer, W. W., and D. R. Kleintop, "Design, Operation and Maintenance of a 150-Ton Electric Furnace Dust Collection System," Iron Steel Eng., June 1967.

28. Kane, John M., "Equipment for Cupola-Emission Control." Gray and Ductile Iron Founders Society, Cleveland, Ohio, 1968.

29. American Foundrymen's Society, Foundry Air Pollution Control Manual, Des Plaines, Illinois, 1967.
30. Cowen, P. S., ed., Cupola Emission Control, Cleveland, Ohio, Gray and Ductile Iron Founders' Society, 1969.
31. Konopka, A. P., Particulate Control Technology in Primary Nonferrous Smelting, Presented at Third Joint Meeting of AIChE, Denver, Colorado, September 1970.
32. "Restricting Dust Emissions from Copper-Ore Smelters," Ver. Deut. Ing., VD1, 2101 (January 1960).
33. Stern, A. C., Air Pollution, Vol. III, New York, Academic Press, 1968.
34. "Restricting Dust and Sulfur Dioxide Emission from Lead Smelters," Ver. Deut. Ing., VD 1, 2285 (September 1961).
35. Wallis, E., "Atmospheric Pollution and the Zinc Industry," Chem. Ind., October 1955.
36. Johnson, G. A., "Air Pollution Prevention at a Modern Zinc Smelter," Air Repair, February 1954.
37. Roggero, C. E., "High Temperature Fluid Bed Roasting of Zinc Concentrates," Trans. Metallurgical Society of AIME, February 1963.
38. Hargrave, J. H. D., "Recovery of Fume and Dust from Metallurgical Gases at Trail, B. C.," Can. Mining Met. Bull., June 1959.
39. Bainbridge, R., "New Developments in Smoke Control at Cominco," J. Metals, November 1956.
40. Air Pollution from the Primary Aluminum Industry, A Report to Washington Air Pollution Control Board, Office of Air Quality Control, Washington State Department of Health, Seattle, Washington, October 1969.
41. Doherty, R. E., "Current Status and Future Prospects--Cement Mill Air Pollution Control," Proc. National Conference on Air Pollution, Washington, D. C. (1966).

42. Burke, E., "Dust Arrestment in the Cement Industry," Chem. Ind., October 1955, 1312-1319.
43. Kreichelt, T. E., D. A. Kemnitz, and S. T. Cuffe, "Atmospheric Emissions from the Manufacture of Portland Cement," U. S. Public Health Service Publication No. 999 AP-17, 1967.
44. Lewis, C. J., and B. B. Crocker, "The Lime Industry's Problem of Airborne Dust," J. Air Pollution Control Assoc., 19, 31 (1969).
45. Gallaer, C. A., "Fine Aggregate Recovery and Dust Collection," Roads and Streets, October 1956, pp. 112-117.
46. Schell, T. W., "Cyclone/Scrubber System Quickly Eliminates Dust Problems," Rock Products, July 1968, pp. 66-69.
47. Personal communication.
48. "Guide for Air Pollution Control of Hot Mix Asphalt Plants," National Asphalt Pavement Association. Riverdale, Maryland.
49. "Atmospheric Emissions from Sulfuric Acid Manufacturing Processes," Public Health Service Publication No. 999-AP-13, Cincinnati, Ohio, 1965.
50. "Teflon Monofilament Cleans Up Acid Stack Gases," Chem. Eng., 112, October 25, 1965.
51. Stastny, E. P , "Electrostatic Precipitation," Chem. Eng. Prog., 62(4), 47-50 (April 1966).
52. "Atmospheric Emissions from Thermal-Process Phosphoric Acid Manufacture," NAPCA Publication No. AP-48, Durham, North Carolina, 1968.
53. Strauss, W., Industrial Gas Cleaning, Pergamon Press, New York, 1966, p.326.
54. Brink, J. A., "Air Pollution Control with Fiber Mist Eliminators," Can. J. Chem. Eng., 41, 135-38 (June 1963).

55. Anon., "Collector 99.9% Efficient: Pressure Drop Moderate," Chemical Process., Mid-November 1966, 48-50.

AUTHOR INDEX

Numbers in parentheses are reference numbers and indicate that an author's work is referred to although his name is not cited in the text. Underlined numbers give the page on which the complete reference is listed.

Y

Z

SUBJECT INDEX